Stem Cell Therapy for Organ Failure

Indumathi Somasundaram
Editor

Stem Cell Therapy for Organ Failure

Springer

Editor
Indumathi Somasundaram
Department of Stem Cell and Regenerative Medicine
Centre for Interdisciplinary Research
D.Y. Patil University
Kolhapur, India

ISBN 978-81-322-2109-8 ISBN 978-81-322-2110-4 (eBook)
DOI 10.1007/978-81-322-2110-4

Library of Congress Control Number: 2014959098

Springer New Delhi Heidelberg New York Dordrecht London

Printed on acid-free paper

Springer (India) Pvt. Ltd. is part of Springer Science+Business Media (www.springer.com)

Foreword

The term "stem cell" was first proposed by Alexander Maksimov in 1908 at the Congress of Hematologic Society in Berlin where he postulated the existence of hematopoietic stem cells. Since then, there has been much research on the origin of stem cells, the effects of stem cells, the pluripotent aspect of stem cells, and their utility in regenerative medicine. Stem cells are residing in almost all tissues/organs, and the ubiquity of stem cells in all organs may be one reason why they have been of interest in the area of tissue repair and regeneration. The book Stem Cell Therapy for Organ Failure describes valuable aspects of a toolbox of the in vitro and in vivo promises of stem cells in research and therapeutics of various organ systems from bench to bedside. This is the most challenging, fascinating, and necessary task for the scientist who is in the component aspect of stem cells and degenerative diseases.

Dr. Indumathi Somasundaram must be complemented to have developed this book, Stem Cell Therapy for Organ Failure, and for preparing this volume on the massively expanding field of stem cells in organ systems. She collaborated with other authors and contributed chapters into six major sections on various organ systems, ranging from the nervous system to the reproductive system. The topics are well chosen keeping in mind those who treat the diseases of these systems. The book helps to understand how to tackle clinical problems and presents treatment concepts for disorders that have not had effective therapy in the past. I congratulate the editor of this volume for her most prestigious effort to bring together all of the problems and their possible solutions under a single roof. It will undoubtedly be of great help in establishing ideal therapeutic decisions with respect to stem cells in the near future. I am convinced that this outstanding work will meet with success.

Tustin, CA, USA Melvin A. Shiffman, M.D., J.D.

Foreword

The field of stem cells has become one of the most intensely studied areas in the last decade. Currently there are no suitable medical therapies to cure disease associated with cell or tissue loss. Stem cells have the innate property to replicate themselves and to differentiate into other mature cell types. This potential leads to many possible clinical applications including tissue regeneration and transplant in humans.

A major issue for cell transplant therapies is the source of the cells to be used. Three sources of stem cells can be employed for transplant: embryonic stem cells, pluripotent stem cells and adult stem cells. These cells have different properties and differential advantages for use in regenerative medicine. Adult stem cells have the advantage of low tumorogenicity and lack of ethical controversy surrounding their use. Adult stem cells are already in clinical application.

This book summarizes our current understanding of stem cells and recent progress in our ability to cure disease using stem cells. Ever since the first bone marrow transplant, scientists have been slowly exploring the field which has now exploded with new and exciting findings. Our bodies contain a wide variety of stem cells that play a critical role in generation and regeneration of human tissue. Their potential far exceeds original expectations.

The current volume addresses the use of stem cells in several organ systems including: the nervous system, the cardiovascular system, the musculoskeletal system, the hematopoietic system, the gastrointestinal system and the reproductive system. Moreover, considerations for culturing and preserving adult stem cells for therapeutics are addressed. This text serves as a timely guide to help inform this developing field.

The various sections contained in this publication provide a wide overview of the latest discoveries in stem cell physiology.

Yale University, New Haven, CT, USA Hugh S. Taylor, M.D.

Preface

Stem cells possess a wonder-working potential to understand important aspects of human biology involving tissue repair and regeneration. Stem cells have recently become the most fascinating area of research as they can treat and cure diseases and ultimately save lives. They have the potential to treat not just one but many diseases. Stem cells, based on their potential to regenerate and differentiate into multi-lineages coupled with their immunosuppressive properties, seem to play an important role in treating chronic degenerative diseases. The ubiquitous existences of stem cells annex to be a regenerative tool rendering the replacement of worn-out cells. These intriguing properties have transpired them to be a strong candidate for applicability in regenerative therapeutics for a myriad of diseases.

In this regard, this book explores the past, present, and the futuristic in vitro and in vivo therapeutic strategies and applications of pre- and postnatal stem cells for treating the disorders of different organ systems of our body in a wide perspective. The prime focus of this volume is to unravel the basic, advanced, therapeutic, and translational approaches put forth so far in the field of stem cells and regenerative medicine at research, preclinical, and clinical levels. Stem cells have ushered a widespread interest and exciting possibilities in cell-based therapies, albeit shortcomings do prevail and small uncontrolled phase I/II studies are only signal generating rather than giving a definite proof of concept, thereby limiting their applicability in curative therapeutics. Despite certain constructive initiatives and untiring efforts, bringing this basic benchside research to advanced transitional bedside remains a challenge.

For instance, certain characteristic features of stem cells both in vitro and in vivo are quite vague and remain a hindrance in the path of breakthrough studies. It is essential that only defined cell populations are introduced into patients; this requires careful characterization of the cell populations, extensive culturing, and preservation. In addition, long-term integration of functionally transplanted cells into damaged organs is also a major challenge. To achieve this, a tissue engineering–based approach with appropriate support of bio-scaffolds and growth factors is required. Finally, immunological rejection is a major barrier to successful stem cell transplant where use of a patient's own cells is not possible. The expanding researches in these areas that seem decisive in improvising regenerative medical therapeutics are discussed in this volume, thereby leading to further pathbreaking studies that can cure all health-related challenges facing mankind.

Overall, the book is designed in such a way that readers can get reliable information about the various stem cell sources and their therapeutic implications/applications for the treatment of disorders related to multiple organ systems under a single roof, thereby making the readers industrious in this field. Thus, this book reveals the imperativeness of various stem cell sources and their utility in curative therapeutics. On the whole, it creates hope in the light of existing perplexity to bring forth stem cells as a frontline source of therapeutics in treating a multitude of disorders in the field of regenerative medicine.

Kolhapur, India Indumathi Somasundaram

Contents

Contributors

Kanmani Anandan Technology Ed, Midlothian, VA, USA

Rajanna Ajumeera Department of Stem Cell Research, National Institute of Nutrition (ICMR), Secunderabad, Hyderabad, India

Peter A. Barbuti Clinical and Experimental Neuroscience, School of Clinical Sciences, Université du Lukembourg, Bristol, UK

Padmanav Behera Department of Stem Cell Research, National Institute of Nutrition (ICMR), Secunderabad, Hyderabad, India

Niranjan Bhattacharya Department of Regenerative Medicine and Translational Science, Calcutta School of Tropical Medicine and STM Cord Blood Bank, Kolkatta, India

Sanjukta Bhattacharya Department of International Relations, Jadavpur University, Kolkata, India

Ramesh R. Bhonde School of Regenerative Medicine, Manipal Institute of Regenerative Medicine, Manipal University, Bangalore, India

Konstantinos Dean Boudoulas, MD Department of Medicine, Department of Cardiovascular Medicine, The Ohio State University, Columbus, OH, USA

Simi M. Chacko Dorothy M. Davis Heart and Lung Research Institute, Department of Internal Medicine, The Ohio State University, Columbus, OH, USA

Suvro Chatterjee Department of Biotechnology, Vascular Biology Lab, AU-KBC Research Centre, Anna University, Chennai, India

Centre for Biotechnology, Anna University, Chennai, India

Indrani Datta School of Regenerative Medicine, Manipal Institute of Regenerative Medicine, Manipal University, Bangalore, India

Subadra Dravida Tran-Scell Biologics Pvt. Ltd., Hyderabad, India

Apurva Gadgil Stem Plus Biotech Pvt. Ltd., Sangli, MS, India

Palanivel Gajalakshmi Life Sciences Division, Vascular Biology Lab, AU-KBC Research Centre, Anna University, Chennai, India

Kavina Ganapathy School of Regenerative Medicine, Manipal Institute of Regenerative Medicine, Manipal University, Bangalore, India

Sireesha Ganneru Department of Stem Cell Research, National Institute of Nutrition (ICMR), Secunderabad, Hyderabad, India

Venkata Naga Srikanth Garikipati Center for Translational Medicine, Temple University School of Medicine, Philadelphia, PA, USA

Namitha Haridoss Department of Stem Cell Research, National Institute of Nutrition (ICMR), Secunderabad, Hyderabad, India

Kouichi Hasegawa Institute for Stem Cell Biology and Regenerative Medicine (inStem), National Centre for Biological Sciences (NCBS), Tata Institute of Fundamental Research (TIFR), Bangalore, India

Institute for Integrated Cell-Material Sciences (iCeMS), Kyoto University, Kyoto, Japan

R. Indarapu Curatum Cell Sciences Pvt. Ltd., Hyderabad, India

Meghnad G. Joshi Department of Stem Cells and Regenerative Medicine, D.Y. Patil University, Kolhapur, Maharashtra, India

Mohammad Mahboob Kanafi School of Regenerative Medicine, Manipal Institute of Regenerative Medicine, Manipal University, Bangalore, India

Raj Kishore, Ph.D., FAHA Center for Translational Medicine, Temple University School of Medicine, Philadelphia, PA, USA

Prasanna Krishnamurthy Department of Cardiovascular Sciences, Houston Methodist Research Institute, Houston, TX, USA

Department of Cardiovascular Sciences, Houston Methodist Research Institute, Houston, TX, USA

Sandeep Kumar Kotturu Molecular Biology Unit, National Institute of Nutrition (ICMR), Secunderabad, Hyderabad, India

Periannan Kuppusamy Dorothy M. Davis Heart and Lung Research Institute, Department of Internal Medicine, The Ohio State University, Columbus, OH, USA

EPR Center, Geisel School of Medicine at Dartmouth, Lebanon, NH, USA

Nikhita Kutala SVS Institute of Dental Sciences, Mahabubnagar, AP, India

Vijay Kumar Kutala Department of Clinical Pharmacology and Therapeutics, Nizam's Institute of Medical Sciences, Hyderabad, India

Debanjana Majumdar School of Regenerative Medicine, Manipal Institute of Regenerative Medicine, Manipal University, Bangalore, India

Syamantak Majumder Aab Cardiovascular Research Institute, School of Medicine and Dentistry, University of Rochester, Rochester, NY, USA

Dhanasekaran Marappagounder Stem Cell Banking & Research, Ree Laboratories Private Limited, Mumbai, India

Shaik Mohammad Naushad School of Chemical & Biotechnology, SASTRA University, Thanjavur, Tamil Nadu, India

Paramasivam Nithyanand School of Chemical & Biotechnology, SASTRA University, Thanjavur, Tamil Nadu, India

Revathy Raj Hematopoietic Stem Cell Transplantation Unit, Apollo Speciality Hospitals, Chennai, India

Ravikumar Rajappa Harvard Medical School, Dr. Spector's Lab-VA Medical Center, Jamaica Plain, MA, USA

Uksha Saini, PhD Division of Cardiovascular Medicine, Department of Internal Medicine, Davis Heart and Lung Research Institute, The Ohio State University, Columbus, OH, USA

Subhanwita Sarkar Dey Institute for Stem Cell Biology and Regenerative Medicine (inStem), National Centre for Biological Sciences (NCBS), Tata Institute of Fundamental Research (TIFR), Bangalore, India

Himadri Singh Department of Stem Cell Research, National Institute of Nutrition (Indian Council of Medical Research), Secunderabad, Hyderabad, India

Phillip G. Stubblefield Department of Obstetrics and Gynecology, School of Medicine, Boston University, Boston, MA, USA

Indumathi Somasundaram Assistant Professor, Department of Stem Cell and Regenerative Medicine, Centre for Interdisciplinary Research, D.Y. Patil University, Kolhapur, India

Susinder Sundaram School of Chemical & Biotechnology, SASTRA University, Thanjavur, Tamil Nadu, India

Vijayalakshmi Venkatesan Department of Stem Cell Research, National Institute of Nutrition (ICMR), Secunderabad, Hyderabad, India

Suresh Kumar Verma Center for Translational Medicine, Temple University School of Medicine, Philadelphia, PA, USA

Noriko Yoshida Institute for Integrated Cell-Material Sciences (iCeMS), Kyoto University, Kyoto, Japan

About the Editor

Dr. Indumathi Somasundaram is presently working as an Assistant Professor in the Department of Stem Cell and Regenerative Medicine, Centre for Interdisciplinary Research, D.Y. Patil University, Kolhapur, India. Previously, she was working as a Research Associate at the Department of Stem Cell Research, National Institute of Nutrition (ICMR), Secunderabad, Hyderabad, India, in a DBT funded stem cell research project. She completed her master's degree in 2008, from University of Madras (Loyola College, Chennai, India). After a brief work experience at Lifeline RIGID Hospitals as a Researcher for nearly 2 years, where she got introduced to stem cells, she proceeded to do her Ph.D. in the field of stem cells, and was awarded Ph.D. from University of Madras in early 2014. During her work and Ph.D. tenure, she gained research experience on different kinds of adult stem cells including bone marrow, adipose tissue and endometrium, their in vitro attributes of self-renewal and differentiation, and their therapeutic implications/applications. Her areas of interest and expertise include adult stem cells, regenerative biology, reproductive biology, diabetes, cancer stem cells and so on. She has more than 21 publications (of which 15 are indexed international journals) and 4 book chapters (Springer Verlag) to her credit and few more publications in the pipeline. Her present interest lies on exploring the potentials of human endometrial stem cells under physiology and pathology conditions including infertility. Her present projects involve exploring microarray/microRNA regulatory mechanisms in physiology and pathology of diseases including cancers. Besides, her future projects lie on enhancing the efficiency of stem cells using tissue engineering and nanoparticle based approach for various diseases including wound healing.

Dr. Indumathi Somasundaram

Part I

Stem Cells: Science/Concepts and Database

Promises of Stem Cell Research and Therapeutics

Indumathi Somasundaram, Kanmani Anandan, and Dhanasekaran Marappagounder

Introduction

Stem cells are smart cells of the body. It is the origin of life. As stated by the great pathologist Rudolf Virchow, “All cells come from cells.” Today it is proved with evidence that all cells come from stem cells. The stem cells are derived from embryo, fetal tissues, and adult organs. Explicitly, stem cells can generate daughter cells identical to their mother (self-renewal), and under certain physiologic or experimental condition, depending on the source, they can differentiate into any type of cells such as heart muscle cells, blood cells, or the insulin-producing cells of the pancreas (differentiated cells). As the plant leaves flourishes from the stem, the body nourishes from the stem cells resting in our body. A more complete description of a stem cell includes a consideration of replication capacity, clonality, and potency. Thus, stem cells are considered to be unique cells with special attributes. It is these special attributes that offer the vast potential serving as a repair system of the body. Ever since the discovery of stem cells [1–5], scientists have dreamed of using them to repair damaged tissue or create new organs and has revolutionized the field of medicine.

This chapter will thus, introduce stem cells, the history of stem cell research and its promises in the field of regenerative medicine.

Stem Cell Backgrounder

Aristotle deduced that the embryo was derived from the mother’s menstrual blood. The hypothesis that life did not arise spontaneously, but rather only from preexisting life (*omne vivum ex vivo*) was pronounced by Leydig in 1855. Virchow (1855) then extended this to postulate that all cells in an organism are derived from preexisting cells (*omnis cellula e cellula*), the fertilized egg. According to the principles derived from Leydig, Virchow, and Pasteur, life as we know it neither ends nor begins but is continuous. The adult human, for example, is only one stage in the cycle of human life (Fig. 1) [6].

Thus, during early embryonic development, each stem cell divides and gives rise to two daughter cells with the same potential, which is called symmetric division, whereas during normal tissue renewal in the adult, each stem/progenitor cell gives rise to one daughter cell that remains a stem/progenitor cell and one daughter cell that begins the process of determination

I. Somasundaram (✉)
Assistant Professor, Department of Stem Cell and Regenerative Medicine, Centre for Interdisciplinary Research, D.Y. Patil University, Kolhapur, India
e-mail: indu.stemcell@gmail.com

K. Anandan
Technology Ed, Midlothian, VA 23112, USA
e-mail: Kanmani.a@gmail.com

D. Marappagounder
Stem Cell Banking & Research, Ree Laboratories Private Limited, Mumbai 400053, India
e-mail: dhanasekarbio@gmail.com

I. Somasundaram (ed.), *Stem Cell Therapy for Organ Failure*, DOI 10.1007/978-81-322-2110-4_1, © Springer India 2014

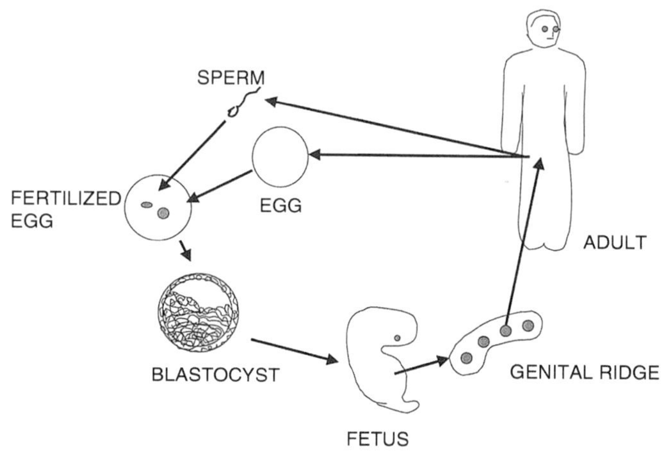

Fig. 1 Cycle of life. The life of an individual begins with formation of a fetus. Totipotent cells in the developing fetus migrate to the genital ridge and in adults produce germinal stem cells in the gonads. Germinal cells give rise to gametes (egg and sperm), resulting in cells containing half the chromosomes of an adult. Genetic reconstitution occurs when the sperm fertilizes the egg. Thus, life is continuous

to a terminally differentiated cell, which is called asymmetric division. The number of cells increases exponentially during early embryogenesis, but the cell number remains constant during normal tissue renewal, as the number of new stem/progenitor cells equals the number of cells destined to die.

History of Stem Cells from the Scratch

The potential of stem cells to revolutionize medicine got a huge boost with the emergence of adult stem cells. This timeline takes us through the ups and downs of the stem cell rollercoaster. The history of stem cell research started in the beginning of 1800s with the discovery that some cells could generate other cells. In the early 1900s, the first real stem cells were discovered when it was found that some cells generate blood cells. Research into adult stem cells in animals and in humans has been ongoing since this time, and bone marrow transplants – actually a transplant of adult stem cells – have in fact been used in patients receiving radiations and chemotherapy since the 1950s [7].

In 1973, a team of physicians performed the first unrelated bone marrow transplant. It required seven transplants to be successful. The 1990s saw rapid expansion and success of the National Marrow Donor Program (NMDP) with more than 38,000 transplants to date for the treatment of immunodeficiency and leukemia [8]. Developments in biotechnology in the 1980s and 1990s saw the introduction of first identity of embryonic stem cells and cloning by Martin Evans and Ian Wilmut, respectively [9–11], which paved way for targeting and altering genetic material and methods for growing human cells in the laboratory. These advances really launched stem cell research into the limelight, establishing the world's first human embryonic stem cell line which still exists today [12].

In the same year, John Gearhart (Johns Hopkins University) derived germ cells from cells in fetal gonadal tissue (primordial germ cells). Pluripotent stem cell "lines" were developed from both sources. The blastocysts used for human stem cell research typically come from in vitro fertilization (IVF) procedures. Since this discovery, a plethora of evidence has emerged to suggest that these embryonic stem cells are capable of becoming almost any of the specialized cells in the body and therefore have the potential to generate replacement cells for a broad array of tissues and organs such as the heart [13–15], liver [16,17], pancreas [18,19], and nervous system [20–22].

Progress in stem cell research is now astounding, with over 21,193 research papers on embryonic, iPS cells and adult stem cells being published globally in the year 2012 (Source: Stem Cell Research: Trends and perspectives on the evolving international landscape. 2013

Elsevier B.V). Embryonic stem cells are still in its infancy in clinics with only 20 reported clinical trials till now; however, adult stem cells are already being used in treatments for over 4,000 clinical trials till date for a number of conditions including spinal cord injury, liver cirrhosis, neurodegenerative diseases, leukemia, cardiac problems, and so on (results from A service of the U.S. National Institutes of Health: clinicaltrials.gov).

Timeline of Stem Cell Research

1956	First successful bone marrow transplant between a related donor and recipient is performed by Dr E Donnall Thomas in New York. The patient, who has leukemia, is given radiotherapy and then treated with healthy bone marrow from an identical twin [7].
1958	Human histocompatibility antigens – Dausset discovers the first human leukocyte antigen [23].
1961	James Till and Ernest McCulloch first published evidence of the existence of stem cells in mouse bone marrow, a population of clonogenic bone marrow (BM) cells was found to generate myeloerythroid colonies in the spleens of lethally irradiated hosts [24].
1968	First bone marrow transplant for noncancer treatment. Dr Robert Good uses a bone marrow transplant to treat an 8-year-old boy with severe combined immunodeficiency syndrome (SCID). The donor is an HLA-matched sister (http://www.fhcrc.org/en/treatment/long-term-follow-up/FAQs/transplantation.html).
1970	Friedenstein and colleagues were the first to report that the rodent bone marrow had fibroblastoid cells, capable of forming colonies on plastic, which was later called a "mesenchymal stem cell" or MSC [25].
1973	First bone marrow transplant between unrelated patients. A 5-year-old SCID patient in New York treated with multiple infusions of bone marrow from a donor in Denmark (http://www.fhcrc.org/en/treatment/long-term-followup/FAQs/transplantation.html).
1978	The first IVF baby is born in England [26].
	Blood stem cells are discovered in human umbilical cord blood [27].
1981	Mouse embryonic stem cells are derived for the first time from the inner cell mass of a mouse blastocyst and grown in vitro [9,10].
1984–1988	Isolated pluripotent cell line of human testicular teratocarcinoma [28], a first example of clonal human embryonal carcinoma cells. When exposed to retinoic acid, these cells differentiate into neuron-like cells and other cell types [29, 30].
1987	National Marrow Donor Program (NMDP) and its Be The Match Registry initiated [31].
1989	Preimplantation genetic diagnosis (PGD) is developed – a method where a single stem cell can be removed from an IVF embryo and tested for inherited diseases. A clonal line of human embryonal carcinoma cells is derived that yields tissues from all three primary germ layers. They have limited replicative and differentiative capacity [32].
1990	Dr Thomas receives the Nobel Prize in Physiology or Medicine for his pioneering work on bone marrow transplants (http://www.nobelprize.org/nobel_prizes/medicine/laureates/1990/).
1992	The first direct evidence of nervous system stem cells came from the identification and isolation of rat neural crest stem cells, clonogenic precursors that give rise to all known neural crest cell types, while self-renewing the neural crest progenitors [33].
1995	Scientists at the University of Wisconsin derive the first embryonic stem cells from nonhuman primates. These cells were reported pluripotent [34].
1996	The first organism ever to be cloned from adult cells is Dolly [11].
1997	Leukemia is shown to originate from a hematopoietic stem cell, the first direct evidence for cancer stem cells [35].
1998	Scientists at the University of Wisconsin, led by James Thompson, isolate and grow the first stem cells from human embryos. The embryos used in these studies were created by IVF [12].
	John Gearhart at Johns Hopkins University derived pluripotent germ cells from cells in fetal gonadal tissue (primordial germ cells) [36].

(continued)

2000	Scientists derive human ES cells from the inner cell mass of blastocysts. They proliferate in vitro for a long time and form all three germ layers and teratomas when injected into immunodeficient mice [37].
	Retinal stem cells (RSCs) isolated from the pigmented ciliary epithelium (CE) of the mouse and differentiate into retinal-specific cell types, including rod photoreceptors, bipolar neurons, and Müller glia [38].
2001	US president George W. Bush prohibits federal funding of research on human embryonic stem cells. The president claims that more than 60 stem cell lines are available for funding (http://georgewbushwhitehouse.archives.gov/news/releases/2001/08/20010809-2.html).
2003	Rare human breast cancer stem cells were identified and isolated [39].
	Cancer stem cells isolated in human brain tumors [40].
2004	California becomes the first state in the USA to provide its own fund for embryonic stem cell research (http://www.cirm.ca.gov/pdf/prop71.pdf).
2005	George W. Bush's restrictions on embryonic stem cell research are loosened (http://query.nytimes.com/gst/fullpage.html?res=9A01E0DE113FF93AA15754C0A9639C8B63&\penalty-\@Msec=health).
	Ernest McCulloch and James Till won the prestigious Albert Lasker Basic Medical Research Award, for first setting stage for all current research on adult and embryonic stem cells (http://laskerfoundation.org/awards/2005_b_description.htm).
2006	Induced pluripotent stem cells from mouse fibroblast cells by defined factors were generated [41].
2007	Evans shares the Nobel prize for medicine with Mario Capecchi and Oliver Smithies for their work on genetics and embryonic stem cells (http://www.nobelprize.org/nobel_prizes/medicine/laureates/2007/).
	Induced pluripotent stem cells from adult fibroblast cells identified [42,43].
	Researchers at Wake Forest University and Harvard University report that stem cells drawn from amniotic fluid donated by pregnant women hold much the same promise as embryonic stem cells [44].
2008	Robert Lanza and colleagues at Advanced Cell Technology create the first human embryonic stem cells lines created without destruction of the embryo [45].
	First trachea transplant using woman's own stem cells [46].
2009	President Barack Obama lifts 2001 restrictions on federal funding for human embryonic stem cell research (http://www.gpo.gov/fdsys/pkg/FR-2009-03-11/pdf/E9-5441.pdf).
2010	First clinical trial of human embryonic-derived stem cells for treatment of spinal cord injury [47].
2011	World's first stem cell-derived synthetic windpipe transplant (http://www.newscientist.com/article/dn20671-man-receives-worlds-first-synthetic-windpipe.html#.UzlZGqiSxaY).
2012	Yamanaka wins a Nobel prize for creating induced pluripotent stem cells, which he shares with John Gurdon of the University of Cambridge (http://www.nobelprize.org/nobel_prizes/medicine/laureates/2012/).
	Blindness eased by historic stem cell treatment [48].
2013	Scientists generate pluripotent stem cells derived from cloned human embryos. Reprogramming human skin cells, using a cloning technique called "somatic cell nuclear transfer" (SCNT) (http://www.nature.com/news/human-stem-cells-created-by-cloning-1.12983).
2014	Charles Vacanti of Harvard Medical School together with Haruko Obokata at the Riken Center for Developmental Biology in Kobe, Japan, and colleagues announced a revolutionary discovery that any cell can potentially be rewound to a pre-embryonic state using a simple, 30-min technique (http://www.newscientist.com/article/mg22129542.500-stem-cell-power-unleashed-after-30-min-dip-in-acid.html#.Uzi-p6iSxaY).
	Masayo Takahashi at the same Riken Center is due to select patients for what promises to be the world's first trial of a therapy based on induced pluripotent stem cells, to treat a form of age-related blindness [49].

Stem Cells and Its Promises

Despite the biomedical research advances of the last 50 years, much is still left to be discovered in human biology and millions of people still suffer from devastating diseases. Stem cell research started in the mid-1800s, and it is viewed from the researcher's point of view that stem cells are the key to understand the biology of various diseases and can lead to treatment and cures, and ultimately saves lives. Stem cells have the ability to literally develop into every cell of the body and thereby have the potential to replace damaged or diseased tissues and to treat disease. Stem cells are considered to have tremendous potential in repairing damaged organs, including the spinal cord, which normally does not undergo regeneration [50–52]. This ability to repair human tissue and to someday provide actual organ systems would provide hope to many with debilitating diseases [53–58].

The most obvious application of human stem cells and the one that receives the most attention are replacement therapies: to replace diseased or degenerating tissues or to replace cell populations, such as those of the hematopoietic system (e.g., blood), that have been destroyed by chemotherapy [59–63]. Stem cells could additionally provide an unlimited supply of specific cell types for transplantation. To date, stem cell-derived cardiomyocyte, neural precursors, and hematopoietic precursors have been transplanted into recipient animals [64–72]. Although the analyses of the long-term outcome of such experiments are limited, the findings suggest that the transplanted cells were able to function in the host animal. Stem cells have a high therapeutic value in the last few years to repair damaged spinal cords [73–76]; cure Crohn's disease and liver cirrhosis [77–84], Alzheimer's, and Parkinson's [85–91]; regrow arteries around a blockage [92–97]; regrow limbs [98–100]; replace failed kidneys and hearts [101–110]; cure diabetes by replacing nonfunctional cells in the pancreas [111–114]; restore vision and hearing [115–123]; and treat leukemia and lymphoma that are nonresponsive to normal therapy [124–129]. These are merely a few of the potential applications of this phenomenal science. In fact most of the treatments listed above have already been studied, and with promising results. However, by latest, miraculous cures that supposedly occurred in faraway lands, the use of stem cells to treat certain degenerative diseases has not advanced beyond mice and needs extensive research.

Other applications of stem cells are in the study of development in both human and animal model systems. This approach includes the identification and isolation of novel precursor cells and of medically important genes. Such genes might encode proteins that have direct therapeutic applications, such as novel growth factors, or genes that would be important targets for drug development [130–135]. Human stem cells will also be valuable as a test system for evaluating the toxicity and efficacy of new medicines or chemicals. The wide ranges of cell types and tissues that may develop from stem cells represent a biological system that mimics many of the complex interactions of the cells and tissues of the body and, as such, provides an attractive and valuable screening tool. This type of assay could have wide applications in the pharmaceutical, chemical, cosmetics, and agrochemical industries. It has the potential to reduce the need for animal testing and to increase efficiency and safety and reduces the costs of developing safe and effective drugs and chemicals [136–140].

Hence, it is clear that stem cell technology continues to revolutionize modern biology and provide unique opportunities in understanding of both the mechanisms that control basic biological processes and in the treatment of diseases. Additional research will be necessary to apply the full therapeutic potential of this technology, but the resulting novel therapies and approaches should more than justify the effort.

Conclusion

Stem cell research provides a useful tool for unraveling the molecular mechanisms that determine the differentiation fate of a pluripotent cell and for understanding the gene expression properties and epigenetic modifications essential to maintain the pluripotent state. There are many ways in which human stem cells can be used in basic and translational research. However, there are a number of hurdles in the path of stem cell research that are preventing the routine application of the technology in regenerative medicine. This will only be overcome by continuous extensive research. Currently, stem cell research is nearing transplantation therapies with high efficiency whereby a specific cell population compromised by disease is replaced with new, functional cells. The greatest promise of stem cell research may lie in an area not yet imagined.

References

1. Bruder SP, Fink DJ, Caplan AI (1994) Mesenchymal stem cells in bone development, bone repair, and skeletal regeneration therapy. J Cell Biochem 56(3):283–294
2. Kolb HJ, Holler E (1997) Hematopoietic transplantation: state of the art. Stem Cells 15(Suppl 1):151–157, discussion 158
3. Prockop DJ (1997) Marrow stromal cells as stem cells for nonhematopoietic tissues. Science 276: 71–74
4. Pittenger MF, Mackay AM, Beck SC, Jaiswal RK, Douglas R et al (1999) Multilineage potential of adult human mesenchymal stem cells. Science 284:143–147
5. Weissman IL (2000) Stem cells: units of development, units of regeneration, and units in evolution. Cell 100:157–168
6. Gilbert SF (2000) Developmental biology, 6th edn. Sinauer Associates, Sunderland
7. Thomas ED, Lochte HL Jr, Lu WC, Ferrebee J (1957) Intravenous infusion of bone marrow in patients receiving radiation and chemotherapy. N Engl J Med 257:491
8. Theresa H, McCarthy PL Jr, Hassebroek A, Bredeson C, Gajewski JL et al (2013) Significant improvement in survival after allogeneic hematopoietic cell transplantation during a period of significantly increased use, older recipient age, and use of unrelated donors. J Clin Oncol 31(19): 2437–2449
9. Evans MJ, Kaufman MH (1981) Establishment in culture of pluripotential cells from mouse embryos. Nature 292:154–156
10. Martin GR (1981) Isolation of a pluripotent cell line from early mouse embryos cultured in medium conditioned by teratocarcinoma stem cells. Proc Natl Acad Sci U S A 78:7634–7638
11. Wilmut I, Schnieke AE, McWhir J et al (1997) Viable offspring derived from fetal adult mammalian cells. Nature 385:810
12. Thomson JA, Itskovitz-Eldor J, Shapiro SS, Waknitz MA, Swiergiel JJ, Marshall VS, Jones JM (1998) Embryonic stem cell lines derived from human blastocysts. Science 282:1145–1147
13. Leor J, Gerecht S, Cohen S, Miller L, Holbova R et al (2007) Human embryonic stem cell transplantation to repair the infarcted myocardium. Heart 93(10):1278–1284
14. Ardehali R, Ali SR, Inlay MA, Abilez OJ, Chen MQ et al (2013) Prospective isolation of human embryonic stem cell-derived cardiovascular progenitors that integrate into human fetal heart tissue. Proc Natl Acad Sci U S A 110(9):3405–3410
15. Karakikes I, Senyei GD, Hansen J, Kong CW, Azeloglu EU et al (2014) Small molecule-mediated directed differentiation of human embryonic stem cells toward ventricular cardiomyocytes. Stem Cells Transl Med 3(1):18–31
16. Hay DC, Zhao D, Ross A, Mandalam R, Lebkowski J et al (2007) Direct differentiation of human embryonic stem cells to hepatocyte-like cells exhibiting functional activities. Cloning Stem Cells 9(1):51–62
17. Pal R, Mamidi MK, Das AK, Gupta PK, Bhonde R (2012) A simple and economical route to generate functional hepatocyte-like cells from hESCs and their application in evaluating alcohol induced liver damage. J Cell Biochem 113(1):19–30
18. Shim JH, Kim SE, Woo DH, Kim SK, Oh CH et al (2007) Directed differentiation of human embryonic stem cells towards a pancreatic cell fate. Diabetologia 50(6):1228–1238
19. Bose B, Shenoy SP, Konda S, Wangikar P (2012) Human embryonic stem cell differentiation into insulin secreting β-cells for diabetes. Cell Biol Int 36(11):1013–1020
20. Zhang SC, Wernig M, Duncan ID, Brüstle O, Thomson JA (2001) In vitro differentiation of transplantable neural precursors from human embryonic stem cells. Nat Biotechnol 19(12):1129–1133
21. Kriks S, Shim JW, Piao J, Ganat YM, Wakeman DR et al (2011) Dopamine neurons derived from human ES cells efficiently engraft in animal models of Parkinson's disease. Nature 480(7378):547–551
22. Maroof AM, Keros S, Tyson JA, Ying SW, Ganat YM et al (2013) Directed differentiation and functional maturation of cortical interneurons from human embryonic stem cells. Cell Stem Cell 12(5):559–572
23. Dausset J (1958) Iso-leuco-anticorps. Acta Haematol 20:156
24. Till J, McCulloch E (1961) A direct measurement of the radiation sensitivity of normal mouse bone marrow cells. Radiat Res 14:213–224
25. Friedenstein AJ, Chailakhjan RK, Lalykina KS (1970) The development of fibroblast colonies in monolayer cultures of guinea-pig bone marrow and spleen cells. Cell Tissue Kinet 3(4):393–403

26. Trounson AO, Gardner DK, Baker G, Barnes FL, Bongso A et al (2000) Handbook of in vitro fertilization. CRC Press, Boca Raton/London/New York/Washington, DC
27. Prindull G, Prindull B, Meulen N (1978) Haematopoietic stem cells (CFUc) in human cord blood. Acta Paediatr Scand 67(4):413–416
28. Andrews PW, Damjanov I, Simon D, Banting GS, Carlin C et al (1984) Pluripotent embryonal carcinoma clones derived from the human teratocarcinoma cell line Tera-2. Differentiation in vivo and in vitro. Lab Invest 50(2):147–162
29. Thompson S, Stern PL, Webb M, Walsh FS, Engstrom W et al (1984) Cloned human teratoma cells differentiate into neuron-like cells and other cell types in retinoic acid. J Cell Sci 72:37–64
30. Andrews PW (1988) Human teratocarcinomas. Biochim Biophys Acta 948:17–36
31. NIH MedlinePlus Magazine. Bone Marrow Transplants: The National Marrow Donor Program and Be The Match Registry. Summer 2011. 6(2):17
32. Handyside AH, Pattinson JK, Penketh RJ, Delhanty JD, Winston RM et al (1989) Biopsy of human preimplantation embryos and sexing by DNA amplification. Lancet 1(8634):347–349
33. Stemple DL, Anderson DJ (1992) Isolation of a stem cell for neurons and glia from the mammalian neural crest. Cell 71(6):973–985
34. Thomson JA, Kalishman J, Golos TG, Durning M, Harris CP et al (1995) Isolation of a primate embryonic stem cell line. Proc Natl Acad Sci USA 92:7844–7848
35. Bonnet D, Dick JE (1997) Human acute myeloid leukemia is organized as a hierarchy that originates from a primitive hematopoietic cell. Nat Med 3:730–737
36. Shamblott MJ, Axelman J, Wang S, Bugg EM, Littlefield J et al (1998) Derivation of pluripotent stem cells from cultured human primordial germ cells. Proc Natl Acad Sci U S A 95(23): 13726–13731
37. Reubinoff BE, Pera MF, Fong CY, Trounson A, Bongso A (2000) Embryonic stem cell lines from human blastocysts: somatic differentiation in vitro. Nat Biotechnol 18(4):399–404
38. Tropepe V, Coles BLK, Chiasson BJ, Horsford DJ, Elia AJ et al (2000) Retinal stem cells in the adult mammalian eye. Science 287:2032–2203
39. Al-Hajj M, Wicha MS, Benito-Hernandez A, Morrison SJ, Clarke MF (2003) Prospective identification of tumorogenic breast cancer cells. Proc Natl Acad Sci U S A 100:3983–3988
40. Singh SK, Clarke ID, Terasaki M, Bonn VE, Hawkins C, Squire J, Dirks PB (2003) Identification of a cancer stem cell in human brain tumors. Cancer Res 63:5821–5828
41. Takahashi K, Yamanaka S (2006) Induction of pluripotent stem cells from mouse embryonic and adult fibroblast cultures by defined factors. Cell 126:663–676
42. Yu J, Vodyanik MA et al (2007) Induced pluripotent stem cell lines derived from human somatic cells. Science 318(5858):1917–1920
43. Takahashi K et al (2007) Induction of pluripotent stem cells from adult human fibroblasts by defined factors. Cell 131(5):861–872
44. De Coppi P, Jr Bartsch G, Minhaj Siddiqui M, Tao X, Santos CC et al (2007) Isolation of amniotic stem cell lines with potential for therapy. Nat Biotechnol 25:100–106
45. Advanced cell technology. Human embryonic stem cells lines created without destruction of the embryos. ScienceDaily 2008
46. Macchiarini P, Jungebluth P, Go T, Asnaghi MA, Rees LE et al (2008) Clinical transplantation of a tissue-engineered airway. Lancet 372(9655): 2023–2030
47. Bretzner F, Gilbert F, Baylis F, Brownstone RM (2011) Target populations for first-in-human embryonic stem cell research in spinal cord injury. Cell Stem Cell 8(5):468–475
48. Schwartz SD, Hubschman J-P, Heilwell G, Franco-Cardenas V et al (2012) Embryonic stem cell trials for macular degeneration: a preliminary report. Lancet 379(9817):713–720
49. Kamao H, Mandai M, Okamoto S, Sakai N, Suga A et al (2014) Characterization of human induced pluripotent stem cell-derived retinal pigment epithelium cell sheets aiming for clinical application. Stem Cell Rep 2:1–14
50. Yoon SH, Shim YS, Park YH, Chung JK, Nam JH et al (2007) Complete spinal cord injury treatment using autologous bone marrow cell transplantation and bone marrow stimulation with granulocyte macrophage-colony stimulating factor: Phase I/II clinical trial. Stem Cells 25(8): 2066–2073
51. Chernykh ER, Stupak VV, Muradov GM, Sizikov MY, Shevela EY et al (2007) Application of autologous bone marrow stem cells in the therapy of spinal cord injury patients. Bull Exp Biol Med 143(4): 543–547
52. Cristante AF, Barros-Filho TE, Tatsui N, Mendrone A, Caldas JG et al (2009) Stem cells in the treatment of chronic spinal cord injury: evaluation of somatosensitive evoked potentials in 39 patients. Spinal Cord 47(10):733–738
53. Vieira NM, Brandalise V, Zucconi E, Jazedje T, Secco M et al (2008) Human multipotent adipose-derived stem cells restore dystrophin expression of Duchenne skeletal-muscle cells in vitro. Biol Cell 100(4):231–241
54. Bahat-Stroomza M, Barhum Y, Levy YS, Karpov O, Bulvik S et al (2009) Induction of adult human bone marrow mesenchymal stromal cells into functional astrocyte-like cells: potential for restorative treatment in Parkinson's disease. J Mol Neurosci 39(1–2):199–210
55. González MA, Gonzalez-Rey E, Rico L, Büscher D, Delgado M (2009) Treatment of experimental

arthritis by inducing immune tolerance with human adipose-derived mesenchymal stem cells. Arthritis Rheum 60(4):1006–1019
56. Connick P, Kolappan M, Crawley C, Webber DJ, Patani R et al (2012) Autologous mesenchymal stem cells for the treatment of secondary progressive multiple sclerosis: an open-label phase 2a proof-of-concept study. Lancet Neurol 11(2):150–156
57. Glass JD, Boulis NM, Johe K, Rutkove SB, Federici T et al (2012) Lumbar intraspinal injection of neural stem cells in patients with amyotrophic lateral sclerosis: results of a phase I trial in 12 patients. Stem Cells 30(6):1144–1151
58. Paracchini V, Carbone A, Colombo F, Castellani S, Mazzucchelli S et al (2012) Amniotic mesenchymal stem cells: a new source for hepatocyte-like cells and induction of CFTR expression by coculture with cystic fibrosis airway epithelial cells. J Biomed Biotechnol 2012:575471
59. Intragumtornchai T, Jootar S, Unganon A, Swasdikul D, Udomprasertgul V (1999) Quality of life in Thai patients after bone marrow and peripheral blood stem cell transplantation: a comparison study with patients treated with conventional chemotherapy. Int J Hematol 70(3):181–189
60. Michallet M, Philip T, Philip I, Godinot H, Sebban C et al (2000) Transplantation with selected autologous peripheral blood CD34 + Thy1+ hematopoietic stem cells (HSCs) in multiple myeloma: impact of HSC dose on engraftment, safety, and immune reconstitution. Exp Hematol 28(7):858–870
61. Müller AM, Kohrt HE, Cha S, Laport G, Klein J et al (2012) Long-term outcome of patients with metastatic breast cancer treated with high dose chemotherapy and transplantation of purified autologous hematopoietic stem cells. Biol Blood Marrow Transplant 18(1):125–133
62. Wallet HL, Sobh M, Morisset S, Robin M, Fegueux N et al (2013) Double umbilical cord blood transplantation for hematological malignancies: a long-term analysis from the SFGM-TC registry. Exp Hematol 41(11):924–933
63. Fouillard L, Francois S, Bouchet S, Bensidhoum M, Elm'selmi A, Chapel A (2014) Innovative cell therapy in the treatment of serious adverse events related to both chemo-radiotherapy protocol and acute myeloid leukemia syndrome: the infusion of mesenchymal stem cells post-treatment reduces hematopoietic toxicity and promotes hematopoietic reconstitution. Curr Pharm Biotechnol 14(9): 842–848
64. Smits AM, van Laake LW, den Ouden K, Schreurs C, Szuhai K et al (2009) Human cardiomyocyte progenitor cell transplantation preserves long-term function of the infarcted mouse myocardium. Cardiovasc Res 83(3):527–535
65. Okura H, Matsuyama A, Lee CM, Saga A, Kakuta-Yamamoto A et al (2010) Cardiomyoblast-like cells differentiated from human adipose tissue-derived mesenchymal stem cells improve left ventricular dysfunction and survival in a rat myocardial infarction model. Tissue Eng Part C Methods 16(3): 417–425
66. Kawamura M, Miyagawa S, Fukushima S, Saito A, Miki K et al (2013) Enhanced survival of transplanted human induced pluripotent stem cell-derived cardiomyocytes by the combination of cell sheets with the pedicled omental flap technique in a porcine heart. Circulation 128(11 Suppl 1):S87–S94
67. Zhilai Z, Hui Z, Yinhai C, Zhong C, Shaoxiong M et al (2011) Combination of NEP 1-40 infusion and bone marrow-derived neurospheres transplantation inhibit glial scar formation and promote functional recovery after rat spinal cord injury. Neurol India 59(4):579–585
68. Drury-Stewart D, Song M, Mohamad O, Guo Y, Gu X et al (2013) Highly efficient differentiation of neural precursors from human embryonic stem cells and benefits of transplantation after ischemic stroke in mice. Stem Cell Res Ther 4(4):93
69. Fujiwara N, Shimizu J, Takai K, Arimitsu N, Saito A et al (2013) Restoration of spatial memory dysfunction of human APP transgenic mice by transplantation of neuronal precursor derived from human iPS cells. Neurosci Lett pii(13):00947-6
70. Benjamin BP, Galy A, Kyoizumi S, Namikawa R, Scarborough J et al (1994) Engraftment of human hematopoietic precursor cells with secondary transfer potential in SCID-hu mice. Blood 84(8): 2497–2505
71. Liu Y, Yi L, Zhang X, Gao L, Zhang C et al (2011) Cotransplantation of human umbilical cord blood derived stromal cells enhances hematopoietic reconstitution and engraftment in irradiated BABL/c mice. Cancer Biol Ther 11(1):84–94
72. Shi Q, Kuether EL, Chen Y, Schroeder JA, Fahs SA et al (2014) Platelet gene therapy corrects the hemophilic phenotype in immunocompromised hemophilia A mice transplanted with genetically manipulated human cord blood stem cells. Blood 123(3):395–403
73. Mazzini L, Ferrero I, Luparello V, Rustichelli D, Gunetti M et al (2010) Mesenchymal stem cell transplantation in amyotrophic lateral sclerosis: a Phase I clinical trial. Exp Neurol 223(1):229–237
74. Karamouzian S, Nematollahi-Mahani SN, Nakhaee N, Eskandary H (2012) Clinical safety and primary efficacy of bone marrow mesenchymal cell transplantation in sub acute spinal cord injured patients. Clin Neurol Neurosurg 114(7):935–939
75. Liu J, Chen J, Liu B, Yang C, Xie D et al (2013) Acellular spinal cord scaffold seeded with mesenchymal stem cells promotes long-distance axon regeneration and functional recovery in spinal cord injured rats. J Neurol Sci 325(1–2):127–136
76. Amr SM, Gouda A, Koptan WT, Galal AA, Abdel-Fattah DS et al (2014) Bridging defects in chronic spinal cord injury using peripheral nerve grafts combined with a chitosan-laminin scaffold and enhancing regeneration through them by co-transplantation

with bone-marrow-derived mesenchymal stem cells: Case series of 14 patients. J Spinal Cord Med 37(1):54–71

77. Goldman O, Han S, Sourrisseau M, Dziedzic N, Hamou W et al (2013) KDR identifies a conserved human and murine hepatic progenitor and instructs early liver development. Cell Stem Cell 12(6):748
78. Jang YO, Kim YJ, Baik SK, Kim MY, Eom YW et al (2014) Histological improvement following administration of autologous bone marrow-derived mesenchymal stem cells for alcoholic cirrhosis: a pilot study. Liver Int 34(1):33–41
79. Liao X, AnCheng JY, Zhou QJ, Liao C (2013) Therapeutic effect of autologous bone marrow-derived liver stem cells transplantation in hepatitis B virus-induced liver cirrhosis. Hepatogastroenterology 60(123):406–409
80. Hong J, Jin H, Han J, Hu H, Liu J et al (2014) Infusion of human umbilical cord-derived mesenchymal stem cells effectively relieves liver cirrhosis in DEN-induced rats. Mol Med Rep 9(4):1103–1111
81. de la Portilla F, Alba F, García-Olmo D, Herrerías JM, González FX, Galindo A (2013) Expanded allogeneic adipose-derived stem cells (eASCs) for the treatment of complex perianal fistula in Crohn's disease: results from a multicenter phase I/IIa clinical trial. Int J Colorectal Dis 28(3):313–323
82. Ciccocioppo R, Bernardo ME, Sgarella A, Maccario R, Avanzini MA et al (2011) Autologous bone marrow-derived mesenchymal stromal cells in the treatment of fistulising Crohn's disease. Gut 60(6):788–798
83. Burt RK, Craig RM, Milanetti F, Quigley K, Gozdziak P et al (2010) Autologous nonmyeloablative hematopoietic stem cell transplantation in patients with severe anti-TNF refractory Crohn disease: long-term follow-up. Blood 116(26): 6123–6132
84. González MA, Gonzalez-Rey E, Rico L, Büscher D, Delgado M (2000) Adipose-derived mesenchymal stem cells alleviate experimental colitis by inhibiting inflammatory and autoimmune responses. Gastroenterology 136(3):978–989
85. Doi D, Samata B, Katsukawa M, Kikuchi T, Morizane A, Ono Y et al (2014) Isolation of human induced pluripotent stem cell-derived dopaminergic progenitors by cell sorting for successful transplantation. Stem Cells Transl Med 2(3):337, -50
86. Battista D, Ganat Y, El Maarouf A, Studer L, Rutishauser U (2014) Enhancement of polysialic acid expression improves function of embryonic stem-derived dopamine neuron grafts in Parkinsonian mice. Stem Cells Transl Med 3(1):108–113
87. Yan M, Sun M, Zhou Y, Wang W, He Z et al (2013) Conversion of human umbilical cord mesenchymal stem cells in Wharton's jelly to dopamine neurons mediated by the Lmx1a and neurturin in vitro: potential therapeutic application for Parkinson's disease in a rhesus monkey model. PLoS One 8(5):e64000
88. Sundberg M, Bogetofte H, Lawson T, Jansson J, Smith G, Astradsson A et al (2013) Improved cell therapy protocols for Parkinson's disease based on differentiation efficiency and safety of hESC-, hiPSC-, and non-human primate iPSC-derived dopaminergic neurons. Stem Cells 31(8):1548–1562
89. Bae JS, Jin HK, Lee JK, Richardson JC, Carter JE (2013) Bone marrow-derived mesenchymal stem cells contribute to the reduction of amyloid-β deposits and the improvement of synaptic transmission in a mouse model of pre dementia Alzheimer's disease. Curr Alzheimer Res 10(5):524–531
90. Lee HJ, Lee JK, Lee H, Shin JW, Carter JE et al (2010) The therapeutic potential of human umbilical cord blood-derived mesenchymal stem cells in Alzheimer's disease. Neurosci Lett 481(1):30–35
91. Wicklund L, Leão RN, Strömberg AM, Mousavi M, Hovatta O et al (2010) Β-amyloid 1-42 oligomers impair function of human embryonic stem cell-derived forebrain cholinergic neurons. PLoS One 5(12):e15600
92. Mutirangura P, Ruangsetakit C, Wongwanit C, Chinsakchai K, Porat Y et al (2009) Enhancing limb salvage by non-mobilized peripheral blood angiogenic cell precursors therapy in patients with critical limb ischemia. J Med Assoc Thai 92(3):320–327
93. Kim H, Cho HJ, Kim SW, Liu B, Choi YJ et al (2010) CD31+ cells represent highly angiogenic and vasculogenic cells in bone marrow: novel role of nonendothelial CD31+ cells in neovascularization and their therapeutic effects on ischemic vascular disease. Circ Res 107(5):602–614
94. Rahnemai-Azar A, D'Ippolito G, Gomez LA, Reiner T, Vazquez-Padron RI et al (2011) Human marrow-isolated adult multilineage-inducible (MIAMI) cells protect against peripheral vascular ischemia in a mouse model. Cytotherapy 13(2): 179–192
95. Padfield GJ, Tura-Ceide O, Freyer E, Barclay GR, Turner M et al (2013) Endothelial progenitor cells, atheroma burden and clinical outcome in patients with coronary artery disease. Heart 99(11): 791–798
96. Katare R, Riu F, Rowlinson J, Lewis A, Holden R et al (2013) Perivascular delivery of encapsulated mesenchymal stem cells improves post ischemic angiogenesis via paracrine activation of VEGF-A. Arterioscler Thromb Vasc Biol 33(8):1872–1880
97. Katare R, Stroemer P, Hicks C, Stevanato L, Patel S et al (2014) Clinical-grade human neural stem cells promote reparative neovascularization in mouse models of hindlimb ischemia. Arterioscler Thromb Vasc Biol 34(2):408–418
98. Rinkevich Y, Lindau P, Ueno H, Longaker MT, Weissman IL (2011) Germ-layer and lineage-restricted stem/progenitors regenerate the mouse digit tip. Nature 476(7361):409–413
99. Lehoczky JA, Robert B, Tabin CJ (2011) Mouse digit tip regeneration is mediated by fate-restricted progenitor cells. Proc Natl Acad Sci U S A 108(51):20609–20614

100. Fernando WA, Leininger E, Simkin J, Li N, Malcom CA et al (2011) Wound healing and blastema formation in regenerating digit tips of adult mice. Dev Biol 350(2):301–310
101. Wagner L, Lengyel L, Mikala G, Reményi P, Piros L et al (2013) Successful treatment of renal failure caused by multiple myeloma with HLA-identical living kidney and bone marrow transplantation: a case report. Transplant Proc 45(10):3705–3707
102. Narayanan K, Schumacher KM, Tasnim F, Kandasamy K, Schumacher A et al (2013) Human embryonic stem cells differentiate into functional renal proximal tubular–like cells. Kidney Int 83: 593–603
103. Harari-Steinberg O, Metsuyanim S, Omer D, Gnatek Y, Gershon R et al (2013) Identification of human nephron progenitors capable of generation of kidney structures and functional repair of chronic renal disease. EMBO Mol Med 5(10):1556–1568
104. Perico N, Casiraghi F, Gotti E, Introna M, Todeschini M et al (2013) Mesenchymal stromal cells and kidney transplantation: pretransplant infusion protects from graft dysfunction while fostering immunoregulation. Transpl Int 26(9):867–878
105. Bukulmez H, Bilgin A, Caplan AI, Jones O (2014) Prevention of late stage renal failure in BXSB SLE mouse model with human bone marrow derived mesenchymal stem cell treatment. Arthritis Rheumatol 66(Suppl 11):S149
106. Leistner DM, Fischer-Rasokat U, Honold J, Seeger FH, Schächinger V et al (2011) Transplantation of progenitor cells and regeneration enhancement in acute myocardial infarction (TOPCARE-AMI): final 5-year results suggest long-term safety and efficacy. Clin Res Cardiol 100(10):925–934
107. Bolli R, Chugh AR, D'Amario D, Loughran JH, Stoddard MF et al (2011) Cardiac stem cells in patients with ischaemic cardiomyopathy (SCIPIO): initial results of a randomised phase 1 trial. Lancet 378(9806):1847–1857
108. Jianqin Y, Boyle A, Shih H, Sievers RE, Zhang Y et al (2012) Sca-1^{+} cardiosphere-derived cells are enriched for Isl1-expressing cardiac precursors and improve cardiac function after myocardial injury. PLoS One 7(1):e30329
109. Martinez EC, Vu DT, Wang J, Lilyanna S, Ling LH et al (2013) Grafts enriched with sub amnion-cord-lining mesenchymal stem cell angiogenic spheroids induce post-ischemic myocardial revascularization and preserve cardiac function in failing rat hearts. Stem Cells Dev 22(23):3087–3099
110. Heldman AW, DiFede DL, Fishman JE, Zambrano JP, Trachtenberg BH (2014) Transendocardial mesenchymal stem cells and mononuclear bone marrow cells for ischemic cardiomyopathy: the TAC-HFT randomized trial. JAMA 311(1):62–73
111. Rezania A, Bruin JE, Riedel MJ, Mojibian M, Asadi A et al (2012) Maturation of human embryonic stem cell–derived pancreatic progenitors into functional islets capable of treating pre-existing diabetes in mice. Diabetes 61(8):2016–2029
112. Zhao Y, Jiang Z, Zhao T, Ye M, Hu C et al (2012) Reversal of type 1 diabetes via islet β cell regeneration following immune modulation by cord blood-derived multipotent stem cells. BMC Med 10:3
113. Zhao Y, Jiang Z, Zhao T, Ye M et al (2013) Targeting insulin resistance in type 2 diabetes via immune modulation of cord blood-derived multipotent stem cells (CB-SCs) in stem cell educator therapy: phase I/II clinical trial. BMC Med 11
114. Dave SD, Trivedi HL, Chooramani SG, Chandra T (2013) Management of type 1 diabetes mellitus using in vitro autologous adipose tissue trans-differentiated insulin-making cells. BMJ Case Rep
115. Siqueira RC, Messias A, Voltarelli JC, Scott IU, Jorge R (2011) Intravitreal injection of autologous bone marrow-derived mononuclear cells for hereditary retinal dystrophy: a phase I trial. Retina 31(6):1207–1214
116. Marchini G, Pedrotti E, Pedrotti M, Barbaro V, Di Iorio E, Ferrari S et al (2012) Long-term effectiveness of autologous cultured limbal stem cell grafts in patients with limbal stem cell deficiency due to chemical burns. Clin Experiment Ophthalmol 40(3):255–267
117. Basu S, Ali H, Sangwan VS (2012) Clinical outcomes of repeat autologous cultivated limbal epithelial transplantation for ocular surface burns. Am J Ophthalmol 153(4):643–650
118. Hambright D, Park KY, Brooks M, McKay R, Swaroop A, Nasonkin IO (2012) Long-term survival and differentiation of retinal neurons derived from human embryonic stem cell lines in un-immunosuppressed mouse retina. Mol Vis 18: 920–936
119. Li T, Lewallen M, Chen S, Yu W, Zhang N, Xie T (2013) Multipotent stem cells isolated from the adult mouse retina are capable of producing functional photoreceptorcells. Cell Res 23(6):788–802
120. Kamao H, Mandai M, Okamoto S, Sakai N, Suga A et al (2014) Characterization of human induced pluripotent stem cell-derived retinal pigment epithelium cell sheets aiming for clinical application. Stem Cell Rep 2(2):205–218
121. Oshima K, Shin K, Diensthuber M, Peng AW et al (2010) Mechanosensitive hair cell-like cells from embryonic and induced pluripotent stem cells. Cell 141(4):704–716.
122. Koehler KR, Mikosz AM, Molosh AI, Patel D et al (2013) Generation of inner ear sensory epithelia from pluripotent stem cells in 3D culture. Nature 500(7461):217–221
123. Liu Z, Fang J, Dearman J, Zhang L, Zuo J (2014) In vivo generation of immature inner hair cells in neonatal mouse cochleae by ectopic atoh1 expression. PLoS One 9(2):e89377
124. Kobos R, Steinherz PG, Kernan NA, Prockop SE, Scaradavou A et al (2012) Allogeneic hematopoietic stem cell transplantation for pediatric patients with treatment-related myelodysplastic syndrome or acute myelogenous leukemia. Biol Blood Marrow Transplant 18(3):473–480

125. Hale GA, Shrestha S, Le-Rademacher J, Burns LJ, Gibson J et al (2012) Alternate donor hematopoietic cell transplantation (HCT) in non Hodgkin lymphoma using lower intensity conditioning: a report from the CIBMTR. Biol Blood Marrow Transplant 18(7):1036–1043
126. Bhatnagar B, Rapoport AP, Fang HB, Ilyas C, Marangoz D et al (2014) Single center experience with total body irradiation and melphalan (TBI-MEL) myeloablative conditioning regimen for allogeneic stem cell transplantation (SCT) in patients with refractory hematologic malignancies. Ann Hematol 93(4):653–660
127. Han Q, Sun M, Wu L, Chen J, Wang W et al (2014) Primary distal femur T-cell lymphoma after allogeneic haematopoietic stem cell transplantation for chronic myeloid leukaemia: a rare case report and literature review. J Int Med Res 42(2): 598–605
128. Cohen JB, Geyer SM, Lozanski G, Zhao W, Heerema NA et al (2014) Complete response to induction therapy in patients with Myc-positive and double-hit non-Hodgkin lymphoma is associated with prolonged progression-free survival. Cancer 120(11):1677–1685
129. McClune BL, Ahn KW, Wang HL, Antin JH, Artz AS et al (2014) Allotransplantation for patients age 40 years and greater with non-Hodgkin Lymphoma (NHL): encouraging progression-free survival. Biol Blood Marrow Transplant pii:S1083–S8791
130. Lee G, Papapetrou EP, Kim H, Chambers SM et al (2009) Modelling pathogenesis and treatment of familial dysautonomia using patient-specific iPSCs. Nature 461:402–406
131. Itzhaki I, Maizels L, Huber I, Zwi-Dantsis L, Caspi O et al (2011) Modelling the long QT syndrome with induced pluripotent stem cells. Nature 471:225–229
132. Boone N, Bergon A, Loriod B, Devèze A, Nguyen C et al (2012) Genome-wide analysis of familial dysautonomia and kinetin target genes with patient olfactory ecto-mesenchymal stem cells. Hum Mutat 33(3):530–540
133. Robinton DA, Daley GQ (2012) The promise of induced pluripotent stem cells in research and therapy. Nature 481(7381):295–305
134. Xu XH, Zhong Z (2013) Disease modeling and drug screening for neurological diseases using human induced pluripotent stem cells. Acta Pharmacol Sin 34:755–764
135. Musunuru K, Domian IJ, Chien KR (2010) Stem cell models of cardiac development and disease. Annu Rev Cell Dev Biol 26:667–687
136. Zeng X, Chen J, Deng X, Liu Y, Rao MS et al (2006) An in vitro model of human dopaminergic neurons derived from embryonic stem cells: MPP + toxicity and GDNF neuroprotection. Neuropsychopharmacology 31:2708–2715
137. Wobus AM, Löser P (2011) Present state and future perspectives of using pluripotent stem cells in toxicology research. Arch Toxicol 85(2):79–117
138. Marx U, Walles H, Hoffmann S, Lindner G, Horland R et al (2012) 'Human-on-a-chip' developments: a translational cutting-edge alternative to systemic safety assessment and efficiency evaluation of substances in laboratory animals and man? Altern Lab Anim 40(5):235–257
139. Kostadinova R, Boess F, Applegate D, Suter L, Weiser T et al (2013) A long-term three dimensional liver co-culture system for improved prediction of clinically relevant drug-induced hepatotoxicity. Toxicol Appl Pharmacol 268(1):1–16
140. Sciancalepore AG, Sallustio F, Girardo S, Gioia Passione L, Camposeo A et al (2014) A bioartificial renal tubule device embedding human renal stem/progenitor cells. PLoS One 9(1):87496

Considerations for Culturing and Preserving Adult Stem Cells for Therapeutics

Simi M. Chacko and Periannan Kuppusamy

Introduction

Stem-cell research is currently being intensely fuelled with the expectancy of developing a therapeutic strategy that could become a viable alternative to short-term medication and surgical procedures to treat several degenerative disease conditions [1–7]. At present several nonclinical studies on stem cells are being done using the established protocols which have been prevalent since several decades of cell culture, expansion, and storage. Over the recent past years, these laboratory protocols are being challenged because of their disparity from physiological conditions and the inability to reproduce the effect observed in vitro, in vivo, and human models. One of the most important factors that will matter specifically while establishing stem-cell protocols in the in vitro conditions with the intention of clinical translation is that contamination or modification of these cells by the external environment such as surrounding oxygen [8–10] or growth medium [11, 12] will bear significant consequence on transplantation, as these cells are inherently capable of differentiating and grafting to the host tissue.

In the human body, the stem cell is surrounded by a complex microenvironment consisting of the parenchymal cells, the extracellular matrix (ECM) that supports the tissue, signaling molecules, etc., all of which simultaneously appear to influence and play a role in the development and regulation of these stem cells [13, 14]. Identifying the requirements for cultivating these cells outside of the body initially appeared to be focused on cell survival. The nutrients supplied by the fetal calf serum and a medium prepared at physiological pH, supplemented with growth factors and essential amino acids, at room air conditions are still sufficient for culturing and keeping cells alive in laboratory conditions. However, now, the focus is shifting to bring reevaluation of existing protocol for stem-cell culture such that stem cells can be expanded and stored with minimal alteration of its biological characteristic and function which is of paramount significance to be preserved for clinical use.

In this review, we focus on three important culture conditions that have potential to influence the effectiveness of adult stem-cell therapy. The first one is cryopreservation of cells which is the most commonly used storage techniques for cells and tissues. Although cells are able to endure extremely low temperatures, cryoinjury

S.M. Chacko
Dorothy M. Davis Heart and Lung Research Institute, Department of Internal Medicine, The Ohio State University, 420 W. 12th Avenue, Room 114, Columbus, OH 43210, USA

P. Kuppusamy (✉)
Dorothy M. Davis Heart and Lung Research Institute, Department of Internal Medicine, The Ohio State University, 420 W. 12th Avenue, Room 114, Columbus, OH 43210, USA

EPR Center, Geisel School of Medicine at Dartmouth, 48 Lafayette Street, Lebanon, NH 03766, USA
e-mail: kuppu@darmouth.edu

I. Somasundaram (ed.), *Stem Cell Therapy for Organ Failure*,
DOI 10.1007/978-81-322-2110-4_2, © Springer India 2014

could result as a consequence to three major factors associated with freezing – cooling rate [15], type and concentration of cryoprotective [16, 17] used and the warming rate of the cells [18, 19]. We address here the effects of cryopreservation on different stem-cell types, the different cooling rates and methods currently used to achieve this, and also attend to the concern about DMSO toxicity especially in the context of clinical translation of stem-cell therapy. Another variable of cell culture discussed in this article is serum conditions and growth factors which have been traditionally prepared from animal products. We have gathered the efforts by several research groups currently underway to determine serum-free solutions and the use of human supplements for culturing especially human stem cells, to not only improve its efficacy but also to preserve its integrity before transplantation by preventing contamination from the use of animal products. Finally, we highlight probably the most important yet most complex cell-culture variable – the oxygen concentration – that pose immense practical challenges to be controlled in a laboratory environment but failure of which affect the outcome of studies done to understand stem-cell processes and also stem-cell preparations for cell therapy. Oxygen, a key regulatory signaling molecule [20], is found in much lower concentration in the human body, especially in the stem-cell niches [21, 22]. When adult stem cells are introduced for therapy especially to repair and replace ischemic tissue, which is severely hypoxic, there is a breakdown of the expected process of stem cells to differentiate and graft to the host tissue. These cells fail to survive in such hostile hypoxic conditions [23, 24]. Hypoxic preconditioning strategies, along with attempts to culture cells in a constant hypoxic environment, are being investigated and the results of some of their efforts are also presented in this article. In summary, this review has been an attempt to direct attention to some of the primary culture conditions that is in need of further modification and optimization in order to advance adult stem-cell research for the purpose of effective clinical translation.

Cryopreservation

The scarcity of stem cells in the body of a healthy individual poses a limitation for obtaining sufficient autologous stem cells for the purpose of cell therapy [25, 26]. Autologous stem cells obtained from the same individual will bring down the incidence of graft-vs-host disease (GVHD) a common problem in allogenic transplant [27] and foster-safe transplantation. However, these cells are harder to procure from patients who are older [28] or suffering from diseases such as diabetics [29] or osteoporosis [30] because in such conditions it has been observed that there is a reduced number of circulating stem cells and also impaired functionality of the existent ones. This is not a universal observation in the case of all kinds of stem cells or in all diseases requiring stem-cell transplantation, but it does raise the need for optimal storage of multipotent stem cells for facilitating a constant supply of cells especially in the case where healthy autologous stem cells are unavailable in sufficient number for therapy. This issue of long-term preservation is also of vital importance in the case of stem cells obtained from umbilical cord blood (UCB) which is typically isolated for use at a later period.

Freezing has been used as an effective method of preservation for biological tissues for long-term storage in animal studies [31]. In order that stem cells can be readily made available for clinical use, it is important to optimize methods of storage of stem cells without compromising the properties of self-renewal and differentiation. Cryopreservation is based on the principle that at extreme low temperatures when liquid water cease to exist (~ -130 °C) and further below -196 °C when all thermal reactions are also suspended, all biochemical mechanisms that drive cellular activities cease and the cells enter into a state of dormancy, which later can be revived to an active state on increasing the temperature [32, 33]. Mammalian cells have been reported to be capable of enduring extreme low temperatures without accumulating chromosomal damage over time [34]. However, the process of freezing itself has been known to inflict injury to the cellular organelles and its function especially during the transition phase of freezing and thawing between extreme temperatures [35, 36]. Slow cooling rates result in dehydration and an increase in solute concentration in the cytosol (called "toxicity" or "solution" effects) [37]. On the other hand, fast cooling rates of a cell cause intracellular ice formation due to insufficient loss of water [38]. Both of these conditions can be lethal to the cell [39–41]. The lethal effects of cryopreservation occur primarily during the transition temperature between −15 and −60 °C which a cell has to cross during the freezing and thawing process and the amount of injury will depend on the cooling rate [38]. In order to achieve an optimum cooling rate that is neither too slow (dehydration) nor too fast (intracellular ice formation), cryopreservatives like DMSO are currently being used. The cryopreservation stabilizes the cell membrane and also reduces the nucleation rate that determines formation of ice and its growth by increasing the viscosity of the intra- and extracellular solution [42]. The exact mechanism of how cryopreservatives function is still an area of disagreement.

Stem cells have been cryopreserved and used successfully for research purposes without significant issues of loss of stem-cell properties such as self-renewal and differentiation. For research studies in laboratories, it is necessary to have an ongoing supply of cells and often cells are experimented as an allograft transplant in animal models. Intracellular ice formation in stem cells reduces proliferation of stem cells and is known to damage cell membrane in suspension [43]. In clinical use for treating humans, the focus of acquiring and preparing stem cells needs to be emphasized more on maximizing the survival of stem cells without eliciting immune response and preventing long-term detrimental effects due to cryopreservant toxicity on stem cells rather than long-term storage and availability whenever stem cells are available, albeit the latter is of critical importance especially when autologous stem cells are difficult to procure in sufficient number because of age or disease.

Freshly Isolated Versus Culture-Expanded Cells

The effect of cryopreservation is not the same on all cell types. Some population of progenitor cells reacts more adversely to the stresses of the storage procedure than others. Placental/umbilical cord blood (PCB) which is a source of hematopoietic/mesenchymal stem cells and progenitor cells requires longer periods of cryopreservation compared to cells used in bone-marrow transplantation. It was noticed that although the overall recovery of cells after cryopreservation was on an average 82.8 %, CFU-Meg, a megakaryocytic progenitor cells which produce platelets, had a slower recovery rate compared to other population of progenitors. This was attributed as one of the reasons why thrombocytopenia was observed in patients who got PCB transplantation [44]. In an another study on cord blood cells by Lu et al., EPCs differentiated from cryopreserved samples also showed impaired differentiation and a higher incidence of apoptosis in cryopreserved cells compared to freshly isolated cells [45]. However, another study reported EPCs obtained from bone-marrow-derived EPC from cryopreserved mononuclear cells (MNCs) showed comparable proliferation and migration activity between fresh and cryopreserved EPCs, although their freezing protocol resulted in 46 % loss of MNCs [46].

Freshly isolated stem cells have been shown to have comparable beneficial effects as culture-expanded cells showing that cryopreservation is not a necessary step in the protocol for cell therapy [47, 48]. In a study to treat renal ischemia, freshly isolated adipose tissue-derived stem and regenerative cells (ADRCs) improved renal function by reducing tissue necrosis and inflammatory cytokines similar to culture-expanded cells [49]. A comparison of the results of the two groups in this study shows a better performance of freshly isolated cells at improving survival and decreasing serum creatinine levels post-ischemia than cultured cells [49]. Culture-expanded stem cells loose proliferative potential and more importantly pluripotency on repeated culture [50, 51]. However, to achieve sufficient number of cells for treatment, cells have to be passaged repeatedly, for example, a single injection to treat a myocardial infarct patient requires ~4 million mesenchymal stem cells [52]. It is, however, a debatable topic as to how many stem cells is sufficient to start repair and engraftment process in the tissue, if it will be a question of quantity or quality of cells that favors survival.

Some studies have reported the issue of culture-expanded cells becoming an obstruction in blood vessels during cell delivery because of cell-size enlargement. Attention to cell density and passage of stem cell while preparing samples prior to transplantation are also important factors that affect the quality of therapy [51, 53, 54]. Culture expansion of stem cells can result in a phenotype of cells that does not suit the method of delivery because of which all injected cells are not available for tissue repair. The size of freshly isolated ADRC cells is ~11 μM which is about the diameter of capillaries, while cultured cells are ~20 μM in size which obstruct the microvascular vessels during arterial infusions [49, 55]. The authors of the study also indicated that positive effect of freshly isolated ADRC could be attributed to their increased expression of VEGF-A and IGF observed in freshly isolated cells as observed in a similar study by Zhu et al. [49, 56].

A good number of reports have claimed in vitro and in vivo that cryopreserved stem cells and freshly isolated and expanded stem cells do not differ from each other in their morphological and functional characteristics [46, 57, 58]. In our study, we observed that cryopreserved rat MSCs in the immediate passage after being thawed failed to show ability to differentiate and express survival proteins in response to a hypoxic stimulus, however, in subsequent early passages (passage 3–4) they showed significant recovery [59]. In studies done on human teeth like dental pulp stem cells (DPSC) [60] and stem cells from apical papilla (SCAP) [61], it has been reported that these stem cells retain their multilineage differentiation and morphofunctional properties after cryopreservation. In SCAP although cryopreserved cells showed a slight reduction in cell viability, proliferation, and colony-forming efficiency compared to fresh cells, the two groups grew to become similar in rate of growth, surface molecules, and multidifferentiation capability [61]. In equine peripheral blood MSCs, a comparison of cryopreserved with fresh cells for up to eight passages showed no difference in growth and morphology, immunophenotype expression pattern, and plasticity of stem cells [62]. In an in vivo study, Bai et al. in their work with cultured and freshly isolated adipose tissue-derived stem cells to treat acute myocardial infarction found both groups gave similar beneficial effects [63]. They, however, observed that the freshly isolated cells had heterogenous cell-surface markers compared to the homogenous profile of cultured cells. Over a period of culture, stem cells become more of a homogenous population from being initially a heterogenous population. Although they lost some of the surface markers, the cultured stem cells have comparable morphofunctional characteristics as the freshly isolated cells. In an in vitro study comparing the influence of freezing and thawing on the characteristics and function of hMSCs, similar observation has also been made. The rate of proliferation and functional ability to aid in angiogenesis on being stimulated by VEGF was comparable between the two groups, cultured and frozen hMSCs, obtained from freshly isolated hMSCs [64]. The expression profile of the surface markers followed a similar distribution in the two groups except that freezing and thawing significantly reduced the expression of CD 73, CD 90, and CD 166, the significance of which is not fully understood [65]. This study did not compare the characteristics of the 22–23 days cultured and frozen MSCs used in this study to freshly isolated MSCs [65]. Most of the studies report preservation of stem-cell characteristics and function after cryopreservation compared to freshly isolated stem cells. However, safe cryopreservation of stem cells and its efficacy for use in cell therapy depends on several factors such as the temperature of freezing, rate of freezing, duration of freezing, the cryopreservant used, thawing, and removal of cryopreservant.

Temperature and Rate of Cooling

The method of cryopreservation involves cooling the cells down using the help of cryopreservatives [66] to prevent ice formation and then thawing and removing the cryopreservatives when cells are ready to be used. However, cells can encounter injury at different stages of the cryopreservation process [41]. Rate of cooling is critical in cell freezing to minimize cryoinjury because if the cooling rate is too slow than the rate of water efflux from inside, the cell results in dehydration of the cell. However, if the rate is too fast, then the temperature can drop before sufficient dehydration resulting in intracellular ice formation [41]. More specifically at around 4 °C when cells thaw from freezing, heat is liberated at this temperature and this heat can be detrimental to the cell [67]. Therefore, although cells can endure extreme low temperatures, improper freezing and thawing can afflict injury to the cell.

Stem cells have been frozen at a controlled [15] and uncontrolled rate [68]. Although the controlled rate freezing using programmed freezers is considered to be a standard, it is expensive, complex, and requires special resources and people to execute it. The uncontrolled method is more common and has been studied to show similar or comparable effects to controlled rate freezing. In the uncontrolled method the cells are cooled to 4 °C, after which it is further cooled at −80 °C following which it is placed in liquid nitrogen for long-term storage [67]. A newer controlled rate method called the bob-in-box, a low-cost, simple device which provides consistent and controlled freezing, has been introduced. The cryopreserved hematopoietic stem cells in comparison with other methods showed more consistent viability and CFU-GM (colony-forming unit granulocyte-monocyte) [69]. Clinical studies are in need of more reliable and cost-effective method to preserve cells at low temperature using controlled rate for freezing so that the transplanted stem cells infused in patients can be viable and functional cells. Also different stem cells show optimal preservation at different cooling rates. HSCs have been reported to prefer 1 °C/min while cord blood MNCs a cooling rate of 10 °C/min showed the highest recovery rate for cells compared to 5° and 1 °C/min [15].

DMSO Toxicity

The routine cryoprotective agent (CPA) currently used to prevent freezing injury to cells is 10 % DMSO along with serum or autologous plasma [70, 71]. The existing protocol has been adopted from previous study for the lack of consensus of a safe cryopreservant solution. Although DMSO was introduced as a safe CPA for stem cells [72], it has since then been reported to be toxic at non-cryogenic temperatures in several clinical studies [73]. DMSO-specific side effects such as nausea, vomiting, and abdominal cramps were observed in many patients who received cryopreserved circulating progenitor cells [74]. In several other studies acute injury to patient's vital organs was recorded including fatal cardiac arrhythmia [75], respiratory depression [76], neurotoxicity [77], etc., on receiving cryopreserved stem-cell infusions. Focuses are now being shifted to determine if alternate cryoprotectant could be used other than DMSO or combine it with a lower percentage of DMSO to reduce toxicity and cryoinjury. A concentration of DMSO lesser than the 10 % has been determined to be safe in some stem-cell transplantation studies [78, 79]. Other CPAs being experimented are ethylene glycol [80], propylene glycol [81], trehalose [82], dextran [83], etc. The earliest CPA used was glycerol [31] whose major advantage is that it is nontoxic to cells [84] even at high concentrations and can minimize protein denaturations occurring at low temperatures. Its disadvantage is that it penetrates cells slowly or in some cells fails to penetrate and hence does not serve as a good CPA [84]. Trehalose, a nonreducing disaccharide of glucose, has also been promoted as an effective CPA, on facilitating its impermeability across cell membrane by using porating agents [85, 86]. Trehalose is chemically stable within the cell and has cytoprotective interactions with the cellular proteins and lipid membrane by preventing the damages of dehydration and intracellular ice injury during the freeze-thaw process [87–89]. In contrast, although the more widely used DMSO penetrates cell membrane more rapidly, they are very toxic to cells. To reduce toxicity and enhance recovery, a study on stem cells from adipose tissue was used with DMSO in combination with trehalose, a nonreducing disaccharide of glucose. The solution containing 4 % DMSO + 6 % trehalose + 90 % FBS was determined the optimal cryopreservant for adipose-derived stem cells in terms of improving viability and differentiation [90]. In identifying the protective role of carbohydrates during freezing and the ability of DMSO to reduce intracellular ice formation, they concluded that the best combination for cryopreservation of these stem cells was 4 % DMSO + 6 % trehalose + 90 % FBS which also reduced side effects of DMSO and preserved biological function of the stem cells [90]. In an another study on hematopoietic stem cells obtained from cord blood, effective freezing has been reported using 10 % ethylene glycol + 2 % DMSO using controlled rate freezing method [80].

Cryopreservation is still a cost-effective method to ensure steady supply to meet unpredictable demands for stem cells. Controlled cryopreservation techniques could provide enhanced shelf life, lesser risk of contamination, and genetic modification compared to other tissue storage methods. However, freezing protocols do not function with same efficacy for all types of cells causing low of recovery and loss in function of transplanted cryopreserved cells. Freezing rate, duration of storage, and nontoxic cryopreservant for different stem-cell types have to be optimized to avoid failures in clinical studies on patients due to poor sample preparation.

Animal Serum-Free Culture and Growth Factors

Animal cell cultures are routinely cultivated under standard conditions in the laboratory using a defined basal medium, fetal bovine serum, L-glutamine, and antibiotics. The basal medium contains essential salts, sugars, vitamins, and amino acids required for cell growth. However, in addition, to attain physiological balance hormones, lipoproteins and other essential nutrients are needed for promoting healthy cell growth. Serum from animal source, primarily bovine, has been added traditionally to a defined basal medium along with antibiotics to prepare a crude physiologic mimic to culture stem cells in the laboratory [91, 92]. Stem-cell expansion in addition to the serum sometimes would be exposed to feeder layers and recombinant proteins to increase cell number production [93]. This culture medium is often not static because as the cells grow and metabolites secreted, the medium composition changes. It is this surrounding environment which will dictate the outcome of the cells cultured.

In the case of clinical studies, animal serum and xenoproteins added to culturing cells can be harmful to patients receiving cell therapy. Often for clinical use there is inadequate number of stem cells in the case of both autologous and allogenic stem cells which requires them to be expanded in culture before transplantation [94, 95]. In humans, the adult stem cells isolated for clinical use from sources such as the bone marrow, umbilical cord, and adipose tissue are supported in their niches by the connective stroma and parenchymal tissue [96, 97]. Signaling molecules such as cytokines and colony-stimulating factors (CSF) released through cell-cell interactions from these surrounding connective tissue and the stem cell's self-regulatory mechanism work together to balance between self-renewal and differentiation directing them to the needs of repair and regeneration in tissues [98]. Successful expansion of a subset of stem cells ex vivo will depend on how closely culture conditions can mimic physiological environment to maintain the purity of the expanded stem cells before transfusion/transplantation into tissues to meet its task of regeneration. Efforts are being taken to exclude the use of animal serum and xenoproteins to culture and expand them such that their stem-cell integrity is maintained before transplantation. The use of animal serum for clinical use is of concern for various reasons, primarily due to transmission of diseases, allergic reactions in patients, variability in serum production, etc. [92]. Further they can also affect the cellular functions of the human stem cells as they cover and surround them during culture.

Hematopoietic stem cells (HSCs) from cord blood are obtained in small numbers per unit of cord blood, which limits its use as an allogenic donor for transplantation, although they are easy to harvest and in addition cause lesser instances of graft-versus-host disease than other sources of HSCs [99]. In an attempt to improve expansion of HSCs in vitro, human bone-marrow stroma-derived heparin sulfate (HS), a component isolated from the extracellular matrix, was added along with a cocktail of selected cytokines [100]. Heparin sulfate, which has an important role in many physiological functions, improved the proliferation and maintenance of the committed HSCs in culture, unlike the usual expansion which promotes expansion of the committed progenitor cells but at the cost of the primitive CD34+, CD38+ cells [100]. In another study, purified HS derived from the human stroma was substituted for FGF-2 exogenously added to MSC culture and has been reported to promote its proliferation [101]. However, these studies were done using FBS in their medium.

Several studies are currently investigating the use of serum-free medium to grow stem cells along with defined growth factors to suit the specific cell type. In a study using human dental pulp cell, the use of bFGF in a serum-free medium produced cells that strongly expressed STRO-1, a marker of stem cell, and sustained its self-renewal property [102]. Human MSCs have been reported to possess a rare property of being capable of surviving in low cell densities in the absence of serum in a naïve state without proliferating or differentiation [103], although after being supported in the initial 2 passages by FCS. Using this to their advantage, Solemsky et al. found that retinoic acid (RA) in a serum-free medium inhibits ERK activated pathways in hMSCs, while EGF and bFGF promoted these pathways. MSCs were observed to lose their properties of migration, adhesion, proliferation, and differentiation by cell-signaling molecule RA, but EGF and bFGF promoted these properties of MSCs [104].

Another alternative to eliminate animal products in human cell culture was to examine the use of human serum. Both autologous [105] and allogenic mediums have been studied [106]. In a study isolating CD271+ selected MSC population, the cells were successfully grown in 10 % pooled allogenic human serum without loss of MSC phenotype or differentiation potential [106]. However, there have been reports contradicting the safe use of allogenic human serum to culture hMSCs [107]. Incidence of cell arrest and fewer attachments on using allogenic human serum has been observed indicating the difference in composition of the pooled allogenic serum [108]. Autologous serum on the other hand has been reported to increase cell proliferation and maintains the stemness of MSCs [109, 110]. On comparing the genetic and epigenetics on the use of autologous serum to FBS on hMSCs, it was observed that although in both serum there is a possibility for localized genetic alterations, the autologous serum showed more consistent genetic background thereby promoting autologous serum as a viable and better alternative to FBS [111]. However, autologous serum is limited by the paucity in its availability for a large-scale expansion for clinical use. A study comparing serum from whole blood, serum from platelet-rich plasma, and serum from platelet-poor plasma with FBS reported that the serum preparation had differential effect on different types of cells. Fibroblasts proliferation was highest in serum from whole blood; although adipose-derived stem/stromal cells and vascular endothelial cells showed no difference in proliferation rate among the three human serum preparations. This could be a very promising method for cell-based therapies on optimizing conditions based on the cell type used for therapy [112]. Successful studies on hMSCs culture using human platelet lysate have also shown that MSCs preserve their phenotype as defined by the ISCT [113], yield high number of multipotent cells compared to FBS, and are also smaller in size on dividing, which might indicate that they still retain their immaturity [114]. In a study comparing 10 % FBS to pooled human serum, thrombin-activated platelet releasate in plasma (tPRP), and pooled human platelet lysate (pHPL), it was observed that pHPL promoted proliferation of bone-marrow-derived MSCs without altering its phenotype and restricting spontaneous differentiation, although for adipose-derived MSCs tPRP seemed to be better suited for culture. The human platelet lysate differs from FBS as it has a higher concentration of VEGF, bFGF, TGF-b, PDGF, and a different cytokine profile which together plays a role to promote human stem-cell characteristics [115]. Although platelet lysate is gaining a lot of acceptance as a replacement for FBS, further studies are required to optimize based on cell type and purpose of therapy as they show a dose-dependent activity and a low immune-suppressive activity.

Alongside of determining optimal medium for cell culture, efforts are underway to determine efficient monitoring system for bioactive molecules secreted in a dynamic culture system. NMR spectroscopy has been used in a technique called Fermentanomics to evaluate metabolite accumulation and nutrient consumption in culture systems [116]. These methods could further help understand and improve mechanisms that go behind the maintenance and regulation of stem-cell properties in vitro.

Oxygen Conditions

The understanding of several cellular mechanisms and their underlying physiological significance in healthy and disease state is currently based on in vitro studies that have been done under room air conditions. There is a growing consensus that the conclusions drawn from these studies are affected by the disparity in oxygen levels cells experience in the body compared to the level of exposure in laboratory conditions [117]. The physiological oxygen within the human body ranges from 2 to 9 %, although some regions in the bone marrow and kidney medulla could exist in as low as 1 % oxygen [118]. Stem-cell niches like the bone marrow and the developing embryo are marked by severe hypoxic microenvironment compared to atmospheric levels [119, 120]. Maintaining such low levels of oxygen in such regions of the body has been observed to be important for regulating and maintaining the stem-cell homeostasis [22]. However, when these stem cells are isolated for culture, they adapt to their surrounding oxygen environment. Oxygen uptake in culture has been shown to have a negative linear association with cell density. This uptake is estimated to be around 0.03–0.4 nmol/sec/10^6 cells [121]. With stem cells, oxygen concentration chosen for cell culture might have a greater effect than mature somatic cells because of their sensitivity to the surrounding environment which can direct its course to retain its stemness or to enter a differentiated state.

Critical Role of Oxygen in Stem-Cell Functionality

Oxygen plays an influential role in the dynamic state of stem cells from quiescence, proliferation, and differentiation. Stem cells in the bone marrow are mostly found in the G0 phase of the cell cycle maintaining a quiescence state until prompted by external stimuli [122]. Hematopoietic stem cells (HSCs), a population of stem cell found in the hypoxic environment of the bone marrow, maintain quiescence through the stabilization of Hif-1α in response to intracellular hypoxia. The lower oxygen availability preserves the cell under less oxidative stress and lower metabolic state which slows down differentiation of these cells [21]. Hypoxia reduces ROS production in HSCs and this population of HSCs with low ROS shows better self-renewal ability than population of HSCs with high ROS. HSC population that had higher ROS was observed to be depleted of self-renewing HSCs through an increased activation of p-38 mitogen-activated protein kinase (MAPK) and mammalian target of rapamycin (mTOR) [123].

It is interesting to note that stem cells in culture can adapt to hypoxia (physiologic normoxia) and hyperoxia (culture normoxia) when they are sufficiently provided with nutrients to grow. It is, however, the imbalance in the cell machinery that tightly regulates ROS and the surrounding availability of molecular oxygen that leads to excessive and uncontrolled intracellular ROS leading to undesired outcome [124]. In adipose-derived stem cells (ASC), it was observed that these cells showed decreased proliferation in serum-free media under 2 % hypoxia, but when supplemented with serum, proliferation of ASCs significantly increased. The same study also showed that acute hypoxia generated ROS which at a controlled rate acts as secondary messengers promoting proliferation and regeneration in adipose-derived stem cells (ASCs) through the activation of PDGFR-β, followed by phosphorylation of Akt and ERK1/2 signal pathways [125]. We have observed similar results with stem cells in "culture normoxia" but "physiological hyperoxia." Rat MSCs were grown in 20 % oxygen in vitro and they exhibited healthy proliferation and differentiation capability. A comparative study using 0.5 % oxygen hypoxic preconditioning also showed healthy proliferation and differentiation in vitro. However, when these cells, the "physiological hyperoxic" and the "hypoxic preconditioned" cells, were introduced to very low oxygen (~0.1 %), a larger proportion of the hyperoxic-cultured cells failed to survive [59].

The cellular response to different hypoxic levels is not uniform or predictable, especially with stem cells. They vary with stem-cell types, age of the donor, and other factors such as its passages and its environment. To begin with Hif-1α, the primary response transcription factor shows exponentially varying levels at low oxygen levels in cells. A study on He-la cells showed that Hif-1α-DNA binding and its protein accumulation have a negative linear association with oxygen concentration, a maximum response observed at 0.5 % oxygen [126]. It is no surprise that stem cells which are sensitive to their environment show different response when exposed to slightly varying hypoxic oxygen tensions. For example, in a study on hypoxic preconditioning of MSCs, it was observed that even a small change in low oxygen levels influenced the differentiation course of MSCs. The differentiation of human MSCs to the osteogenic lineage was promoted at 3 % oxygen concentration; however, it was inhibited at 1 % oxygen [127]. These results are supported by another study on human MSCs in which on exposing them to ≤1 % oxygen it affected their bone-forming potential but on the contrary promoted VEGF expression in the mRNA and protein level by twofold, which aids in repair and the process of angiogenesis [128]. Lower passages of MSCs and young donors have been observed to deliver favorable response to hypoxic stress by activating the Hif-1α machinery which can be taken advantage of to enhance angiogenesis during ischemic tissue therapy than their higher passage and older donor counterparts [129]. Sometimes even after successful in vitro preconditioning of stem cells, an external stimuli can attenuate the therapeutic response of stem cells. Superoxide induced by glucose at high concentrations in culture was able to downregulate VEGF expressed by mouse MSCs grown in hypoxic culture [130]. This has been speculated as a reason why hypoxic preconditioned stem cells showed a lack of functional improvement in diabetic patients. Recently, a study by Abachi et al. have shown that different types of stem cells show differing cellular responses to different levels of oxygen. The study done at 1, 5, and 20 % oxygen concluded that although all the stem-cell types showed an ability to adapt to hypoxia, their cellular responses can be modulated by monitoring and controlling the oxygen environment [121]. Neuronal differentiation has been reported to favor 2 % oxygen concentration compared to 20, 10, 4, 3, 2, 1, and 0 % oxygen conditions, although these neural stem cells (NSCs) compared to 20 and 0 % showed higher proliferation in 10, 4, 3, 2, and 1 % oxygen conditions [131]. Human cardiac-derived stem cells showed better yield with lesser chromosomal aberration due to lesser oxidative stress on culturing these cells at 5 % oxygen than 20 %, and they showed improved engraftment and function in a mouse model [132]. The trend observed in most hypoxic studies has been that very low levels of physiological hypoxia cause the accumulation of Hif-1α and promote self-renewal and quiescence, but mild hypoxia starts building cell metabolism, causing intracellular degradation of HIF-1α leading to activation of Hif-1α-dependent pathway that promotes proliferation and differentiation. In the study on hNSCs, at 1 % oxygen the cells expressed highest levels of nestin, a primitive stem-cell marker, and promoted quiescence. However, mild hypoxia (2–5 %) enhanced NSC proliferation and differentiation [133]. A similar effect was observed in hMSCs, which showed reduced proliferation in low oxygen conditions compared to 20 % oxygen; however, at very low oxygen (1 %), they showed poor osteogenic differentiation capacity which was restored on increasing the oxygen levels to 3 % [127]. To comprehend the underlying mechanisms and to be able to fine-tune the switches that promote self-renewal, proliferation, and differentiation, each stem-cell type has to be studied in varied environments and culture conditions that simulate physiological niches.

Strategies Used for Hypoxic Culturing

Recognizing the need for a physiological match of oxygen concentration for stem cells cultured for therapeutic purpose, attempts are being made towards developing and optimizing hypoxic preconditioning strategies, especially for treating ischemic tissue which is marked by severe hypoxia. Several studies have reported the use of preconditioned stem cells by either short periods of hypoxia [48, 134, 135] or anoxia [136, 137] for treating ischemic tissue and have reported successful engraftment and significant functional improvement in the function of the damaged tissue. The following are just few examples of the variations in hypoxic preconditioning efforts of stem cell for the purpose of ischemic tissue repair. Impaired adipose-derived stem cells (ADSCs) obtained from aged mouse on hypoxic preconditioning for 48 h at 1 % oxygen showed increased expression of proangiogenic factors [134]. Similarly NSCs on hypoxic preconditioning for 3 h at 0.5 % oxygen showed increased expression of connexin-43 which helped in enhanced engraftment and host communication [135]. MSCs on preconditioning for 24 h at 0.5 % oxygen showed enhanced angiogenesis and survival in an MI rat model [59]. However, in another study, mouse MSCs on preconditioning using 3 % oxygen showed enhanced levels of chemokines such as SDF-1 and CXCR-4 and improved adhesion, migration and survival of MSCs [138]. Although all of these studies agree that hypoxic culturing enhance the survival and functional property of stem cell, the protocol for preconditioning does not match among the different studies. This makes it difficult to compare the results of hypoxic preconditioning among the various studies on preconditioning stem cells.

Oxygen Requirements to Prepare Stem Cells for Clinical Therapy

Stem cell holds potential to cure a range of degenerative diseases such as bone-related disorders, ischemia, cancer, metabolic disorders, immune system disorders, etc. With respect to oxygen requirements, to prepare stem cells for clinical therapy, a major shift is needed in prioritizing research focus to design workstation that can monitor and culture stem cells under physiological normoxic environment. Currently, in vitro and in vivo hypoxic culturing and preconditioning face several practical challenges to execute cell culture in a physiological normoxic environment. One of the major challenges is to obtain equilibrium of oxygen concentration in the gas and liquid medium of the culture [139, 140]. Although hypoxic chambers have been designed to achieve equilibrium with low oxygen inside them in minutes, it takes more than 3 h for the oxygen to diffuse through the liquid. The challenge increases when cells treated in "hypoxic" environment are removed from the chamber for change of medium, or for doing a treatment or conducting assays to study cellular activities [141]. The cells are reperfused with excess oxygen which could cause a lot of cellular instability as the half-life of Hif-1α accumulated in hypoxic cells is around 5 min [142]. To overcome spatial and temporal heterogeneity of oxygen in hypoxic cultures, microfluidic technology is currently being exploited to construct devices that can tightly control oxygen distribution across the gradients [143]. One such proposed is the Boyden chamber, which is an insert within an insert, which claims to equilibrate 1 % oxygen within the chamber in 20 min. Gases are infused into the chamber, and they freely diffuse through the 100 μM gas-permeable PDMS membrane allowing the possibility of studying cellular processes like migration and cell invasion in either a steady state or intermittent hypoxic environment [144]. Finally there exists a bigger challenge to produce hypoxic-cultured stem cells at a clinical scale and deliver them for therapy with minimal damage from reperfusion injuries. Controlled bioreactors [145] are being researched to achieve this goal such that healthy functional stem cells can be cultured in physiologically relevant oxygen and pH environments.

Conclusion

In conclusion, much has been done towards changing the existing cell-culture protocols to suit stem cells and to promote efficacious clinical translation. Cryopreservation is still one of the best known methods to preserve cells; however, different stem cells show varying response to this method of storage. Although mammalian cells are capable of enduring extreme low temperatures, they can suffer injury in the process of transition between temperatures if factors such as cooling rate, duration of freezing, and toxic levels of cryopreservants used for storage are not optimized according to the stem-cell type. Both following isolation and thawing of frozen stem cells, the cell expansion process involves supplying nutrients for cell growth which is now being substituted from animal-derived products to human derivatives to ensure safety and compatibility for clinical use. However, there is still the challenge of not only standardizing the composition of nutrients but also the scarcity of human derivatives for the purpose of cell growth, along with the limitation of obtaining quality stem cells because of age of the donor and restricted use stem-cell passages. Finally, studies show a close to unanimous consensus on improvement of cell therapy on hypoxic preconditioning of adult stem cells to treat diseased tissue, although it still remains probably the biggest challenge to meet physiological standards. Overall, although much has been done to move forwards to improve stem-cell preparations in vitro, much still requires to be done to maximize the potential of stem-cell therapy.

References

1. Safari M (2010) Stem cells therapy for retinal degeneration. Pak J Biol Sci 13(21):1016–1022
2. Bendtsen M et al (2011) Autologous stem cell therapy maintains vertebral blood flow and contrast diffusion through the endplate in experimental intervertebral disc degeneration. Spine (Phila Pa 1976) 36(6):E373–E379
3. Ali H, Bahbahani H (2010) Umbilical cord blood stem cells – potential therapeutic tool for neural injuries and disorders. Acta Neurobiol Exp (Wars) 70(3):316–324
4. Markert CD et al (2009) Mesenchymal stem cells: emerging therapy for Duchenne muscular dystrophy. PM R 1(6):547–559
5. Tedesco FS et al (2010) Repairing skeletal muscle: regenerative potential of skeletal muscle stem cells. J Clin Invest 120(1):11–19
6. Bonaros N et al (2008) Enhanced cell therapy for ischemic heart disease. Transplantation 86(9): 1151–1160
7. Galle J et al (2010) Mesenchymal stem cells in cartilage repair: state of the art and methods to monitor cell growth, differentiation and cartilage regeneration. Curr Med Chem 17(21): 2274–2291
8. Yun S et al (2011) Oxygen tension regulates NK cells differentiation from hematopoietic stem cells in vitro. Immunol Lett 137(1–2):70–77
9. D'Ippolito G et al (2006) Low oxygen tension inhibits osteogenic differentiation and enhances stemness of human MIAMI cells. Bone 39(3):513–522
10. Csete M (2005) Oxygen in the cultivation of stem cells. Ann N Y Acad Sci 1049:1–8
11. Sotiropoulou PA et al (2006) Characterization of the optimal culture conditions for clinical scale production of human mesenchymal stem cells. Stem Cells 24(2):462–471

12. Szymczak P et al (2010) Effect of matrix metalloproteinases inhibition on the proliferation and differentiation of HUCB-NSCs cultured in the presence of adhesive substrates. Acta Neurobiol Exp (Wars) 70(4):325–336
13. Tottey S et al (2011) Extracellular matrix degradation products and low-oxygen conditions enhance the regenerative potential of perivascular stem cells. Tissue Eng Part A 17(1–2):37–44
14. Reing JE et al (2009) Degradation products of extracellular matrix affect cell migration and proliferation. Tissue Eng Part A 15(3):605–614
15. Ketheesan N et al (2004) Effect of cryopreservation on the immunogenicity of umbilical cord blood cells. Transfus Apher Sci 30(1):47–54
16. Klebe RJ, Mancuso MG (1983) Identification of new cryoprotective agents for cultured mammalian cells. In Vitro 19(3 Pt 1):167–170
17. Fahy GM (1986) The relevance of cryoprotectant "toxicity" to cryobiology. Cryobiology 23(1):1–13
18. Seki S, Mazur P (2008) Effect of warming rate on the survival of vitrified mouse oocytes and on the recrystallization of intracellular ice. Biol Reprod 79(4):727–737
19. Armitage WJ (1987) Cryopreservation of animal cells. Symp Soc Exp Biol 41:379–393
20. Acker H (2005) The oxygen sensing signal cascade under the influence of reactive oxygen species. Philos Trans R Soc Lond B Biol Sci 360(1464): 2201–2210
21. Parmar K et al (2007) Distribution of hematopoietic stem cells in the bone marrow according to regional hypoxia. Proc Natl Acad Sci U S A 104(13): 5431–5436
22. Mohyeldin A, Garzon-Muvdi T, Quinones-Hinojosa A (2010) Oxygen in stem cell biology: a critical component of the stem cell niche. Cell Stem Cell 7(2):150–161
23. Nygren JM et al (2004) Bone marrow-derived hematopoietic cells generate cardiomyocytes at a low frequency through cell fusion, but not transdifferentiation. Nat Med 10(5):494–501
24. Toma C et al (2002) Human mesenchymal stem cells differentiate to a cardiomyocyte phenotype in the adult murine heart. Circulation 105(1):93–98
25. Buchstaller J et al (2004) Efficient isolation and gene expression profiling of small numbers of neural crest stem cells and developing Schwann cells. J Neurosci 24(10):2357–2365
26. Eibes G et al (2010) Maximizing the ex vivo expansion of human mesenchymal stem cells using a microcarrier-based stirred culture system. J Biotechnol 146(4):194–197
27. Przepiorka D et al (2001) Chronic graft-versus-host disease after allogeneic blood stem cell transplantation. Blood 98(6):1695–1700
28. Stenderup K et al (2003) Aging is associated with decreased maximal life span and accelerated senescence of bone marrow stromal cells. Bone 33(6):919–926
29. Sibal L et al (2009) Circulating endothelial progenitor cells, endothelial function, carotid intima-media thickness and circulating markers of endothelial dysfunction in people with type 1 diabetes without macrovascular disease or microalbuminuria. Diabetologia 52(8):1464–1473
30. Rodriguez JP et al (2000) Mesenchymal stem cells from osteoporotic patients produce a type I collagen-deficient extracellular matrix favoring adipogenic differentiation. J Cell Biochem 79(4): 557–565
31. Polge C, Smith AU, Parkes AS (1949) Revival of spermatozoa after vitrification and dehydration at low temperatures. Nature 164(4172):666
32. McGee HA Jr, Martin WJ (1962) Cryochemistry. Cryogenics 2:257–267
33. Mazur P (1984) Freezing of living cells: mechanisms and implications. Am J Physiol 247(3 Pt 1):C125–C142
34. Ashwood-Smith MJ, Friedmann GB (1979) Lethal and chromosomal effects of freezing, thawing, storage time, and x-irradiation on mammalian cells preserved at -196 degrees in dimethyl sulfoxide. Cryobiology 16(2):132–140
35. Farrant J (1977) Water transport and cell survival in cryobiological procedures. Philos Trans R Soc Lond B Biol Sci 278(959):191–205
36. Rapatz G, Nath J, Luyet B (1963) Electron microscope study of erythrocytes in rapidly frozen mammalian blood. Biodynamica 9:83–94
37. Mazur P, Leibo SP, Chu EH (1972) A two-factor hypothesis of freezing injury. Evidence from Chinese hamster tissue-culture cells. Exp Cell Res 71(2):345–355
38. Mazur P (1963) Kinetics of water loss from cells at Subzero temperatures and the likelihood of intracellular freezing. J Gen Physiol 47: 347–369
39. Meryman HT (1970) Absence of unfrozen freezable water in rapidly frozen red cells. Cryobiology 7(4):252–255
40. Acker JP, McGann LE (2000) Cell-cell contact affects membrane integrity after intracellular freezing. Cryobiology 40(1):54–63
41. Meryman HT (1974) Freezing injury and its prevention in living cells. Annu Rev Biophys Bioeng 3:341–363
42. Karlsson JM, Cravalho CE, Toner M (1994) A model of diffusion-limited ice growth inside biological cell during freezing. J Appl Phys 75: 4442–4452
43. Zhurova M, Woods EJ, Acker JP (2010) Intracellular ice formation in confluent monolayers of human dental stem cells and membrane damage. Cryobiology 61(1):133–141
44. Xu Y, Kashiwakura I, Takahashi TA (2004) High sensitivity of megakaryocytic progenitor cells contained in placental/umbilical cord blood to the stresses during cryopreservation. Bone Marrow Transplant 34(6):537–543

45. Lu X, Proctor SJ, Dickinson AM (2008) The effect of cryopreservation on umbilical cord blood endothelial progenitor cell differentiation. Cell Transplant 17(12):1423–1428
46. Mieno S et al (2008) Characteristics and function of cryopreserved bone marrow-derived endothelial progenitor cells. Ann Thorac Surg 85(4):1361–1366
47. Schenke-Layland K et al (2009) Adipose tissue-derived cells improve cardiac function following myocardial infarction. J Surg Res 153(2):217–223
48. Alt E et al (2010) Effect of freshly isolated autologous tissue resident stromal cells on cardiac function and perfusion following acute myocardial infarction. Int J Cardiol 144(1):26–35
49. Feng Z et al (2010) Fresh and cryopreserved, uncultured adipose tissue-derived stem and regenerative cells ameliorate ischemia-reperfusion-induced acute kidney injury. Nephrol Dial Transplant 25(12):3874–3884
50. Bruder SP, Jaiswal N, Haynesworth SE (1997) Growth kinetics, self-renewal, and the osteogenic potential of purified human mesenchymal stem cells during extensive subcultivation and following cryopreservation. J Cell Biochem 64(2):278–294
51. Colter DC et al (2000) Rapid expansion of recycling stem cells in cultures of plastic-adherent cells from human bone marrow. Proc Natl Acad Sci U S A 97(7):3213–3218
52. Segers VF, Lee RT (2008) Stem-cell therapy for cardiac disease. Nature 451(7181):937–942
53. Sekiya I et al (2002) Expansion of human adult stem cells from bone marrow stroma: conditions that maximize the yields of early progenitors and evaluate their quality. Stem Cells 20(6):530–541
54. Prockop DJ, Sekiya I, Colter DC (2001) Isolation and characterization of rapidly self-renewing stem cells from cultures of human marrow stromal cells. Cytotherapy 3(5):393–396
55. Moelker AD et al (2007) Intracoronary delivery of umbilical cord blood derived unrestricted somatic stem cells is not suitable to improve LV function after myocardial infarction in swine. J Mol Cell Cardiol 42(4):735–745
56. Zhu M et al (2010) Supplementation of fat grafts with adipose-derived regenerative cells improves long-term graft retention. Ann Plast Surg 64(2):222–228
57. Zhang HT et al (2011) Neural stem cells differentiation ability of human umbilical cord mesenchymal stromal cells is not altered by cryopreservation. Neurosci Lett 487(1):118–122
58. Liu G et al (2008) Evaluation of the viability and osteogenic differentiation of cryopreserved human adipose-derived stem cells. Cryobiology 57(1): 18–24
59. Chacko SM et al (2010) Hypoxic preconditioning induces the expression of prosurvival and proangiogenic markers in mesenchymal stem cells. Am J Physiol Cell Physiol 299(6):C1562–C1570
60. Lee SY et al (2010) Effects of cryopreservation of intact teeth on the isolated dental pulp stem cells. J Endod 36(8):1336–1340
61. Ding G et al (2010) Effect of cryopreservation on biological and immunological properties of stem cells from apical papilla. J Cell Physiol 223(2): 415–422
62. Martinello T et al (2010) Cryopreservation does not affect the stem characteristics of multipotent cells isolated from equine peripheral blood. Tissue Eng Part C Methods 16(4):771–781
63. Bai X et al (2010) Both cultured and freshly isolated adipose tissue-derived stem cells enhance cardiac function after acute myocardial infarction. Eur Heart J 31(4):489–501
64. Lagar'kova MA et al (2006) Characteristics of human bone marrow mesenchymal stem cells isolated by immunomagnetic selection. Bull Exp Biol Med 141(1):112–116
65. Haack-Sorensen M et al (2007) The influence of freezing and storage on the characteristics and functions of human mesenchymal stromal cells isolated for clinical use. Cytotherapy 9(4):328–337
66. Lovelock JE (1954) The protective action of neutral solutes against haemolysis by freezing and thawing. Biochem J 56(2):265–270
67. Berz D et al (2007) Cryopreservation of hematopoietic stem cells. Am J Hematol 82(6):463–472
68. Halle P et al (2001) Uncontrolled-rate freezing and storage at -80 degrees C, with only 3.5-percent DMSO in cryoprotective solution for 109 autologous peripheral blood progenitor cell transplantations. Transfusion 41(5):667–673
69. Zhou X et al (2009) Cryopreservation of peripheral blood stem cells using a box-in-box cooling device. Biopreserv Biobank 7(2):107–114
70. Lovelock JE, Bishop MW (1959) Prevention of freezing damage to living cells by dimethyl sulphoxide. Nature 183(4672):1394–1395
71. Katayama Y et al (1997) The effects of a simplified method for cryopreservation and thawing procedures on peripheral blood stem cells. Bone Marrow Transplant 19(3):283–287
72. Rowley SD, Anderson GL (1993) Effect of DMSO exposure without cryopreservation on hematopoietic progenitor cells. Bone Marrow Transplant 11(5):389–393
73. Donmez A et al (2007) Clinical side effects during peripheral blood progenitor cell infusion. Transfus Apher Sci 36(1):95–101
74. Zambelli A et al (1998) Clinical toxicity of cryopreserved circulating progenitor cells infusion. Anticancer Res 18(6B):4705–4708
75. Zenhausern R et al (2000) Fatal cardiac arrhythmia after infusion of dimethyl sulfoxide-cryopreserved hematopoietic stem cells in a patient with severe primary cardiac amyloidosis and end-stage renal failure. Ann Hematol 79(9): 523–526

76. Caselli D et al (2009) Respiratory depression and somnolence in children receiving dimethylsulfoxide and morphine during hematopoietic stem cells transplantation. Haematologica 94(1): 152–153
77. Mueller LP et al (2007) Neurotoxicity upon infusion of dimethylsulfoxide-cryopreserved peripheral blood stem cells in patients with and without pre-existing cerebral disease. Eur J Haematol 78(6):527–531
78. Tijssen MR et al (2008) Improved postthaw viability and in vitro functionality of peripheral blood hematopoietic progenitor cells after cryopreservation with a theoretically optimized freezing curve. Transfusion 48(5):893–901
79. Liu Y et al (2010) Cryopreservation of human bone marrow-derived mesenchymal stem cells with reduced dimethylsulfoxide and well-defined freezing solutions. Biotechnol Prog 26(6):1635–1643
80. Son JH et al (2010) Optimization of cryopreservation condition for hematopoietic stem cells from umbilical cord blood. Cryobiology 60(3): 287–292
81. Arnaud FG, Pegg DE (1990) Cryopreservation of human platelets with propane-1,2-diol. Cryobiology 27(2):130–136
82. Sasnoor LM, Kale VP, Limaye LS (2003) Supplementation of conventional freezing medium with a combination of catalase and trehalose results in better protection of surface molecules and functionality of hematopoietic cells. J Hematother Stem Cell Res 12(5):553–564
83. Tsang KS et al (2001) Dextran sedimentation in a semi-closed system for the clinical banking of umbilical cord blood. Transfusion 41(3):344–352
84. Meryman HT (2007) Cryopreservation of living cells: principles and practice. Transfusion 47(5):935–945
85. Buchanan SS, Pyatt DW, Carpenter JF (2010) Preservation of differentiation and clonogenic potential of human hematopoietic stem and progenitor cells during lyophilization and ambient storage. PLoS One 5(9):e12518
86. McGann LE (1978) Differing actions of penetrating and nonpenetrating cryoprotective agents. Cryobiology 15(4):382–390
87. Chen T et al (2001) Beneficial effect of intracellular trehalose on the membrane integrity of dried mammalian cells. Cryobiology 43(2):168–181
88. Hino A et al (1990) Trehalose levels and survival ratio of freeze-tolerant versus freeze-sensitive yeasts. Appl Environ Microbiol 56(5):1386–1391
89. Eroglu A et al (2000) Intracellular trehalose improves the survival of cryopreserved mammalian cells. Nat Biotechnol 18(2):163–167
90. De Rosa A et al (2009) A new method for cryopreserving adipose-derived stem cells: an attractive and suitable large-scale and long-term cell banking technology. Tissue Eng Part C Methods 15(4): 659–667
91. Waymouth C (ed) (1984) Preparation and use of serum-free culture media. In: Barnes DW, Sirbasku DA, Sato GH (eds) Methods for preparation of media, supplements, and substrata for serum-free animal cell culture. Liss, New York, pp 23–68
92. Sandstrom CE, Miller WM, Papoutsakis ET (1994) Serum-free media for cultures of primitive and mature hematopoietic cells. Biotechnol Bioeng 43(8):706–733
93. Magin AS et al (2009) Primary cells as feeder cells for coculture expansion of human hematopoietic stem cells from umbilical cord blood–a comparative study. Stem Cells Dev 18(1):173–186
94. Lange C et al (2007) Accelerated and safe expansion of human mesenchymal stromal cells in animal serum-free medium for transplantation and regenerative medicine. J Cell Physiol 213(1):18–26
95. Hu WS, Piret JM (1992) Mammalian cell culture processes. Curr Opin Biotechnol 3(2):110–114
96. Koller MR, Manchel I, Palsson BO (1997) Importance of parenchymal:stromal cell ratio for the ex vivo reconstitution of human hematopoiesis. Stem Cells 15(4):305–313
97. Naughton BA (1995) The importance of stromal cells. In: Bronzino JD (ed) The biomedical engineering handbook. CRC Press, Boca Raton, pp 1710–1727
98. Colvin GA, Quesenberry PJ, Dooner MS (2006) The stem cell continuum: a new model of stem cell regulation. Handb Exp Pharmacol 174: 169–183
99. Rocha V, Gluckman E (2009) Improving outcomes of cord blood transplantation: HLA matching, cell dose and other graft- and transplantation-related factors. Br J Haematol 147(2):262–274
100. Bramono DS et al (2011) The effect of human bone marrow stroma-derived heparan sulfate on the ex vivo expansion of human cord blood hematopoietic stem cells. Pharm Res 28(6):1385–1394
101. Dombrowski C et al (2009) Heparan sulfate mediates the proliferation and differentiation of rat mesenchymal stem cells. Stem Cells Dev 18(4): 661–670
102. Morito A et al (2009) Effects of basic fibroblast growth factor on the development of the stem cell properties of human dental pulp cells. Arch Histol Cytol 72(1):51–64
103. Pochampally RR et al (2004) Serum deprivation of human marrow stromal cells (hMSCs) selects for a subpopulation of early progenitor cells with enhanced expression of OCT-4 and other embryonic genes. Blood 103(5):1647–1652
104. Solmesky L et al (2010) Serum free cultured bone marrow mesenchymal stem cells as a platform to characterize the effects of specific molecules. PLoS One 5(9):e12689
105. Stute N et al (2004) Autologous serum for isolation and expansion of human mesenchymal stem cells for clinical use. Exp Hematol 32(12): 1212–1225

106. Poloni A et al (2009) Selection of CD271(+) cells and human AB serum allows a large expansion of mesenchymal stromal cells from human bone marrow. Cytotherapy 11(2):153–162
107. Shahdadfar A et al (2005) In vitro expansion of human mesenchymal stem cells: choice of serum is a determinant of cell proliferation, differentiation, gene expression, and transcriptome stability. Stem Cells 23(9):1357–1366
108. Spees JL et al (2004) Internalized antigens must be removed to prepare hypoimmunogenic mesenchymal stem cells for cell and gene therapy. Mol Ther 9(5):747–756
109. Nimura A et al (2008) Increased proliferation of human synovial mesenchymal stem cells with autologous human serum: comparisons with bone marrow mesenchymal stem cells and with fetal bovine serum. Arthritis Rheum 58(2):501–510
110. Perez-Ilzarbe M et al (2009) Comparison of ex vivo expansion culture conditions of mesenchymal stem cells for human cell therapy. Transfusion 49(9):1901–1910
111. Dahl JA et al (2008) Genetic and epigenetic instability of human bone marrow mesenchymal stem cells expanded in autologous serum or fetal bovine serum. Int J Dev Biol 52(8):1033–1042
112. Kurita M et al (2008) Differential effects of three preparations of human serum on expansion of various types of human cells. Plast Reconstr Surg 122(2):438–448
113. Dominici M et al (2006) Minimal criteria for defining multipotent mesenchymal stromal cells. The International Society for Cellular Therapy position statement. Cytotherapy 8(4):315–317
114. Haasters F et al (2009) Morphological and immunocytochemical characteristics indicate the yield of early progenitors and represent a quality control for human mesenchymal stem cell culturing. J Anat 214(5):759–767
115. Bieback K et al (2009) Human alternatives to fetal bovine serum for the expansion of mesenchymal stromal cells from bone marrow. Stem Cells 27(9):2331–2341
116. Bradley SA et al (2010) Fermentanomics: monitoring mammalian cell cultures with NMR spectroscopy. J Am Chem Soc 132(28):9531–9533
117. Carreau A et al (2011) Why is the partial oxygen pressure of human tissues a crucial parameter? Small molecules and hypoxia. J Cell Mol Med 15(6):1239–1253
118. Simon MC, Keith B (2008) The role of oxygen availability in embryonic development and stem cell function. Nat Rev Mol Cell Biol 9(4): 285–296
119. Harrison JS et al (2002) Oxygen saturation in the bone marrow of healthy volunteers. Blood 99(1):394
120. Kofoed H et al (1985) Bone marrow circulation after osteotomy. Blood flow, pO2, pCO2, and pressure studied in dogs. Acta Orthop Scand 56(5): 400–403
121. Abaci HE et al (2010) Adaptation to oxygen deprivation in cultures of human pluripotent stem cells, endothelial progenitor cells, and umbilical vein endothelial cells. Am J Physiol Cell Physiol 298(6):C1527–C1537
122. Hermitte F et al (2006) Very low O2 concentration (0.1 %) favors G0 return of dividing CD34+ cells. Stem Cells 24(1):65–73
123. Jang YY, Sharkis SJ (2007) A low level of reactive oxygen species selects for primitive hematopoietic stem cells that may reside in the low-oxygenic niche. Blood 110(8):3056–3063
124. Griendling KK, FitzGerald GA (2003) Oxidative stress and cardiovascular injury: part I: basic mechanisms and in vivo monitoring of ROS. Circulation 108(16):1912–1916
125. Kim JH et al (2011) The pivotal role of reactive oxygen species generation in the hypoxia-induced stimulation of adipose-derived stem cells. Stem Cells Dev 20(10):1753–1761
126. Jiang BH et al (1996) Hypoxia-inducible factor 1 levels vary exponentially over a physiologically relevant range of O2 tension. Am J Physiol 271(4 Pt 1):C1172–C1180
127. Holzwarth C et al (2010) Low physiologic oxygen tensions reduce proliferation and differentiation of human multipotent mesenchymal stromal cells. BMC Cell Biol 11:11
128. Potier E et al (2007) Hypoxia affects mesenchymal stromal cell osteogenic differentiation and angiogenic factor expression. Bone 40(4): 1078–1087
129. Jiang S et al (2008) Transcriptional profiling of young and old mesenchymal stem cells in response to oxygen deprivation and reparability of the infarcted myocardium. J Mol Cell Cardiol 44(3): 582–596
130. Ishizuka T, Hinata T, Watanabe Y (2010) Superoxide induced by a high-glucose concentration attenuates production of angiogenic growth factors in hypoxic mouse mesenchymal stem cells. J Endocrinol 208(2):147–159
131. Horie N et al (2008) Effects of oxygen concentration on the proliferation and differentiation of mouse neural stem cells in vitro. Cell Mol Neurobiol 28(6):833–845
132. Li TS et al (2010) Expansion of human cardiac stem cells in physiological oxygen improves cell production efficiency and potency for myocardial repair. Cardiovasc Res 89(1):157–165
133. Santilli G et al (2010) Mild hypoxia enhances proliferation and multipotency of human neural stem cells. PLoS One 5(1):e8575
134. Efimenko A et al (2011) Angiogenic properties of aged adipose derived mesenchymal stem cells after hypoxic conditioning. J Transl Med 9:10
135. Jaderstad J, Brismar H, Herlenius E (2010) Hypoxic preconditioning increases gap-junctional graft and host communication. Neuroreport 21(17): 1126–1132

136. He A et al (2009) The antiapoptotic effect of mesenchymal stem cell transplantation on ischemic myocardium is enhanced by anoxic preconditioning. Can J Cardiol 25(6):353–358
137. Wang JA et al (2009) Anoxic preconditioning: a way to enhance the cardioprotection of mesenchymal stem cells. Int J Cardiol 133(3):410–412
138. Liu H et al (2010) Hypoxic preconditioning advances CXCR4 and CXCR7 expression by activating HIF-1 alpha in MSCs. Biochem Biophys Res Commun 401(4):509–515
139. Pettersen EO et al (2005) Pericellular oxygen depletion during ordinary tissue culturing, measured with oxygen microsensors. Cell Prolif 38(4):257–267
140. Baumgardner JE, Otto CM (2003) In vitro intermittent hypoxia: challenges for creating hypoxia in cell culture. Respir Physiol Neurobiol 136(2–3):131–139
141. Allen CB, Schneider BK, White CW (2001) Limitations to oxygen diffusion and equilibration in in vitro cell exposure systems in hyperoxia and hypoxia. Am J Physiol Lung Cell Mol Physiol 281(4):L1021–L1027
142. Moroz E et al (2009) Real-time imaging of HIF-1 alpha stabilization and degradation. PLoS One 4(4):e5077
143. Oppegard SC et al (2009) Modulating temporal and spatial oxygenation over adherent cellular cultures. PLoS One 4(9):e6891
144. Oppegard SC et al (2010) Precise control over the oxygen conditions within the Boyden chamber using a microfabricated insert. Lab Chip 10(18):2366–2373
145. Sud D et al (2006) Optical imaging in microfluidic bioreactors enables oxygen monitoring for continuous cell culture. J Biomed Opt 11(5):050504

Bioinformatics Toolbox for Stem Cell Research and Therapeutics

Himadri Singh

Introduction

Stem cell therapy can be used for treatment of neurodegenerative diseases, spinal cord injury, diabetes (type 1 and type 2), and cardiovascular diseases [1]. Molecular mechanisms of regeneration and differentiation can be elucidated by integrating high-throughput "omics" studies at multiple levels of regulation (genetic, epigenetic, interactome, RNA, metabolomic and proteomic levels) [2]. Data-driven biology of these "omics" technologies has ushered a new bioinformatics discipline wherein information technology is employed for creating tools and databases focused on stem cell biology for managing and interpreting biological high-throughput data [3]. Databases focused on stem cell have an edge over general databases since they provide highly relevant information on stem cell specific processes [4]. These databases and tools have helped us address and interpret molecular mechanism(s) associated with stem cell renewal and differentiation, and hold promise(s) to solve current issues associated with stem cell therapy and transplantation [4]. This chapter outlines databases and online tools focused on stem cells, useful for biomedical researchers/clinicians/scientists in the field of stem cell biology and regenerative medicine. The databases are listed accordingly along with their websites (Table 1).

H. Singh (✉)
Department of Stem Cell Research, National Institute of Nutrition (Indian Council of Medical Research), Jamai Osmania, Tarnaka, Secunderabad, Hyderabad, India
e-mail: himadrisrmc@gmail.com

Transcriptomics-Based Databases and Tools

Molecular mechanism of pluripotency and stem cell differentiation can be better understood by establishing gene-expression patterns [5]. As compared to protein, transcript levels (RNA) at large scale is easier to detect; a large number of studies have included gene expression at whole genome levels. This led to generation of huge microarray datasets depicting physiological condition(s) of various stem cell types [5]. Many stem cell-specific databases/tools have been created using a variety of gene-expression data. These databases are able to extract and display the expression-related data for stem cell-specific physiological conditions. Databases/tools include StemCellDB, PluriTest, the Stem Cell Discovery Engine and StemBase.

StemCellDB: The Human Pluripotent Stem Cell Database

StemCellDB (NIH stem cell data management system) is a database created by gene-expression profiling of 21 hESCs and 8 hiPSC lines to

I. Somasundaram (ed.), *Stem Cell Therapy for Organ Failure*,
DOI 10.1007/978-81-322-2110-4_3, © Springer India 2014

Table 1 Stem Cell Resources; the database names, a brief description and the web address are indicated

Databases	Description	URL http://
StemCellDB	Expression level of different genes in hESCs and hiPSC	stemcelldb.nih.gov/
PluriTest	Assay of pluripotency based on gene expression	www.pluritest.org/
StemBase	Expression studies on stem cells	www.stembase.ca/
SCDE	Cancer stem cell profile	discovery.hsci.harvard.edu/
ESCAPE	Build networks based on multiple regulatory layers of stem cells	www.maayanlab.net/ESCAPE/
SyStemCell	Multiple-level experimental data of stem cells	lifecenter.sgst.cn/SyStemCell/
Stem Cell Omics Repository	Quantitative information at multiple levels of measurements	scor.chem.wisc.edu/

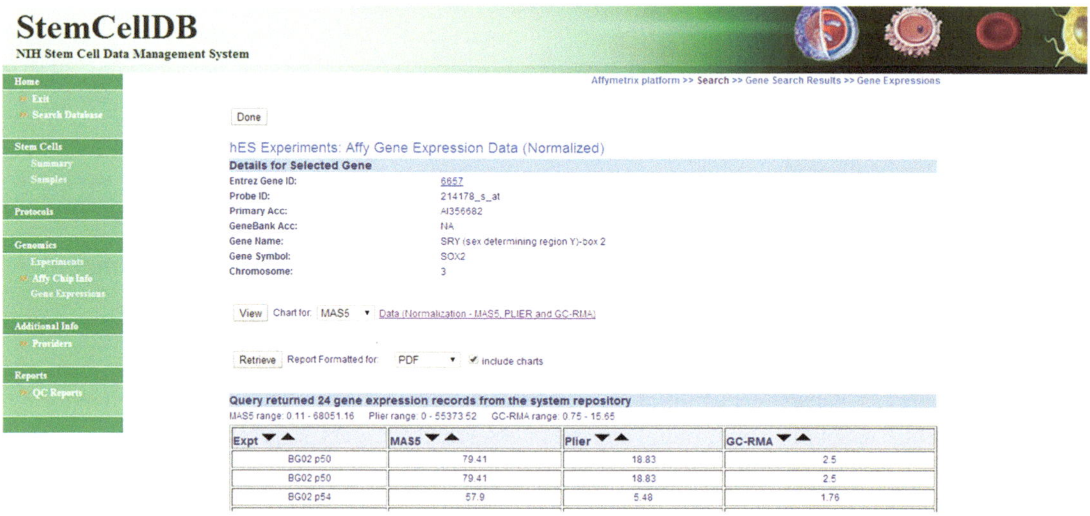

Fig. 1 Snapshot of StemCellDB database

generate a comprehensive snapshot of the undifferentiated state of pluripotent stem cells using Affymetrix microarray system. A gene can be queried for its expression in different studies using different normalization algorithm (MAS5, PLIER, GC-RMA) [6] (Fig. 1).

PluriTest

PluriTest is a robust bioinformatic functional assay of pluripotency in human cells based on their unique gene-expression profiles. Workflow for PluriTest involves (1) global gene-expression analysis of the sample using the Illumina HT12v3 and v4 platform, (2) the raw data (.dat file) is obtained and uploaded to the PluriTest tool, (3) the data is analyzed on the PluriTest server, and (4) the results are interpreted through two related parameters, Pluripotency Score and Novelty Score. Pluripotency Score and Novelty Score are derived from 285 peer-reviewed publications selected from PubMed [7] (Fig. 2).

StemBase

StemBase was initially developed as a collection of gene-expression levels in samples of stem cells (mouse and human) from the data generated within the framework of the Stem Cell Genomics Project (Canadian Stem Cell Network). The database is equipped with analysis tools. StemBase currently contains gene-expression

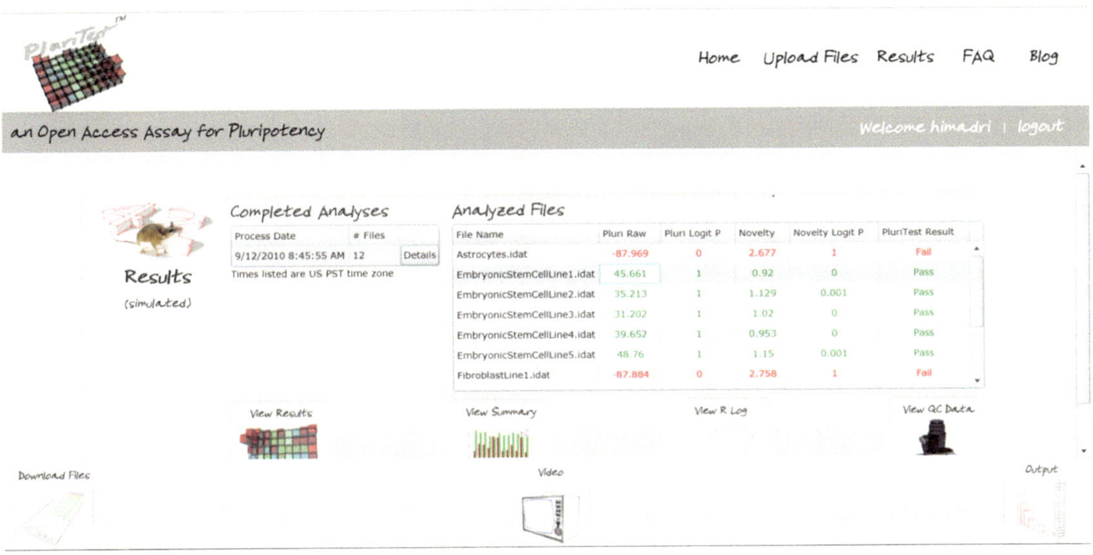

Fig. 2 Snapshot of PluriTest

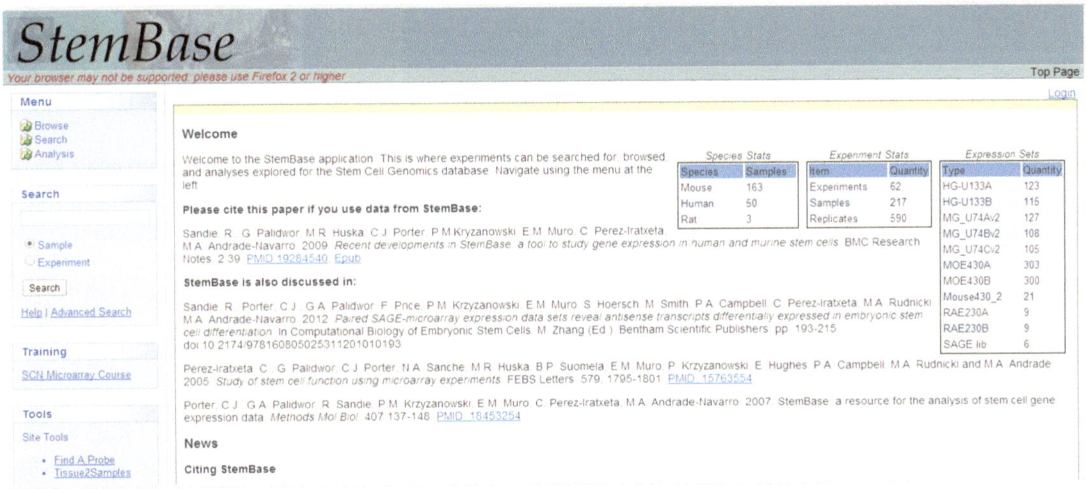

Fig. 3 Snapshot of StemBase database

datasets (microarray and SAGE) from 210 stem cell samples from 60 different experiments [8] (Fig. 3).

The Stem Cell Discovery Engine

Stem cell discovery engine is an online system initiated with the aim to understand cancer stem cells profile from the data generated from the Harvard Stem Cell Institute. The Stem Cell Discovery Engine is a repository of carefully collected cancer stem cell/stem cells-related information from 52 public studies and 1098 assays. Experimental information is stored as multiomics Investigation/Study/Assay (ISA Tab) format which can be queried in the repository [9] (Fig. 4).

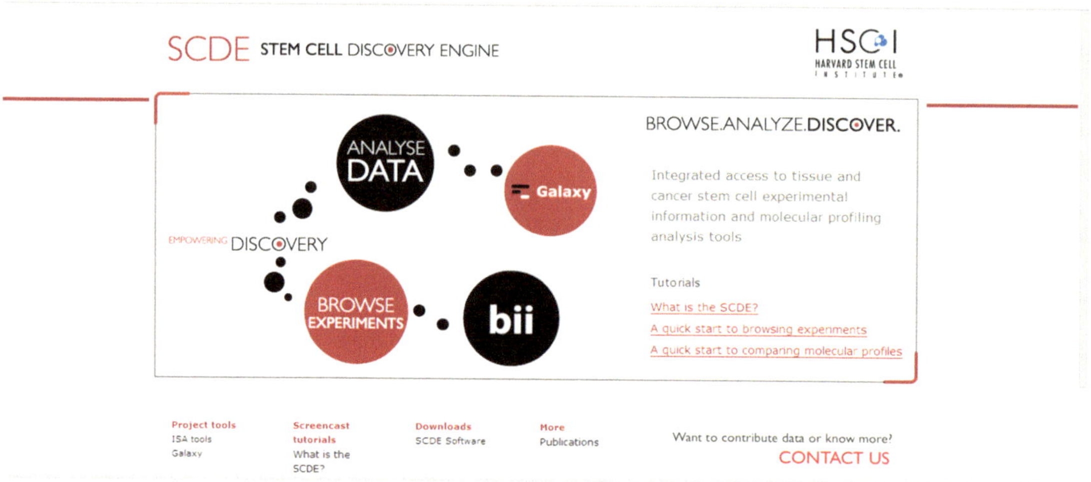

Fig. 4 Snapshot of Stem Cell Discovery Engine

Multiple Level-Based Databases and Tools

Complex stem cell differentiation process is orchestrated by multiple levels of regulatory processes. Single regulatory layer information (transcript levels) suffer from some inherent disadvantages as they overlook other layers as well as cross talk among different layers [2]. Many valuable resources (databases/tool) can integrate multiple levels of regulatory processes including genomics, genetics, epigenetics, transcriptomics, proteomics, and phosphoproteomics. Databases/tools include ESCAPE, SyStemCell, and Stem Cell Omics Repository.

ESCAPE

Embryonic Stem Cell Atlas from Pluripotency Evidence (ESCAPE) is a comprehensive Embryonic Stem cell database aimed at integrating published high-throughput "omics" studies which include deep sequencing, RNA sequencing, knockout or over-expression studies, proteomics and phospho-proteomics. The novel cross talk between different layers can be identified by building subnetwork by processing data sets into gene lists, gene-gene and protein-protein interactions which enable to build subnetworks and thus identify known and novel interactions across various regulatory layers. Information available via rich web-based intuitive search and visualization tools of ESCAPE database will enable us to build subnetworks and also identify known and novel interactions across various regulatory layers (epigenetics, proteomics, genomics) [10] (Fig. 5).

SyStemCell

SyStemCell is an online resource that aims to provide users with comprehensive and curated information on multiple-level experimental data of stem cells and proteins. The database contains 43,434 genes for four organisms (Homo sapiens, Mus musculus, Rattus norvegicus, and Macaca mulatta) and various stem cell sources (e.g., embryonic stem cells, neural stem cells, and induced pluripotent stem cells). Rigorously curated high-quality data from peer-reviewed journals are compiled and can be queried by Entrez Gene ID, gene symbol or stem cell type. An online analysis tool is integrated for assisting data mining to mine potential relation among different regulatory layers [11] (Fig. 6).

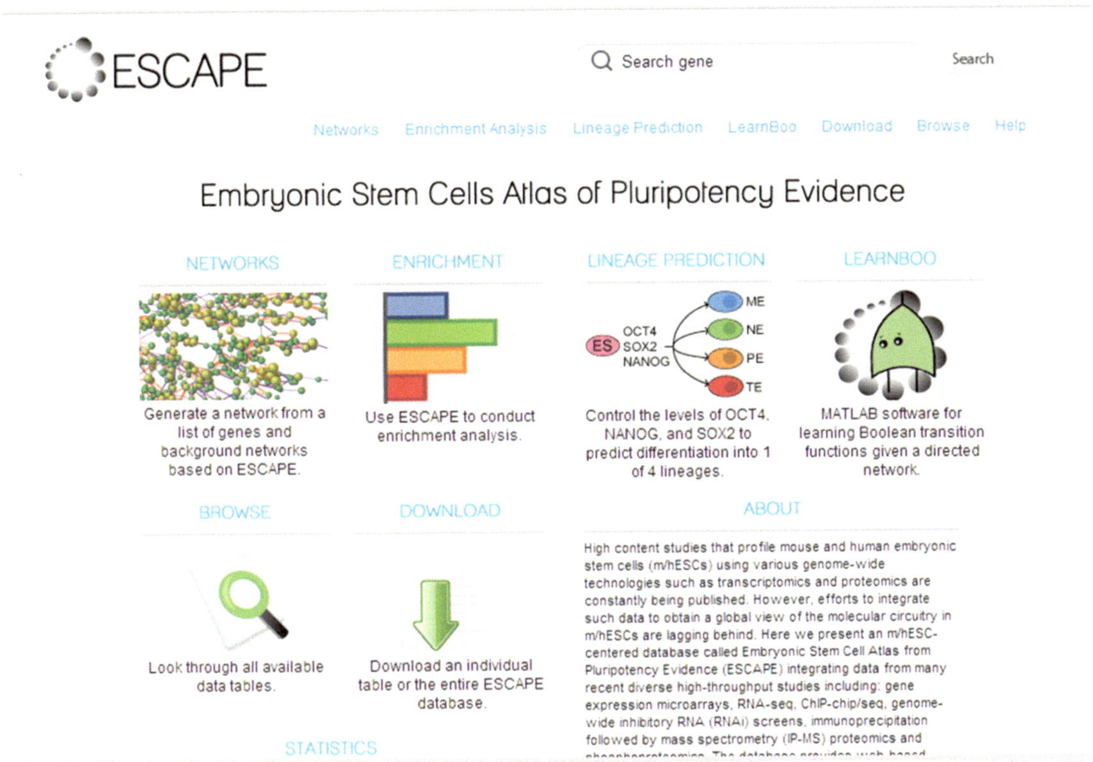

Fig. 5 Snapshot of ESCAPE database

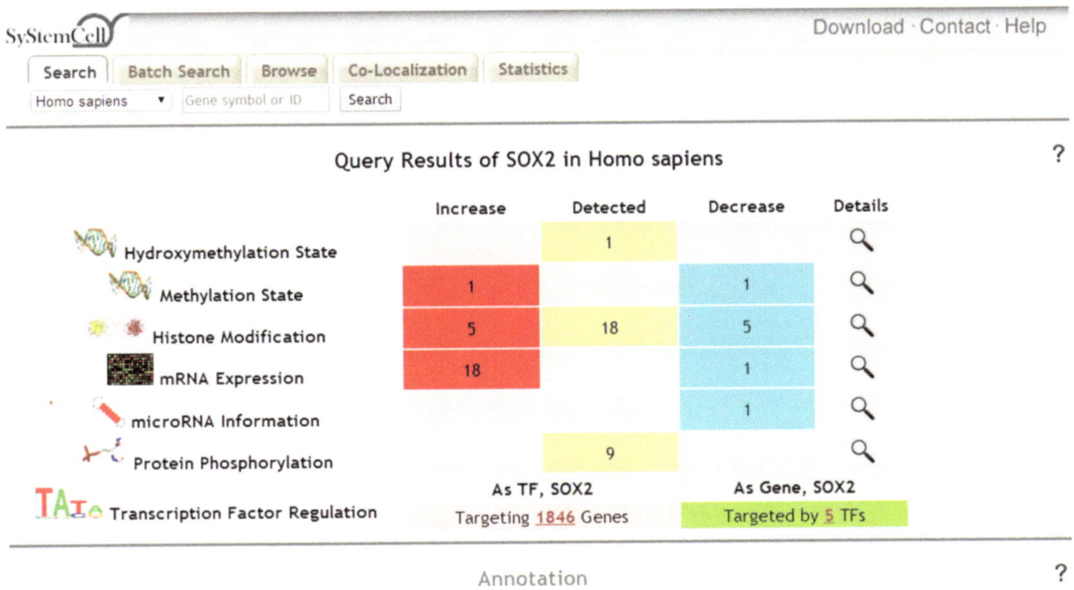

Fig. 6 Snapshot of SyStemCell database

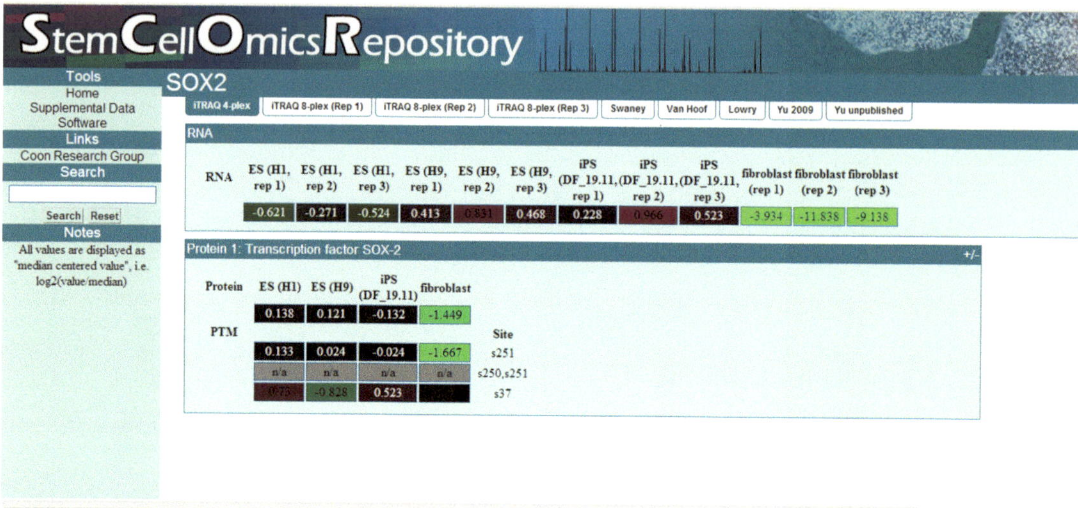

Fig. 7 Snapshot of Stem Cell Omics Repository

Stem Cell Omics Repository

Stem Cell Omics Repository (SCOR) contains quantitative information at multiple levels of measurements (RNA, protein, posttranslational measurements). SCOR is constructed as a centralized and queriable database of large-scale studies regarding embryonic stem cells and induced pluripotent stem cells. SCOR is a data-rich resource wherein data sets can be downloaded as well as gene/protein of interest can be queried in a dedicated search box. Single as well as multiple datasets can be viewed using this repository [12] (Fig. 7).

References

1. Trounson A, Thakar RG, Lomax G, Gibbons D (2011) Clinical trials for stem cell therapies. BMC Med 9:52
2. Macarthur BD, Ma'ayan A, Lemischka IR (2009) Systems biology of stem cell fate and cellular reprogramming. Nat Rev Mol Cell Biol 10:672–681
3. Nestor MW, Noggle SA (2013) Standardization of human stem cell pluripotency using bioinformatics. Stem Cell Res Ther 4(2):37
4. Xu H, Baroukh C, Dannenfelser R, Chen EY, Tan CM, Kou Y, Kim YE, Lemischka IR, Ma'ayan A (2013) ESCAPE: database for integrating high-content published data collected from human and mouse embryonic stem cells. Database (Oxford)
5. Liang Y, Russell I, Walworth C, Chen C (2009) Gene expression in stem cells. Crit Rev Eukaryot Gene Expr 19(4):289–300
6. Mallon BS, Chenoweth JG, Johnson KR, Hamilton RS, Tesar PJ, Yavatkar AS, Tyson LJ, Park K, Chen KG, Fann YC, McKay RD (2013) StemCellDB: the human pluripotent stem cell database at the National Institutes of Health. Stem Cell Res 10(1): 57–66
7. Müller FJ, Schuldt BM, Williams R, Mason D, Altun G, Papapetrou EP, Danner S, Goldmann JE, Herbst A, Schmidt NO, Aldenhoff JB, Laurent LC, Loring JF (2011) A bioinformatic assay for pluripotency in human cells. Nat Methods 8(4):315–317
8. Sandie R, Palidwor GA, Huska MR, Porter CJ, Krzyzanowski PM, Muro EM, Perez-Iratxeta C, Andrade-Navarro MA (2009) Recent developments in StemBase: a tool to study gene expression in human and murine stem cells. BMC Res Notes 2:39
9. Ho Sui SJ, Begley K, Reilly D, Chapman B, McGovern R, Rocca-Sera P, Maguire E, Altschuler GM, Hansen TA, Sompallae R, Krivtsov A, Shivdasani RA, Armstrong SA, Culhane AC, Correll M, Sansone SA, Hofmann O, Hide W (2012) The Stem Cell Discovery Engine: an integrated repository and analysis system for cancer stem cell comparisons. Nucleic Acids Res 40:D984–D991
10. Xu H, Baroukh C, Dannenfelser R, Chen EY, Tan CM, Kou Y, Kim YE, Lemischka IR, Ma'ayan A (2013) ESCAPE: database for integrating high-content published data collected from human and mouse embryonic stem cells. Database (Oxford); bat 045

11. Yu J, Xing X, Zeng L, Sun J, Li W, Sun H, He Y, Li J, Zhang G, Wang C, Li Y, Xie L (2012) SyStemCell: a database populated with multiple levels of experimental data from stem cell differentiation research. PLoSOne 7(7): e35230

12. Phanstiel DH, Brumbaugh J, Wenger CD, Tian S, Probasco MD, Bailey DJ, Swaney DL, Tervo MA, Bolin JM, Ruotti V, Stewart R, Thomson JA, Coon JJ (2011) Proteomic and phosphoproteomic comparison of human ES and iPS cells. Nat Methods 8(10): 821–827

Part II

Stem Cells and Organ Systems: Nervous System

Stem Cell Therapy for Neurological Disorders: From Bench to Bedside

Peter A. Barbuti

Introduction

According to the National Institute of Neurological Disorders and Stroke (NIH), there are found to be 488 neurological disorders present in the world at the time of writing, ranging from acid lipase disease to Zellweger syndrome [1]. No book chapter is equipped to tackle all of these in detail; however, it will focus upon the five major neurological diseases: Alzheimer's disease, Parkinson's disease, Huntington's disease, Multiple Sclerosis and Amyotrophic Lateral Sclerosis. The chapter will begin by summarising the epidemiology, pathology and disease susceptibilities for each of these five major neurological diseases. The chapter will then look in detail at generating patient-specific cell lines for personalised medicine, genome editing and the differentiation protocols necessary for cell replacement therapy related to the aforementioned neurodegenerative diseases. This chapter will then look at the alternative cell sources that have been used as existing and current cell therapeutic strategies before summarising with the advantages and constraints of stem cells in research and clinical translation. Finally, the chapter will conclude on the current research findings with a particular focus on patient-derived research in Parkinson's disease and how different therapeutic strategies can be targeted at different neurological diseases focusing on – Parkinson's disease and multiple sclerosis – before summarising on the challenges for stem cell therapy in neurological disorders: from bench to the bedside.

P.A. Barbuti (✉)
Clinical and Experimental Neuroscience, School of Clinical Sciences, Université du Lukembourg, Dorothy Hodgkin Building, Whitson Street, Bristol BS1~3NY, UK
e-mail: peter.barbuti@uni.lu; peterbarbuti@hotmail.com

Neurological Disorders

Alzheimer's Disease

Epidemiology and Pathology

Alzheimer's disease (AD) is the most common age-related neurodegenerative disease in the world currently estimated to affect 30 million people, a figure expected to quadruple in 40 years [2]. In 2010 alone AD was estimated to have cost the world $604 billion [3]. This staggering figure will only increase.

Pathologically AD is characterised by three cardinal changes in the brain: the presence of amyloid β (Aβ) plaques, intra-neuronal hyper-phosphorylated microtubule-associated protein tau and the loss of specific neurons and synapses, principally pyramidal neurons that are located in the cerebral cortex and cholinergic neurons of the basal forebrain [4]. However, as the disease progresses serotoninergic and noradrenergic neurons are also affected, with post-mortem tissue showing further loss of GABA and somatostatin cell types. Loss of these

I. Somasundaram (ed.), *Stem Cell Therapy for Organ Failure*,
DOI 10.1007/978-81-322-2110-4_4, © Springer India 2014

neuronal subtypes leads multiple atrophy of these affected brain regions, often beginning with the hippocampus and including the entorhinal cortex, frontal, parietal and temporal cortices [5]. For further reading on the progressive AD stages, see Braak and Braak. AD is symptomatically characterised by a progressive loss in learning, memory and cognitive decline.

Disease Susceptibility

AD is largely thought to be sporadic with genetic mutations thought to account for 0.5 % of all AD patients worldwide [1]. Mutations in three autosomal dominant Mendelian risk genes – *APP*, *PSEN1* and *PSEN2* – are highly penetrant and lead to an early onset of this disease [1–4]. Mutations in *APP* result in increased levels of Aβ and also change the ratio of cleaved Aβ peptides from 40 amino acids ($A\beta_{40}$) to increasing levels of the more toxic $A\beta_{42}$ peptides [5]. The $A\beta_{42}$ peptide is more hydrophobic and amyloidogenic than the $A\beta_{40}$ form and leads to increasing amyloid fibrillogenesis [1]. Mutations in *PSEN1* and *PSEN2* have been found to impair the activity of γ-secretase that is involved in the cleavage of *APP*, which also results in an increased $A\beta_{42}$ to $A\beta_{40}$ ratio [6–9]. Mutations in *APP*, *PSEN1* and *PSEN2* account for only 13 % of all early-onset AD patients [5]. *APOE* is a moderately penetrant gene with semi-dominant inheritance [10]. Inheriting the ε4 allele, homozygous *APOE* ε4ε4 and heterozygous *APOE* ε3ε4, increases the risk of AD 15x and 3x, respectively, compared to the most common *APOE* ε3ε3 form of the gene [11].

Parkinson's Disease

Epidemiology and Pathology

Parkinson's disease (PD) is a multifactorial neurodegenerative disease characterised by the loss of A9 dopaminergic neurons in the substantia nigra pars compacta (SNc) of the midbrain. The presence of intra-cytoplasmic inclusions (Lewy bodies) is another characteristic feature in those midbrain dopaminergic neurons (mDA) in the SNc that remain [12]. PD is the second most common neurodegenerative disease worldwide; the results from a 2010 census in the USA found PD to affect 630,000 people in the USA costing $14.4 billion in 2010 alone; as the prevalence to PD is projected to double by 2040 due to an increased elderly population, this cost will only increase [13].

The dopaminergic (DA) neurons project from the SNc to the dorsolateral striatum, caudate and putamen forming the nigrostriatal pathway that releases the neurotransmitter dopamine; it is thus the reduction in the dopamine following the progressive loss of DA neurons that allows the disease to manifest and become symptomatic. The four cardinal motor symptoms that signify the manifestation of the disease remains the primary tool of clinical diagnosis: bradykinesia, resting tremor, rigidity and postural instability. However, there are many other non-motor symptoms that are associated with the disease and can often precede the initial diagnosis, such as depression and gastrointestinal difficulties. As no PD patient presents with a homogenous aetiology, PD remains a difficult disease to diagnose without any of the motor symptoms. At the stage when the PD has manifested into the characteristic motor symptoms that precede diagnosis, up to 70 % of the mDA neurons of the SNc have already degenerated [14].

Disease Susceptibility

Approximately 90 % of all PD is idiopathic and thus of an unknown aetiology [15]. However, there are genetic susceptibility loci and risk genes attributed to PD, known as the PARK genes, of which there are 18 identified to date [16]. Two of these key PARK genes are PARK1/4 (*SNCA*) and PARK8 (*LRRK2*). *SNCA* encodes α-synuclein, a monomeric protein that in its mutated form undergoes a conformational change from an α-helical structure into β-sheets, which aggregate and oligomerise to form toxic protofibrils that then fibrilise [12, 17]. Fibrils of α-synuclein have been identified as the main components of LBs and LNs [13, 14]. Duplications and triplications of *SNCA* is directly correlated to the earlier onset and disease severity of familial PD [18–20], with single nucleotide polymorphisms associated with an increased risk in sporadic PD [21, 22].

The most common (>10 %) autosomal dominant locus of PD is *LRRK2*. *LRRK2* is also the locus most commonly associated with idiopathic PD where spontaneous mutations in this gene accounts for 3.6 % of all idiopathic PD; the *LRRK2* (G2019S) mutation is responsible for 1–2 % of all these cases [23]. Mutations in *LRRK2* have been implicated with impairments of lysosomal packaging and chaperone-mediated autophagy [24], reducing the clearance of α-synuclein and ubiquitin, leading to presynaptic accumulation and subsequent neuronal toxicity [25]. Thus dysfunction in both *SNCA* and *LRRK2* represents two of the main sources of PD; it is unknown, however, how the dysfunction of these genes is triggered.

Recent evidence, however, has shown that tau, a protein heavily associated with Alzheimer's disease (AD), enhances the aggregation and toxicity of α-synuclein [26]. This research article follows a publication [27] that suggests that tau and Aβ interact synergistically with α-synuclein in vivo to promote aggregation and accumulation of each other leading to cognitive dysfunction.

Huntington's Disease

Huntington's disease (HD) is a progressive, fatal, monogenic neurodegenerative disorder. HD is caused by an expansion of a triplet region of polyglutamine (CAG) repeats in the huntingtin (*HTT*) protein; this trinucleotide repeat then results in the addition of a long stretch of glutamines (polyQ) near the N-terminus of the protein [28]. Initially, HD causes loss of medium spiny neurons (MSN) in the neostriatum before progressing to loss of entire cortical structures [29, 30]. Another pathological feature of HD is astrogliosis [31]. HD is autosomal dominant and displays age-dependent penetrance with the increased length of CAG repeats inversely proportional to age of onset with CAG repeat lengths greater than 36 repeats considered a pathological threshold [32, 33]. The prevalence of Huntington's disease is 4–10 per 100,000 in the Western world; the mean age of onset is 40 years, with death occurring 15–20 years from time of onset. HD patients have progressive motor dysfunction, cognitive decline and psychological problems such as suicidal ideation [34].

Multiple Sclerosis

Epidemiology and Pathology

Multiple sclerosis (MS) is a chronic neurodegenerative demyelinating disease. MS affects 2.5 million people worldwide with 80 % of patients developing a progressive disability and costs the EU economy €9 billion per year [35]. There are two disease-specific symptoms of MS: Lhermitte's symptom, an electrical sensation that runs down the spine on neck flexion, and Uhthoff's phenomenon, a worsening of symptoms due to higher than usual temperatures [36]. MS is not localised to any specific region of the brain are affects the cerebellum and cerebellar pathways, optic nerve, brainstem and also the spinal cord. MS is a demyelinating neuronal disease caused by the loss of the oligodendrocytes that normally create and maintain the myelin sheath. MS is in part a disease of the innate immune system, as following inflammation the blood–brain barrier becomes dysregulated, transendothelial leukocytes cross the blood–brain barrier and become autoreactive: attacking oligodendrocytes; this leads to the eventual demyelination of neurons and MS pathology [37].

Disease Susceptibility

The interaction of the Epstein–Barr virus (EBV) is heavily implicated in aetiology of MS with almost 100 % of patients analysed in the study by the Ramagopalan group having previous serological markers of a past EBV infection [38]. The environment plays a hugely significant factor in determining disease susceptibility in MS with the biggest risk factor being latitude [39]. There is also an increasing female to male bias that has markedly increased in 50 years [40]. Although the role of genetics are not as pivotal in MS compared to other highly penetrant Mendelian diseases such as HD, with only a 30 % likelihood of MS if both parents are sufferers [41], there is still a genetic link with heterogeneity on the major histocompatibility antigen (HLA) class II complex gene locus *MHA-DRB1* being associated with increased risk of the disease [42, 43].

Amyotrophic Lateral Sclerosis

Epidemiology and Pathology

Amyotrophic lateral sclerosis (ALS) is the most common adult motor neuron disease, affecting 1:500,000 people worldwide per year [44]. ALS is fatal, is idiopathic and varies from patient to patient with loss of both upper motor neurons (UMN) in the motor cortex and lower motor neurons (LMN) in the brainstem and spinal cord. Loss of neurons leads to muscle atrophy with patients additionally presenting with dysphagia, dysarthria, spasticity and hyperreflexia symptoms and an abnormal reflex commonly called Babinski's sign. The age of onset of ALS is variable but tends to be after 40 years of age; only 4 % of ALS patients survive longer than 10 years [45].

Disease Susceptibility

Approximately 10 % of the cases of ALS are familial [46]. Of those familial cases, mutations in three genes – superoxide dismutase *SOD1* [47, 48], TAR DNA-binding protein-43 (TDP-43) *TARDBP* [49] and fused in sarcoma *FUS* [28, 29] – account for 30 %.

Mutations in TDP-43, encoded by *TARDBP*, enhance protein aggregation, fibril formation and neurotoxicity in ALS [30]. TDP-43 also behaves as a prion with intracellular TDP-43 exhibiting seed-dependent and self-templating aggregation, with propagation of TDP-43 aggregates via the exosome [31]. The *SOD1* mutations in ALS are fascinating as they show that the motor neurons are selectively degenerated via astrocytic- and microglial-mediated toxicity [32, 33]. The original study by the Przedborski group and follow-up papers have shown that when the mutant *SOD1* is carried on motor neurons, fibroblasts, cortical neurons and myocytes, they do not cause toxicity; similarly *SOD1* mutated astrocytes and glial are not toxic to spinal GABAergic, dorsal root ganglion neurons or hESC-derived interneurons, therefore implicating ALS as a non-autonomous neurodegenerative disease [32–37].

Stem Cell Therapy for Neurological Disease – Introduction

An advantageous route in which to research and subsequently treat neurological disorders and neurodegenerative diseases is to recapitulate in vitro the endogenous, patient-derived cell type where the disease is present and prevalent in vivo, thus determining the reason(s) for their specific vulnerability and selective degeneration.

Existing ways to elucidate and research the disease phenotype typically involve post-mortem tissue sections, neuroblastoma cell lines, non-human animal models, including small rodents, yeast, drosophila and zebrafish and non-human primary culture cell lines. All of these various research sources have been useful and valuable in studying the disease further; however, no research strategy is without limitation. The increasing use of stem cell therapy, particularly from patient-derived induced pluripotent stem cells (iPSCs) in countless groups around the world, should lead to further understanding and hopefully better treatment of these complex progressive chronic neurological diseases.

The reprogramming of the patients' somatic cells, typically fibroblasts from a skin biopsy, to a pluripotent, neural precursor or terminally differentiated cell type, or the differentiation of the pluripotent or neural precursors cells into the terminally differentiated cell type.

Once the terminally differentiated cells such as mature, electrophysiologically active neurons are generated, they can be used in drug screening to elucidate the efficacy of novel or pre-existing drugs or neurotrophins that can be subsequently used in patient therapy. These de novo neurons or neural precursor cells have the long-term potential to be utilised as cell replacement therapies. Fundamentally, however, the generation of these patient-derived disease lines has the ability to enhance research and understanding of the disease aetiology, pathogenesis and manifestation; to ascertain the selective vulnerability and thus disease mechanism; and to halt the disease progression and prevent further cell loss. An alternative strategy to modelling the disease by generating patient-derived in vitro de novo neurons could be using adult stem cells, such as in the bone marrow, to mobilise endogenous protection to treat the selective vulnerabilities of the disease.

Generating iPS Cell Lines to Study Neurological Disease

Induced pluripotent stem cells were first derived in 2006 in a seminal paper in Cell by Takahashi and Yamanaka [50]; this ground-breaking work was based upon Sir John Gurdon's work on frogs in 1962 that challenged the dogma that mature cells are irreversibly committed to their fate [51]. Yamanaka and Takahashi proved this in a mouse model using a retrovirus encoding *Oct4*, *Sox2*, *Klf4* and *c-Myc* to induce pluripotency from dermal fibroblasts. One caveat however with the paper was that the cells failed to produce a viable chimera, a hallmark of pluripotent stem cells [50]. However, 1 year later Yamanaka and Takahashi were the first to generate iPS cells using the same four factors in human cells, this time being able to produce viable chimeras [52]. A month later using lentiviruses encoding the four factors of *Oct4*, *Sox2*, *Nanog* and *Lin28*, James Thomson's group also generated iPS cells including a viable chimera from human fibroblasts [53]. What these two landmark papers proved and repeated was that pluripotency can be induced from terminally differentiated mature somatic cells, developing the possibility of personalised medicine: using the patient-specific cells to treat his individual disease. The pioneering work of iPS generation based on the vectors discovered by Yamanaka and Thomson has been replicated in many laboratories throughout the world using different tissue sources to generate iPS cell lines including the amnion [54], dental pulp [55], adipose tissue [56], blood [57–59] and urine [60].

There are caveats with the original Yamanaka and Thomson studies, however, such as by using integrating viruses to reprogram somatic cells; both the vector backbone and transgenes are integrated into the genome. These integrating vectors have the potential to create mutations upon genome insertion that interfere with the normal function of the cell. The integrating vectors can also result in residual transgene expression that can influence and affect the differentiation propensity of cells to specific lineages [53]. Also two of the Yamanaka factors – *Klf4* and c-*Myc* – are oncogenic and have previously resulted in tumourigenicity due to reactivation of the *c-Myc* oncogene [61]. Consequently, as summarised in Table 1, research was undertaken to either use nonintegrating transduction strategies [62, 63, 65–68], integrating vectors that can be excised out of the genome [69–71], DNA-free delivery of vectors in the form of RNA [64], proteins [72], mRNAs [73], microRNA (miRNA) [74] and a chemical only induction of pluripotency [75]. However, all of these strategies have their own specific constraint of using each method; in particular the reprogramming efficiencies of the non-integrating adenoviral and episomal methods are very low.

The Sendai virus being an RNA virus will never produce DNA that integrates into the host genome during transduction; it can easily be removed by antibody-mediated negative selection and generates iPS cells with a high efficiency [64]; as a result the Sendai virus method of iPS reprogramming is commonly used. Unfortunately the transgenes from the Sendai virus can only be removed by diluting, i.e. multiple passages. In the original research paper by Fusaki et al., the transgenes however were still there after 20 passages [64]. A caveat in using Sendai viruses to reprogram to pluripotency is the time the iPS cell line will need to spend in in vitro culture to facilitate the removal of these transgenes. Consequently reprogramming to pluripotency using this technique becomes a long process. The consequence of an extended time in in vitro culture is that there is greater chance of losing genomic stability and acquiring karyotypic abnormalities such as amplifications or trisomy on chromosome 8,12, 17q, 20q and X found in both iPSCs and hESCs [71, 76–78]. Genes found on these chromosomes such as the pluripotency gene *Nanog* and anti-apoptosis gene *Survivin* is encoded in 12 and 17q, respectively, where increasing dosage of these genes can confer a selective advantage [76].

Another strategy used to generate iPS cell for clinical therapy was the excision of the already integrated transgenes. The two systems available

Table 1 Strategies for generating iPS cells

Method	Species	Vectors	Days to generate iPS	Reprogramming efficiency (%)	References
Integrating vectors					
Retrovirus	Human	OSKcM	30	2×10^{-4}	[52]
Lentivirus	Human	OSNL	20	2.2×10^{-4}	[53]
Nonintegrating vectors					
Episomal	Human	OSNLKcM40	30–35	$3–6 \times 10^{-6}$	[62]
Adenovirus	Human	OSKcM	25–30	2×10^{-6}	[63]
Sendai	Human	OSKcM	7+ p5–p20[a]	$10^{-3} - 10^{-2}$	[64]
Minicircle	Human	OSNL	28	5×10^{-5}	[65]
Episomal plasmids	Human	OSKLlM	26–32	$1 \times 10^{-5} - 3 \times 10^{-4}$	[66]
Liposomal magnetofection	Mouse	OSKcM	7	4×10^{-2}	[67]
Doxycycline inducible	Human	OSKcM	7	92 %	[68]
Excised integrated vectors					
PiggyBac	Human	OSKcM	20–30	3×10^{-4}	[69, 70]
Cre/loxP (lentiviral)	Human	OSKcM	15–20	0.5 %	[71]
Viral-free delivery					
Proteins	Human	OSKcM	56	1×10^{-5}[b]	[72]
mRNA	Human	OSKcML	17–24	2.9 % (4.4 % – Hypoxia)	[73]
miRNA	Human	mir-200c, 302 s, 369 s	20	5×10^{-5}[c]	[74]
Small molecule-only induction					
Chemicals	Mouse	VC6TFZ	36–48	0.2 %	[75]

Abbreviations: *O* Oct4, *S* Sox2, *K* Klf4, *cM* c-Myc, *N* Nanog, *L* Lin28, *40* SV40LT, *lM* L-Myc, *VC6TFZ* small molecules taken from [75]

[a]The numbers of passages (p) required before transgenes are silenced

[b]Transduction repeated 6 times

[c]Transduction repeated every 48 h 4 times

that use this are Cre/loxP recombination and *piggyBac* transposons. In the Cre/loxP system even though the viral cassette containing the Yamanaka factors were removed following reprogramming using transfection of the Cre recombinase, the initial vector sequences still remained integrated into the genome; thus, the Cre/loxP system still has a risk of insertional mutagenesis [71]. In the mouse, *piggyBac* has been shown to be a viable strategy for iPS generation without leaving a genomic footprint [69, 70]. Unfortunately, there is a lack of information to date regarding the removal of the *piggyBac* insertions in the human suggesting more work needs to be done before it can be used routinely as the strategy for personalised medicine.

Three other alternative strategies that avoid the introduction of genetically modifying DNA into the genome include the use of proteins, mRNAs and microRNAs (miRNA). The protein-based method takes the longest out of the all methods summarised in Table 1; it is also the most labour intensive requiring 6 repeated transductions [72]. The use of mRNAs seems to be an attractive way to generate iPS cells for translational research: a high efficiency of up to 4.4 % under hypoxic conditions plus no molecular footprint such as integrating vectors or transposons to remove [73]. The miRNA method is another attractive nonviral solution that can be applied in translation therapy. miRNAs leave no molecular footprint, are reasonably efficient and are quick: can reprogram in under 3 weeks [74]. MicroRNA therapy is already deemed to be safe and is undergoing stage 2b clinical trials in the treatment of Hepatitis C under the brand name Miravirsen [79, 80].

Recently the group led by Hongkui Deng has been able to generate iPS cells from murine cells using only small molecules. Provided this can be replicated in human cells, this is another strategy to generate iPS cells with no footprint [75].

In October 2013, another landmark paper in the field of cellular reprogramming has been published in Nature by the Israeli group led by Jacob Hanna showing the generation of iPS cells at an efficiency of close to 100 % in both human and murine cells in 7 days due to knock-down of the nucleosome repressor complex Mbd3 [68]. The importance of this research is the advancement in the understanding at a molecular level with a huge gain at a practical level. By generating iPS cells at a 92 % efficiency in approximately a week, there is a huge cost benefit by reducing the laboratory hours needed to make iPS cells; also with an efficiency of 92 %, it is the first strategy that could lend itself to automation and scaleup and bring patient-specific medicine a step closer.

Since 2007 when Yamanaka and Thomson showed that somatic human skin could be reprogrammed to pluripotency, iPS cells became instantly more advantageous than hES cells for one significant clinical reason. The genome of the in vitro derived cells matches the in vivo cells; consequently for any future cell replacement therapy, there would be a significant reduction in risk of immune rejection compared to the allogeneic hES cells, which will express the human leukocyte antigen minor histocompatibility complex (HLA/mHC) and low levels of the class I major histocompatibility antigen (HLA/MHC class I) [81]. There are logistical benefits from using iPS cells in that they are easier to derive than hES and are not subject to the ethical concerns or strict financial constraints of federal funding in the USA.

However, although Yamanaka and later Thomson pioneered iPS cell research, enabling the possibility of personalised medicine and a patient-specific clinical resource of iPS lines, there are still many constraints. Many of the strategies developed and in use require integrating vectors and transgenes, with evidence of insertional mutagenesis and transgene reactivation [82, 83]. There is also an increase in copy-number variants and increase in point mutations in protein coding genes as opposed to hES cells [84, 85]. Finally the last constraint that needs rectifying prior to cell therapy is epigenetic memory [86, 87], which not only suggests that these cells are not truly pluripotent but also implies that there is lack of current understanding on epigenetic memory which needs to be fully understood before it is assumed a fact as it is a variable and may influence the results of future studies.

Genome Editing

Yamanaka and Thomson have brought cell-replacement therapy a step closer by allowing it to be patient-specific, provided however that the cells do not need genetic modification. In the cases of a genetic mutation, the cells once derived will also retain that mutation and the disease phenotype, for example, a triplication in the *SNCA* gene is present not only in the dopaminergic neurons but also in the fibroblasts and the iPS cell line [88, 89]. A way to avoid this is by genome editing homologous recombination; the first study on this was by the Jaenisch research group where zinc finger nuclease (ZFN) mediated genome editing was used to edit the genome to correct the A53T point mutation and create an isogenic iPS cell line [90].

There are many methods currently published that have the ability to perform genome editing homologous recombination (Table 2): ZFN [97], bacterial artificial chromosomes (BACs) [98], transcription activator-like effector nucleases (TALENs) [99] and clustered, regularly interspaced, short palindromic repeat (CRISPR) [100]. Helper-dependent adenoviral vectors (HDAdV) have also been used in genome editing being able to edit both the transcriptionally active and inactive loci [92, 94, 95]. HDAdV also have a benefit in that they accommodate up to 35 kb of DNA, as opposed to ~4.7 kb that traditional adeno-associated virus (AAV) system. There are a few concerns however with HDAdV: a possibility of in vivo toxicity from the adenoviral capsid proteins, random integration sites and a genomic footprint [101].

Table 2 Technologies for genomic editing (corrective and inductive) of disease mutations in neurological disease models

Disease	Gene	Defection	Corrective method	References
Parkinson's disease	SNCA (α-synuclein)	A53T point mutation	ZFN	[90]
	LRRK2	G2019S	ZFN[a]	[91]
	LRRK2	G2019S	HDAdV	[92]
Huntington's disease	HTT (huntingtin)	CAG repeats	BAC	[93]
HGPS	LMNA (lamin A)	C1824T point mutation	HDAdV	[94]
AWS	LMNA	A1733T	HDAdV	[95]
Gyrate atrophy	OAT	Base-pair mutation	BAC	[96]

Abbreviations: *HGPS* Hutchinson–Gilford progeria syndrome, *AWS* atypical Werner syndrome
[a]Induction of disease mutation as well as deletion

The biggest concern about genome editing is the potential for off-target double-stranded DNA breaks being introduced into the genome, thus creating non-specific genome variants that are not truly isogenic. There is rationale for this concern as a ZFN study by Hockemeyer there was 1 off-target event per 184 clones analysed [97]. However, considering that the human diploid genome is six billion base pairs, that statistical probability of generating an off-target event is far too high. ZFN are also difficult for non-specialists to design and are associated with high rates of failure [102].

TALENs are cheaper and easier to use than ZFN; however, they are larger molecules, so it can be difficult for them to be efficiently delivered [99]. Although, TALEN technology has been used to correct for mutations in the β-globin gene in the blood disorder disease β-thalassemia [103].

CRISPRs have an advantage over ZRN and TALEN in that using a single vector to guide RNAs in series which can be then processed into individual RNAs allows for simultaneous, multiplexed targeting of multiple sites of the genome in the same cell [100]. The big concern with CRISPR is an increased inherent risk of off-targeting due to the guide RNA being shorter [104]. However, a recent publication by the Zhang group have detailed using 2 guide RNA's in a 'double-nickase' strategy has been able to facilitate a double-strand break and reduce off-target mutagenesis up to 1,000 fold [105].

The use of genome editing in science is a recently emerging field, and it is particularly valuable when the iPS cells derived from the patient also contain the genetic disease. Therefore, by editing out the disease and similarly editing in the disease [91, 106], researchers will understand a lot more about each disease phenotype and how it manifests. Also by editing in a disease to a non-disease control line, it effectively creates a positive disease model that could be of particular use in compound screening (Fig. 1) by being able to distinguish phenotype of the mutation from the genetic background. Additionally depending on the research considered this could minimise the use of animals in research. To be able to correct neurodegenerative diseases using genome editing is a very exciting area of research and could even be used in future clinical therapy in correcting the genetic mutation prior to potential transplantation using the patients' isogenic cells.

Direct Conversion

Pioneered by Marius Wernig and colleagues in their seminal paper published in Nature in 2010, they showed that non-neural adult somatic fibroblasts can be directly reprogrammed to terminally differentiated functional and electrically active neurons without going through a pluripotent intermediary state [108]. These induced neurons (iN) were reprogrammed in 13 days from MEFs using three factors – *Brn2*, *Ascl1* and *Myt1l* (BAM factors) – with an efficiency of 19.5 %, with the iN exhibiting positive immunocytochemical staining for the neuronal markers of Tuj1, NeuN and MAP2.

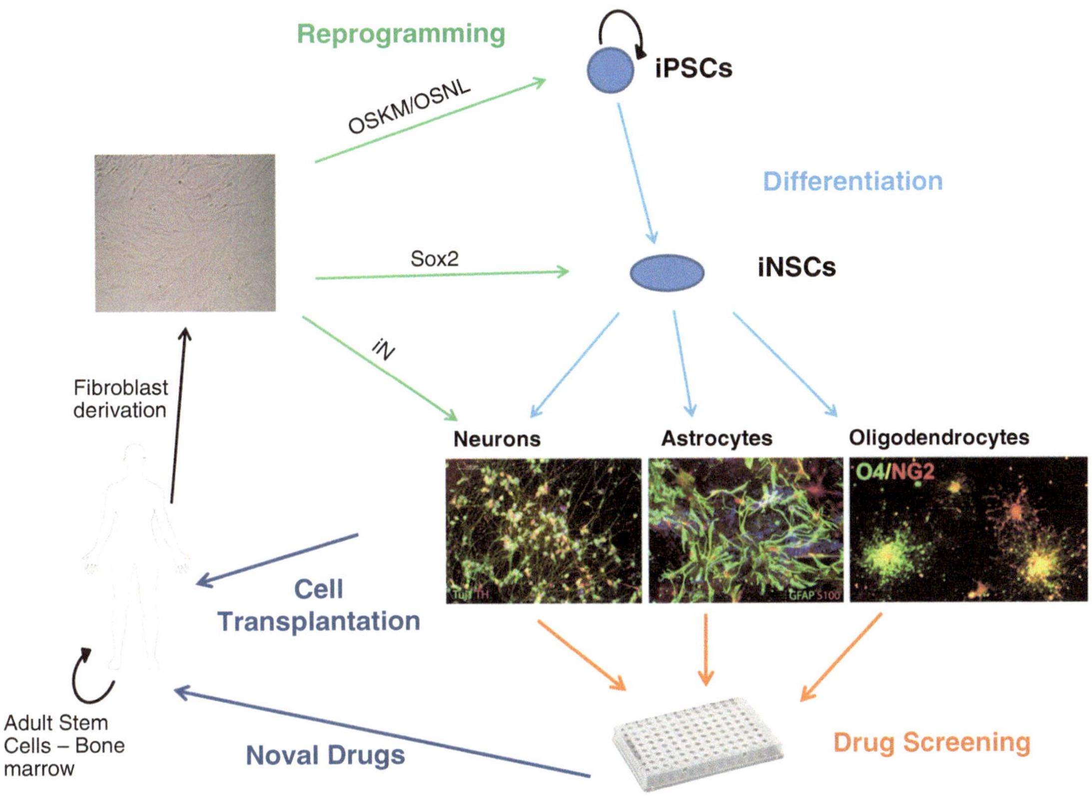

Fig. 1 A schematic representation of the processes that can be undertaken in personalised medicine. Following a patient biopsy, fibroblasts can be derived and expanded into a cell line. The fibroblasts can then be reprogrammed into induced pluripotent stem cells (iPSCs) [52, 53] or an induced neural-progenitor stem cell (iNSCs) state [107] before being terminally differentiated towards a cell type, such as neurons, astrocytes or oligodendrocytes. Alternatively these cells can be directly reprogrammed from the patient somatic cell to the terminally differentiated cell type [108]. These cells can then be used in disease modelling, drug screening and the development on novel drugs or in cell transplantation. Images of fibroblasts and Tuj1^{+}/TH^{+} dopaminergic neurons are provided by the author. The image of the GFAP^{+}/S100^{+} astrocytes is kindly donated by Dr Federica Rinaldi (University of Oxford). The image of the O4^{+}/NG2^{+} oligodendrocytes is used with permission from Professor Nada Zecevic (University of Connecticut) and has previously been published [109]

53 % of the mouse embryonic fibroblasts (MEF)-derived iN expressed Tbr1, a marker of excitatory cortical neurons with both excitatory glutamatergic and inhibitory GABAergic neurons generated. There has been an increase in the number of groups that have used the BAM factors amongst other factors, to directly convert human fibroblasts into terminally differentiated neuronal cell phenotypes which is summarised in Table 3. This includes cell populations specific for neuronal disease such as dopaminergic neurons for PD [114–118], retinal-like ganglion neurons for glaucoma [119] and spinal motor neurons for amyloid lateral sclerosis (ALS) and spinal muscular atrophy (SMA) [120]. All iN for PD exhibited functional dopaminergic neuronal properties: positive expression of midbrain markers, functionally active electrophysiological properties and dopamine release and uptake. In the study led by Abeliovich's group in 2011, the fibroblasts used for the direct conversion were patient-derived from a familial Alzheimer's disease (FAD) patients containing mutations in presenilin-1 and presenilin-2. These FAD-iN were found to show phenotype of AD showing altered processing and localization of amyloid precursor protein (APP) and increased production of Aβ [113].

Table 3 Summary of iN methods following direct conversion

iN-derived subtype	Reprogramming factors	Reprogramming efficiency	Neuronal efficacy	Duration (days)	References
Glutamatergic/ GABAergic neurons	BAM	19.5 % [a]		13	[108]
	BAMN[1]	2–4 %	60 % Tuj1[b]	18	[110]
	BM, miR-124	4–11 %	55 % MAP2[b]	18	[111]
	AM, miR-124, miR9/9	5 %	50 % MAP2[b]	30	[112]
	BAMOZ	7.1–8.9 %	28.4–36.1 % MAP2[b]	21	[113]
Dopaminergic neurons	ANL	3–6 % TH[b] 5–10 % Tuj1[b]		18–24	[114]
	BAMLF	16 %	<95 % MAP2[b]	24	[115]
	AN(Lb)	18.2 %[b]	35.1 % Tuj1[b]	14	[116]
	ANLFEP	9.19 %[a]	2 % Pitx3[b]	18	[117]
	ANPSNg	1–2 %	40 % DDC[b]	20	[118]
Retinal-like ganglion neurons	ANgB[3]	3.5 % Tuj1[b]		14	[119]
Spinal motor neurons	BAMNgHIL[3]	5–10 %[a]		10	[120]

Abbreviations: *B* Brn2, *A* Ascl1, *M* Mytl1, *N[1]* NeuroD1, *O* Oligo2, *Z* Zic1, *L* Lmx1a, *N* Nurr1, *F* FoxA2, *(Lb)* Lmx1b, *E* En1, *P* Pitx3, *S* Sox2, *Ng* Ngn2, *B[3]* Brn3b, *H* Hbx9, *I* Isl1, *L[3]* Lhx3

[a]Relates to reprogramming efficiency in MEFs

[b]Reprogramming efficiency from human astrocytes

Table 4 Summary of various reprogramming methods for iNSC generation

Initial cells	Transcription factors	References
Mouse fibroblasts	(*Oct4, Sox2, Klf4, c-Myc*) initially	[121]
	Oct4 (initially) *Sox2, Klf4, c-Myc*	[122]
	Brn2, Sox2, Klf4, c-Myc	[123]
	Brn2, Sox2, FoxG1	[124]
Mouse and human fibroblasts	*Sox2*	[107]
Murine sertoli cells	*Sox2, Klf4, c-Myc, Brn2, Ascl1, Ngn2, Pax6, Hes1, Id1*	[125]

The field of iN via direct conversion is very exciting and although it does hint at the possibility of far quicker method to reprogram neuronal cells from somatic cells without a pluripotent intermediate, it is still very inefficient. Fibroblasts are not an expandable immortal cell type, with a decreasing capacity to generate iN with every additional passage being exhausted by passage 8 [115]. Therefore, it is questionable if the quantities of cells generated from iN would be sufficient for cell replacement therapy strategies. It is for this reason of scalability and proliferative potential that generating multipotent neuronal progenitors instead of post-mitotic neurons would be a more sustainable long-term strategy in modelling PD.

The multipotent progenitors have been termed induced neural stem cells (iNSCs) and have been generated from murine and human fibroblasts and more recently Sertoli cells (Table 4). After initial research in mouse fibroblasts, there was no consensus on what transcription factors were necessary to generate iNSCs with the Yamanaka pluripotency factors initially used [121, 122]. Han and colleagues then swapped *Oct4* for another POU homeodomain transcription factor in

Brn4 that is normally expressed in the neural tube [126]. Han and colleagues generated stable iNSCs for over 130 passages that could form astrocytes, neurons and oligodendrocytes, although the efficiency for oligodendrocyte generation was extremely poor at approximately 10 % compared to ~100 % efficiency of both neuronal and astrocytic differentiation [123]. The Wernig group found that just *Sox2* and *FoxG1* were necessary to generate iNSCs; however, those iNSCs were bi-potent with the potential to form neurons and astrocytes only, oligodendrocytes were not generated unless the transcription factor *Brn2* was included in the transduction [124]. The first study and only current study published in humans is from the Huang group where transduction with *Sox2* alone could generate stable iNSCs with the potential to differentiate into neurons of multiple subtypes, astrocytes and oligodendrocytes after 2–4 weeks of culture in permissive differentiation conditions. The human iNSC were also able to survive and integrate in mouse brains without any tumourigenicity [107]. The only other published study of iNSC generation is from the Qi Zhou Chinese group using mesoderm-derived Sertoli cells. This study however used 9 transcription factors and needed 1 month of propagation to generate the sufficient amount of cells necessary for analysis [125].

Differentiation Protocols to Model Selected Neurogenerative Diseases

Generation of iPS cells is the strategy typically used by most research groups as a source of pluripotency prior to the terminal differentiation of the required cell type necessary to the model of the disease. Table 5 shows the most efficient protocol currently published for the generation of the neuronal subtypes lost in the 5 aforementioned most common neurodegenerative diseases: basal-forebrain cholinergic neurons for AD, striatal medium-sized spiny neurons for HD, midbrain dopaminergic neurons for PD, motor neurons for ALS and oligodendrocytes for MS.

The Crompton et al. protocol for AD is long at 90 days, but it has an excellent efficiency. 83 % of the cells generated are Tuj1+ neurons, with the remaining being GFAP+ astrocytes. Of those 83 % of neurons, between 91 and 92.4 % of the hES and iPS cell lines taken through

Table 5 Summary of protocols efficiencies used to generate terminally differentiated cell types for disease modelling

Disease	Desired subtype	Efficiency		Duration	References
		Neurons	Regional identity		
AD	Basal-forebrain-derived cholinergic neurons	83 % Tuj1$^+$	97 % Nestin$^+$ (d30)	90 days	[127]
		17 % GFAP$^+$	**92.3 % Nkx2.1$^+$/nestin$^+$**		
		91–92.4 % ChAT$^+$/Tuj1$^+$			
HD	Striatal medium-sized spiny neurons	51 % MAP2$^+$, 80 % Tuj1$^+$,	**70.6 % CTIP2$^+$/calbindin$^+$**	90 days	[128, 129]
		78 % GABA$^+$/MAP2$^+$			
		60.3 % CTIP2$^+$/MAP2$^+$			
		53 % Calbindin$^+$/MAP2$^+$			
		20 % DARPP32$^+$/CTIP2$^+$			
PD	Midbrain dopaminergic neurons	75 % TH$^+$	80 % FoxA2$^+$	50 days	[130]
		20 % TH$^+$/FoxA2$^+$	60 % Lmx1a$^+$/FoxA2$^+$		
			80 % Otx2$^+$/FoxA2$^+$		
			(counts at d25)		
ALS	Motor neurons	**53 % HB9$^+$/ISL1$^+$**	55–70 % FOXP1	31 days	[131]
			63 % FOXP1$^+$/LHX3$^-$		
MS	Oligodendrocytes	**60–80 % O4$^+$**	40–60 % NG2	77 days	[132]
		70 % A2B5$^+$			

Abbreviations: *AD* Alzheimer's disease, *HD* Huntington's disease, *PD* Parkinson's disease, *ALS* amyotrophic lateral sclerosis, *MS* multiple sclerosis

3 replications of this protocol are positive for ChAT: cholinergic acetyl transferase, a marker of cholinergic neurons [127]. The regional identities of the neurons are correct for an AD model being 92.3 % Nkx2.1, a marker of the basal forebrain.

In the HD model study from the Italian group of Elena Cattaneo, striatal medium-sized spiny neurons (MSN) were generated in a very well-described research article that deconstructed each neuronal subtype generated from this differentiation protocol, not all of which are shown in Table 5. A commonly used marker for MSN is DARPP32; however, DARPP32^{+} neurons are also found outside the striatum. By co-staining with CTIP2, a post-mitotic striatal marker, the regional identity of the MSN can be confirmed. 70.6 % of the calbindin + neurons co-expressed CTIP2, thus confirming the general acquisition of an MSN fate and not of an interneuron identity [128]. The Kriks et al. protocol for PD showed the generation of 75 % of dopaminergic neurons; 80 % of cells differentiated through the protocol were positive for the midbrain marker FoxA2. Although of those 75 % dopaminergic neurons, only 20 % were midbrain dopaminergic neurons [130], suggesting that the regional identity of the majority of those neurons was not specific to the midbrain.

A follow-up dopaminergic differentiation study by Kirkeby et al. trialled the GSK3β inhibitor, CHIR99021, that was used in the Kriks protocol; what was found was at high levels of drug, >1 μM, the floor plate begun to caudalise away from the midbrain fate towards the hindbrain showed by an upregulation in *LEF1*, a gene found at the midbrain–hindbrain border, and also hindbrain genes *HoxA2*, *HoxA4*, *IRX3* and *GBX2* [133, 134]. At 4 μM there was over a 100-fold upregulation in *GBX2*, a key gene expressed in the anterior hindbrain that shares a border in development with the MHB. In the study published in Nature by Millet, *GBX2* was shown to not only repress *Otx2*, a key midbrain gene regulator, but also shift and reposition the MHB, creating a smaller midbrain and larger hindbrain region [135]. Furthermore, shifting of the MHB has a direct consequence on increasing the number of serotoninergic neurons to the detriment of the midbrain dopaminergic neurons [136]. As the Kriks et al. protocol used a very high CHIR concentration of 3 μM, this may explain that of the 75 % of dopaminergic neurons produced by the protocol, 45 % of these neuron did not have a positive identity for the ventral midbrain marker *FoxA2* [130]. Unfortunately, with *Lmx1a* and *Lmx1b* now being shown to have overlap in the formation of the anterior hindbrain roof plate [137], it is quite possible that these remaining 45 % TH positive cells are of an anterior hindbrain identity and not the midbrain, although this is yet to be proven.

However, even though the Kriks protocol may have generated a population of dopaminergic neurons of a midbrain/anterior forebrain identity, it still possess 20 % FoxA2^{+}/TH^{+} neurons which is currently the best in the literature. Also the issue of non-midbrain contamination is not insurmountable as this can be selected out by methods such as FACS sorting. In addition to this, the neurons from the Kriks et al. protocol have been grafted in mice, rat and monkey models of PD, in all cases showing evidence of survival, integration and behavioural recovery [130].

The diseases of ALS and MS are both nonautonomous diseases; consequently the generation of MN or oligodendrocytes may not necessarily be the most effective long-term strategy for cell replacement therapy. However, it is still advantageous for researchers to obtain regionally specific terminally differentiated cell types to ascertain the specific susceptibility of MNs and oligodendrocytes to astrocytic/microglial disease transmission in ALS and lymphocytes in MS and then try to prevent this susceptibility.

For the study of ALS, the Wichterle group generates limb-innervating lateral motor column motor neurons. The MN quantification shows by immunocytochemistry the generation of 30 % HB9^{+}-specific motor neurons; of those neurons 53 % are HB9^{+}/ISL1^{+}, with ISL1^{+} specifying spinal motor neurons [131], although ISL1^{+} neurons are also specific for the cranial ganglia [138]. A FOXP1^{+} neuronal identity is specific for neurons that innervate with limb muscles [139, 140]. LHX3^{+} expression determines if a medial motor column that innervates axial muscles with $LHX3^-$ specifying a lateral motor column with limb-innervating muscles [139]. Of the MN identified there was a 70 % $FOXP1^+/LHX3^-$ identity [131]. Thus, the successful generation of limb-innervating lateral motor column motor neurons for ALS research.

Differentiating to oligodendrocytes for the study of MS the protocol by Sundberg et al. generates 60–80 % oligodendrocytes, specified by the commonly used oligodendrocyte specific marker O4 [132]. This paper also confirmed this identity and the prior establishment of oligodendrocyte precursors cells (OPCs) through this 77-day protocol using the markers NG2 and A2B5 to confirm identity. $NG2^+$ positive cells account for the identity of OPCs with NG2 being responsible for directional migration of the OPC through cell polarity [141]. $A2B5^+$ is a marker of the oligodendrocyte–astrocyte shared lineage progenitor cell, shown to be expressed on both OPC and astrocytes; an induction of BMP2 and other BMPs are required for the astroglial switch [142].

In all of these protocols that are used for the treatment of the five major neurodegenerative diseases, the initial recapitulation of development as shown by Crompton et al., is crucial for more efficient differentiation of the terminal cell type of choice. It is of note that the time, necessary generate specific terminally differentiated cells through this protocol is very long. However, if one considers that the gestational period of a human is 266 days, and a mouse is 20 days, a 90-day terminal differentiation protocols that recapitulate the developing embryo would seem accurate. Another important aspect of the differentiation protocols would be to establish maturity of the cells with functional synapses and electrical activity. Consequently it is essential that for disease modelling an accurate recapitulation of the desired cell type must be achieved.

Existing and Future Therapies Using Adult Stem Cells

Since the onset of hES [143] and iPS [52, 53] cells, the field and profile of stem cells has risen exponentially with the potential clinical therapy for every disease ever closer. Unfortunately the progress has been slow, with diseases having increasing layers of complexity and sophistication necessary to first understand before being able to treat; this is in addition to the ethical, logistical and scientific challenges that these hES and iPS cells bring.

Adult stem cells (ASCs) has been used as bone marrow transplants since 1963 [144], with the discovery of what is now known as haematopoietic stem cells (HSCs). Since then there has been the discovery of many more ASC populations: muscle-derived stem cells (MDSCs) [145, 146], mesenchymal stem cells (MSCs) [147], cord blood-derived multipotent stem cells (CB-SCs) [148, 149], neural stem cells (NSCs) [150], adipose stem cells (AdSCs) [151, 152] and most recently amniotic stem cells (AmSCs) [153].

ASCs in the form of CB-SCs are currently undergoing clinical trials for type I diabetes based on studies in mice reversing the disease [154, 155]. It remains to be seen if CB-SCs to treat disease will be an effective and feasible long-term strategy for the treatment of neurological disease. However, whilst the other methods of cell therapy detailed in this chapter are being advanced and refined for future therapeutic use, adult stem cell therapies appear attractive to patients and clinicians alike. Possibly the greatest benefit of ASCs is their availability, with bone marrow, blood and adipose tissue being readily available sources.

ASCs also have an advantage over iPSCs as they do not have to be reprogrammed. However in the brain, the multipotent adult progenitor cell populations are found in the subgranular zone in the dentate gyrus of the hippocampus and the subventricular zone of the lateral ventricles [38, 39]; these are not readily accessible areas. Additionally, a constraint of using ASCs compared to hESC or iPSC research is that ASCs are multipotent; however, this dogma is recently being challenged, particularly in AdSCs [40–43].

However, ASCs in the form of MSCs and bone marrow-derived neural crest stem cells (NCSCs) were used in a recent publication that looked at the efficacy of these cell populations when injected into the striatum of MPTP lesioned mice. At 7 days post ASC intervention, only 3 % of starting 5×10^4 MSCs survived; by 14 days there were no surviving MSCs. The survival rate of the NCSCs was 10 % after 7 days; this lowered to 1 % after 28 days with the same starting cell population. The results of the study showed that those cells that did survive were not able to integrate and migrate through the brain tissue and therefore were unable to modify their initial phenotype and no recovery of any type was observed [156]. The most recent study from the Wislet-Gendebien group has confirmed an earlier implantation study that showed partial and transient survival, poor integration and no neurogenesis or recovery [157]. This study also showed activation of the innate immune system with positive markers for microglial and astrocytes [158]. The most concerning article however is that in vitro expanded neural crest stem cells led to in vivo tumourigenesis in an animal model [159].

ASCs however have also shown to be safe and efficacious; in a study using CB-SCs, the integration and behavioural recovery was shown in rotenone-induced rat models of PD [159]. Furthermore, ASCs have shown efficacy in neuronal protection by secreting protective growth factors such as GDNF and BDNF [160–162]. In October 2013, a publication by a Swedish group has shown that the secretion of BDNF, GDNF, VEGF-A and angiopoietin-1 proteins resulted in axon regeneration, increased vascularity and decreased apoptosis [163]. Consequently it can be seen that CB-SCs show promise as a potential therapy, however more research is necessary to determine its potential use as a bench to bedside therapy.

Foetal Cells – A Case Study from Parkinson's Disease

Foetal cells were first used as a cell therapy in PD. In work pioneered at Lund University which showed in rat models of PD significant functional recovery following foetal nigral transplantations into the host striatum [164–168]. Shortly after, Lindvall and colleagues performed the first transplants of human ventral mesencephalic tissue from tissue collected from elective terminations of pregnancy at 8–10 weeks gestation [169]. In the subsequent paper by the same group six patients had transplanted mesencephalic foetal tissue into the putamen. These foetal cells were able to survive in the brain and produce a significant symptomatic relief, restoring dopamine synthesis and reducing bradykinesia and rigidity [170].

As a consequence of the success results found by Lindvall and colleagues, foetal cell transplantation studies were then replicated with several open-labelled studies taking place across the world [171–176]. Positive results and lack of adverse side effects from these trials led to USA's National Institute of Health (NIH) funding of two double-blind placebo-controlled clinical trials [177, 178]. However, although double-blinded clinical trials were undertaken, there were still concerns that needed to be addressed from the open-labelled studies. In that, although some of the patients showed significant benefit from the foetal transplants, many more did not, with a considerable variation in both intra- and inter-study [179]. Barker and colleagues recently reviewed these open-labelled studies in which variations between these studies were highlighted these included: the age of donor tissue, number of donors, target site for transplantation, immunotherapy and measurable endpoints of each study [179].

Nevertheless, due to positive results in some patients, including a lower Unified Parkinson's disease rating scale (UPDRS), across these open-labelled studies the NIH decided to support two double-blind sham-surgery trials. In 2001, transplantations by Freed and co-workers of mesencephalic foetal tissue from four embryos bilaterally into the putamen were used [177]; what they found was that the younger patients transplanted (<60 years) had a significantly lower UPDRS and an increased ^{18}F-fluorodopa signifying neuronal outgrowth [177]. Unfortunately the older patients (>61 year) had no significant symptomatic improvement over the control group with dyskinesias and dystonias recurring in 15 % of the post-transplanted patients [177].

The second NIH study in 2003 by Olanow and colleagues 1 and 4 foetal donors per transplant were trialled. Despite an increase in striatal ^{18}F-dopa uptake via PET scans and evidence of graft survival and innervation in autopsy studies, there was ultimately no significance between the foetal transplants and the placebo control, with the transplanted patients from four donors marginally failing to show significance ($P = 0.096$) [178]. The withdrawal of the immunosuppressant cyclosporine and subsequent graft rejection was raised by Olanow and co-workers as a possible reason why the foetal transplants ultimately failed. However, equally if not more concerning was that half of the transplanted patients began to develop graft-induced dyskinesias [178].

The two NIH trials had brought a disappointing conclusion to foetal transplantations as a symptomatic treatment for PD after the initial optimism following the earlier open-labelled studies. Consequently, many questions were raised concerning the future of PD and foetal transplantation such as the PD severity and patient selection, with the potential benefit of the transplant weighed up against the side effects of the post-surgery dyskinesias – the patients with more severe PD (i.e. a higher UPDRS score) possibly being more predisposed to dyskinesias post foetal transplantation. The optimum transplanted area and method of transplant were other questions to be determined with donor quality, quantity, age, storage time and conditions and transplanted region of foetal tissue were other important variables that needed addressing, lest not forgetting that increasing numbers of foetal tissue should be accompanied by an increasing need for immunosuppression.

In 2008 three independent research groups revisited their previous work and released their post-mortem results of PD patients with foetal midbrain transplants. Warren Olanow's research group analysed patient data from an open-labelled study from a PD patient 14 years post transplantation [174]. For 11 years the patient had an improved UPDRS, motor function and less dyskinesias; however, for the last 3 years of her life, her PD symptoms had increased [180]. At the post-mortem the grafted neurons were also found to have positive staining for α-synuclein and ubiquitin, reduced dopamine transporter (DAT) staining and more significantly a Lewy body pathology. Although grafted midbrain did improve function and patient quality of life, it did not arrest the disease and worryingly the disease pathology had spread from the endogenous host cells to the grafted striatum [180].

The earlier Lund trials were revisited by Li and colleagues who found that in patient autopsy's 11 and 16 years post-foetal transplant, the transplanted tissues had survived and provided symptomatic relief. However, what was also observed was α-synuclein and ubiquitin positively stained Lewy bodies and neurites were present in the grafted neurons suggesting host-to-graft disease prion-like propagation [181]. However, as these grafted neurons had provided 16 years of symptomatic relief with an unrelated death, it is debatable if an α-synuclein-driven prion disease propagation significantly affected the patient quality of life. The third set of PD patients post-mortems analysed from Isacson trial was the most successful. After receiving intracerebral transplantation of foetal ventral midbrain grafts, there were no Lewy body of Lewy neurites present [182]. However, serotoninergic neurons were found to be transplanted and thus a mixed non-specific cell population. Using PD patient scans, the group led by Paola Piccini showed that it is the serotoninergic neurons that are responsible for the dyskinesias following neural transplantation [183], confirming prior rat model studies [184].

Recently, a large international consortium headed by Roger Barker from the University of Cambridge has been set up with the aim of determining the efficacy of foetal transplantation in PD as a replacement for dopaminergic neurons [185]. This consortium, TRANSEURO, will attempt to rectify the variability and dyskinesias shown in previous trials with strict controls on patient selection, number, age and storage of donor foetuses, surgical techniques, graft size and placement, the use of immunosuppression and length of time on immunosuppression all being addressed [177, 178, 186]. The results of this trial (NCT 01898390) are due in 2017. The success and long-term viability of the foetal transplantation studies are important not only in PD but in other neurodegenerative diseases such as HD where foetal transplants have also had mixed success [187–189]. If the foetal transplantations are successful, it could act as a proof of principle and be a precursor for the use of stem cell therapy in neurodegenerative diseases, such as PD. How the TRANSEURO study will address the propagation of the prion-like disease pathology from the host cells to the transplanted tissue however is still to be determined.

Advantages and Disadvantages of Using Stem Cell Therapy in the Treatment of Neurological Disorders

Advantages

There are significant advantages of using stem cell therapy for the treatment of neurological disorders, the ability to generate patient-derived disease lines and then study those degenerated cells of that disease is of unquestionable benefit the researchers aiding the progression and understanding of the disease. The ability to generate in vitro de novo neurons, oligodendrocytes and astrocytes of a specific regional identity enables researchers to look at specific vulnerabilities to specific cell types, also increasing the understanding of the ameliorating ability of the cell types to aid the disease propagation.

Studies from patient-derived iPS cells have furthermore exemplified some of the limitations of animal models to accurately model disease, with rodent models of PD failing to recapitulate human disease pathology such as the selective loss of midbrain dopaminergic neurons and accumulation of α-synuclein [190]. In addition to this studies have shown that in in vitro neuronal disease models that the disease phenotype is only apparent in the differentiated neurons, not the fibroblasts or the pluripotent iPS cells [89, 191] consequently to study the disease progression and manifestation disease-specific iPS terminally differentiated neurons and also neural progenitors have become an essential tool in research, having the additional potential to minimise the extent of the animal models used in research. Animal models however are still a vital tool for researchers with particular relevance in development and embryology; behavioural studies; ascertaining the effect and side effects of new drugs and treatments on a systemic level; and is also the first step necessary for testing the validity cell replacement therapies prior to human clinical trials and eventually the bedside.

The validity and understanding gauged from these terminally differentiated neurons will become more apparent as the protocols necessary to develop these terminally differentiated diseased cell types are generated with greater specificity and efficacy. The sheer quantity and availability of neurons and neuronal precursor cells that can be differentiated from stem cells lends itself perfectly as a model for toxicity testing via high-content screening and automation in screening new and existing therapies to slow/halt/potentially recover disease manifestation, progression and development of pathology. The ability of an accurate recapitulation of the disease cell type coupled with the exciting potential of genome editing technology such as CRIPSR can be a significant step forward in treating neurodegenerative diseases. In diseases such as HD where a high penetrance monogenic mutation is responsible for aetiology, disease correction by removal of the abnormal CAG repeats, without leaving a genomic footprint can effectively cure the disease in de novo isogenic cells. However to be an effective therapy it can only be used provided 100 % certainty of no off-targeting of the genome, genome editing cannot currently give this guarantee. In AD, PD and ALS, the majority of the cases are idiopathic with no abnormal mutations to correct; thus, it is only by further understanding of the disease aetiology that makes it evidently more treatable.

Challenges

There are many challenges associated with using hES cells for research, such as the ethical issues; legislative funding concerns, particularly in USA; and a risk of immune rejection. Therefore there has been a gravitation towards using patient-derived iPS cells as the primary tool of research as it enables not only disease modelling but also the onset of personalised medicine. However, there are, if not more challenges associated with iPSCs as there were with hESCs.

The first of these challenges unique to iPSCs is the method of reprogramming; it is tantamount to have a methodology in which miRNA is not randomly integrated into the genome, where reprogramming should be transgene-free leaving behind no genomic footprint. Concerns however still remain regarding the heterogeneity of the iPS cell lines generated compared to the hES cell lines and the variable clones that they produce [89]. There is also the matter that the most common somatic cell type used for reprogramming is skin from patients of all ages. The skin is one of the somatic cells that is more likely to have inherent mutations due to constant exposure to UV damaging sunlight with at least half of the mutations seen from reprogramming already present in the pre-existing fibroblasts [85].

Another more recent concern regarding both iPS-based and direct somatic cell reprogramming of both iN and iNSCs is the phenomenon of epigenetic memory [86, 192], which not only exemplifies the amount of unknown variables in this area of research but also changes any assumption that iPS/somatic cell differentiation is a constant, with epigenetic memory known to vary DNA methylation which in turn can influence differentiation capabilities of directing cell fate in addition to molecular and functional properties of iPS cells [86, 193]. This phenomenon also stresses the importance of having multiple control lines and the use of hESCs as non-iPS control lines. However, hESC lines are also variable, having different characteristics and different lineage differentiation propensities [194]. An additional problem of iNSCs and iN compared to iPSCs is the length of time in culture, lack of cell numbers generated and the variable cell population generated [115, 125].

Consequently, this raises another concern regarding the time and thus number of in vitro passages that cells receive which lead to an increasing genotypic abnormality and karyotypic instability. This raises an important point that needs to be addressed if ever patient-specific personalised medicine is able translate to the bedside, which is cell maintenance in an in vitro environment. The longer cells are cultured in vitro; they are more likely to acquire selective adaptations such as alterations in DNA methylation, X-chromosome instability, imprinting instabilities and partial and full chromosomal aberrations such as trisomy of chromosomes 8, 12 and 17 [195].

Inactivated mouse embryonic fibroblast (iMEF) cells are still used as a supporting matrix for both iPS and hES cell maintenance. However, the complications of using these murine layers is that non-human sialic acid N-glycolylneuraminic acid (Neu5Gc) has been detected on the surface of hESCs maintained on MEF feeder layer which is potentially immunogenic [196, 197]. Subsequently feeder-free iPS/hES cell culture has become more apparent using Matrigel™ and CELLstart™; however, chromosomal abnormalities have still been reported [198, 199] and Matrigel is still derived from mouse sarcoma. An alternative of using murine feeders and non-xeno-free matrices is using autologous feeders [200]. Unfortunately after an increasing number of passages, fibroblasts become less viable and increasingly prone to senescence compared to younger samples [201]; also unlike iPSCs or hESCs, fibroblasts are not immortal and have a limited number of passages. Recently xeno-free clinical grade iPS cell lines have been described, with mandatory procedural guidelines necessary for quality control and good manufacturing practice also published [202, 203].

Results From the Bench

Due to the advent of patient-specific disease cells, there is now a plethora of information being released into the literature from disease modelling. The terminal differentiation of iPS and neural precursors into regionally relevant specific neurons, oligodendrocytes and astrocytes to recapitulate the disease conferring selective vulnerability and disease phenotype in response to toxicity has led to a variety of candidate drugs to be further researched for clinical therapy.

Alzheimer's Disease

In AD, both β- and γ-secretases had led to a reduction of Aβ in patient-derived familial *APP-*, *PSEN1-* and *PSEN2*-mutated diseased neurons, with β-secretase inhibitors giving a partial reduction in activated GSK3β and phosphorylated tau [204, 205]. Docosahexaenoic acid (DHA) was found to lower the reactive oxygen species (ROS) in addition to a decreased in cell death in an *APP* patient-derived cell line [206]. The use of DHA was previously described in a mouse model to protect against amyloid and dendritic pathology [207].

Parkinson's Disease

In PD there have been three research articles with a familial triplication in *SNCA* (A53T) [88, 89, 106]. In all three articles, phenotype was shown, with a significant increase in the gene and protein expression levels and secretion levels of α-synuclein [88, 89] and a significant increase in stress gene expression levels, including *UCHL1*, one of the PARK genes (PARK5) [88]. The most recent paper looked at nitrosative stress in the *SNCA* A53T line before and after correction by genome editing [106]. Using yeast as an initial platform to model the disease, two markers of ER stress, BIP and PD1, were found to have significant elevation in A53T. ER-associated dysfunction (ERAD) substrates, such as sensitivity to glucocerebrosidase (GBA), another PD-associated risk gene [106, 114], were reversed using a compound called synoviolin, an E3 ubiquitin ligase. Using a small molecule screen based on yeast, a compound called NAB2 was found to increase post-ER forms and ameliorate ER accumulation of ERAD substrates in A53T PD-iPS neurons [106].

In the *LRRK2* mutation in PD, there have been six papers published from patient-derived neurons [91, 92, 191, 208–210]. Overall, it has been shown that there is a link between *LRRK2* dysfunction and α-synuclein accumulation [91, 191, 209]; tau and phosphorylated tau [91]; autophagy impairment [209]; mitochondrial DNA mutations [210]; significantly lower mitochondrial consumption rate than controls [208]; and a decrease in neurite length [91, 191]. After previous suggestions that *LRRK2* (G2019S) is a toxic gain-of-function kinase [211, 212], a small molecular kinase inhibitor significantly lowered numbers caspase-3^{+}/TH^{+} neurons [91] and also rescued defects in nuclear architecture [92].

Huntington's Disease

In the first patient-derived HD-iPS study in both heterozygous and homozygous patients, there was no change in CAG repeat length following reprogramming, which also did not change either after 40 passages of the culture [213]. CAG instability has been previously shown in HD gametes [214], lymphoblasts [215], post-mitotic murine neurons [216] and human striatal neurons [217]. Camnasio et al. also found HD-iPS neurons did not affect neuronal differentiation and had a significant increase in lysosomal activity over control lines [213]. Another HD paper showed that following transplantation of GABAergic neurons of the forebrain LGE identity from a 72 CAG-repeat HD-iPS line into a rat containing a unilateral excitotoxic striatal lesion, a significant behavioural recovery was obtained [218]. However, in this study the in vivo transplanted neurons developed HD pathology 33 weeks after transplantation. The in vivo neurons when exposed to MG132 and thus proteasomal stress also developed HD pathology [218].

Using genome editing via BACS a 72-CAG repeat was replaced with a normal 21-CAG repeat *HTT* gene [93]. The corrected HD line reversed the analysed HD pathology of significantly reduced mRNA levels of BDNF, TGF-β1 and N-cadherin; significantly reduced maximum respiration levels; and significantly increased activated caspase-3. The corrected HD cell line was able to differentiate in vitro into DARPP32^{+} neurons and integrate and survive in a mouse model. In a study of multiple patient-derived HD-iPS lines, the HD Consortium has also extensively showed that an increased HD pathology of relative intracellular ATP/ADP ratio and cleaved caspase-3 was associated with the number of CAG repeats [219].

Amyotrophic Lateral Sclerosis

In ALS a patient-derived iPS line from TDP-43 mutated neurons displayed phenotype of significantly shorter neurite length and increased insoluble protein of TDP-43 compared to non-disease controls [220]. There was also an upregulation in RNA transcription genes and downregulation in cytoskeletal protein genes, with TDP-43 being involved in RNA metabolism [221, 222]. This study also found that anacardic acid was able to prevent cytotoxicity by arsenite, reversing both neurite length and insoluble protein fraction pathology; it was postulated that this mechanism of action was either via redox reduction or suppression of NF-κB protein complex [220].

Consequently using earlier research done in lower-order mammals and yeast, new pharmacological therapies can quickly be ascertained for their efficacy and therapeutic potential using patient-derived disease-specific cell lines, which have the potential ability to be a causative treatment, reverse pathogenesis, enable better presymptomatic treatment, halt pathogenesis and facilitate better symptomatic treatment and thus improvement in patient quality of life.

Stem Cell Therapy: Readiness to the Bedside I

Parkinson's Disease

PD is a neurological disease where stem cell replacement therapy is a possible strategy for the treatment of the disease in which the precedent has been set transplantation of foetal mesencephalic tissue into the midbrain. Yet PD is not just a motor disorder affecting the dopaminergic neuronal network. PD is a complex and progressive neurodegenerative disease, and PD patients have many non-motor symptoms and often go on to develop dementia. A cell replacement therapy of dopaminergic neurons will not prevent this. Consequently to treat PD, one has to ask what is the aim of the treatment. If the answer is to cure the disease and prevent it, then the field of research is a long way off, as although the disease mechanisms are further being elucidated, such as the increasing number of PARK genes and risk loci from GWAS studies, the accumulation of the α-synuclein and dysfunction of *LRRK2* are still the most common causes. Fundamentally there is still a clear lack of definitive understanding of the PD aetiology and the step-by-step manifestation of the complex and patient-variable disease pathology. Also for much improved outcomes of PD treatment, scientists and clinicians would need to be able to get access to and treat the patient whilst the patient is presymptomatic and has greater than 30 % of the dopaminergic neurons remaining and still functional.

If the purpose of the treatment however is to improve the quality of life, then existing cell therapies have shown to be effective as revealed in the earlier foetal transplant trials [174, 180, 181] with the patient having lower UPDRS scoring indicating improvements in quality of life. The widespread use of foetal transplants in PD however is not a sustainable long-term strategy; therefore, to replace these foetal transplants, stem cell-derived neuronal precursors could be transplanted. The current protocol that produces the greatest amount of TH^+ $FoxA2^+$ neurons is only at approximately 20 % [130], with, the remaining 80 % of cells generated being non-specific. Consequently the current research literature has not advanced sufficiently for iPS-based cell replacement therapy to become a realistic therapeutic strategy yet.

The dangers of transplanting a mixed population of cells have been previously shown in the off-stake dyskinesias in the foetal transplants [177, 223] which has later been confirmed to be a result of contaminating serotoninergic neurons in the transplanted tissue [183]. Although recent evidence has now indicated that it is the loss of cholinergic neurons not dopaminergic neurons that correlates to gait difficulty: one of the cardinal motor symptoms in PD [224]. Therefore, with further research it is becoming more apparent that transplantation of a pure 100 % dopaminergic neuronal population may not in fact be the best possible treatment. The original therapeutic theory was the cell replacement of the degenerated cells with de novo in vitro generated cells; however, more research and understanding is necessary to determine what all of those cells are. It is currently known that the loss of midbrain A9 dopaminergic neurons of the SNc causes PD, yet PD is implicated in non-dopaminergic multiple pathways: noradrenergic, serotoninergic, glutamatergic and cholinergic within the regions of the cortex, brainstem and basal ganglia [225]. Also with 40 % of all PD patients going on to get dementia [226], the replacement of the dopaminergic neurons would not prevent this.

In addition, cell-replacement therapies do not treat the disease aetiology. Eventually the disease will degenerate the de novo cells as it did the endogenous cells to the prion-like propagation of α-synuclein. There is increasingly evidence that α-synuclein, TDP-43, tau and Aβ are prion diseases [181, 227–233, 240]. This will create a problem for any cell replacement therapies in those diseases in which those toxic proteins take place, PD, ALS and AD. Therefore, for any future cell replacement therapies, the propagation of the disease from host to grafted tissue must be prevented. Recently a group headed by Giovanna Mallucci has published that treatment of a prion-infected mouse model with an inhibitor of the protein kinase PERK (protein kinase RNA-like endoplasmic reticulum kinase) prevented over-activation of the unfolded protein response (UPR) system that otherwise leads to protein synthesis accumulation and prion replication [234]. Provided this result can be replicated in human models, this drug could be used in conjugation with any cell replacement therapy and also given routinely as a therapeutic therapy for these diseases.

Each neurological disorder is different; consequently each requires different treatment. In diseases such as AD and ALS, there is systemic loss of multiple cell types; a cell replacement bench to bedside therapy would not necessarily be the easiest and most effective option with a pharmaceutical intervention being a more realistic method of treatment. Also studies in HD have shown diseases can propagate through astrocytes and glial [235]; in addition to this the disease is not localised to a specific area making transplantation to multiple locations difficult.

There are a lot of pathological similarities in neurodegenerative diseases that warrant further research. For the diseases of protein misfolding, such as PD, AD and ALS, this leads to ER stress; therefore, limiting ER stress by reducing protein synthesis could be a strategy for therapeutic treatment [236]. Another example of this and the closer link between PD and ALS is *FIG4*: a phosphatase that regulates intracellular vesicle trafficking along the endosomal–lysosomal pathway, that mutations lead to ALS and also Charcot–Marie–Tooth neuropathy, has been found in Pick bodies, Picks disease, Lewy bodies in PD and dementia with Lewy bodies (DLB) [237]. This adds further scrutiny to the endosomal–lysosomal pathway, pushing it to a central role in many neurodegenerative diseases. As TDP-43 has been found in ALS, FTN, AD [238, 239] and is can mechanisms known to cause TDP-43 dysfunction implicated in other neurological diseases: Lewy body dementia, Down syndrome, hippocampal sclerosis dementia, familial British dementia and spinal cerebellar ataxia [238]. TDP-43 could be a marker of general neurodegeneration with suppression of aggregated TDP-43 propagation a target for therapy [240].

Stem Cell Therapy: Readiness to the Bedside II

Multiple Sclerosis

A previous strategy on treating MS was the generation of oligodendrocytes as a cell therapy to replace the endogenous myelin-sheath-forming oligodendrocytes. However, it is now established that in MS the OPCs have not been ablated by the activated lymphoblasts of the innate immune system and can still be mobilised to myelin-sheath-forming oligodendrocytes [241]. Due to aging however, the OPCs become less efficient in differentiating the myelin-sheath-forming oligodendrocytes [242]. Thus, over time the inherent regenerative process that occurs normally and efficiently becomes chronic and inefficient leading to selective vulnerabilities in the motor neuron and subsequent degeneration. It is currently unknown why the motor neuron is selectively degenerated. MS is not a cell autonomous disease; therefore, replacing the degenerated motor neurons in multiple locations would not be a sustainable or realistic source of therapy. To treat MS a synergistic strategy must be employed of suppressing the immune system and repairing the damaged axons by facilitating the endogenous differentiation of the OPCs.

Bone marrow-derived MSCs have shown to be an effective immunomodulator, more so than both adipose-derived and cord blood-derived MSCs, in the suppression of the in vitro proliferation of mitogen or antigen-stimulated T-cell responses [243]. Bone marrow-derived MSCs have previously been shown to promote endogenous repair and functional recovery in animals of the disease [244]. The results from two of the bone marrow-derived phase I clinical therapies indicated that the delivery of the MSCs were safe with no tumourigenicity detected [245, 246]. In one of the studies, there was also evidence of structural, functional and physiological improvement [245]. There are now 13 phase I or phase II clinical trials taking place using bone marrow-, umbilical cord- or adipose-derived MSCs for the treatment of MS [35]. Therefore, using a strategy specific for a disease, it appears that MSCs may be the best effective therapeutic strategy for the treatment of MS.

Conclusion

In conclusion it can therefore be seen that apart from the hopeful and exciting trials in MS, the bench to bedside stem cell therapy for neurological disorders is still a considerable way off. There still remains a great deal of unknowns regarding these neurological diseases. The aetiology, manifestation and subsequent pathology are patient-variable and idiopathic, with common pathways and processes implicated across the different neurodegenerative diseases. The new technologies of genome editing without a molecular footprint represents an exciting chapter in research by being able to correct, control and better model the disease. The progression in differentiation protocols and the advent of direct reprogramming will hopefully speed up the time taken to conduct the research and enable greater study and understanding of the disease. Thus, by being able to understand these neurological disorders, it allows for a better informed rationale when it comes to treatment. Questions still remain about the safety and reproducibility of the patient-derived iPS-cell technology as a route to cell replacement therapies not withstanding the logistical practical and financial implications. However from these patient-derived iPS based cellular models it is inevitable that greater understanding and mechanisms behind the pathology will be found and then hopefully attenuated. Then next question would be how readily and how efficious will these subsequent therapies be in translating from the localised in vitro terminally differentiated cell model to the systemic in vivo environment in which the disease state is in variable stages of progression and is patient-specific. Once the answers to these questions are ascertained the bench to the bedside approach of using stem cell mediated therapy to treat neurological diseases will be realised.

References

1. http://www.ninds.nih.gov/disorders/disorder_index.htm. Accessed 1 Aug 2013
2. Holtzman DM, Morris JC, Goate AM (2011) Alzheimer's disease: the challenge of the second century. Sci Transl Med 3, 77 sr71. doi:10.1126/scitranslmed.3002369
3. Wimo A, Prince M (2010) World Alzheimer report 2010: the global economic impact of Dementia. Alzheimer's Dis Int 1–56
4. Braak H, Braak E (1991) Neuropathological stageing of Alzheimer-related changes. Acta Neuropathol 82:239–259. doi:10.1007/bf00308809
5. Sherrington R et al (1995) Cloning of a gene bearing missense mutations in early-onset familial Alzheimer's disease. Nature 375(6534):754–760

6. Levy-Lahad E et al (1995) Candidate gene for the chromosome 1 familial Alzheimer's disease locus. Science 269(5226):973–977
7. Goate A et al (1991) Segregation of a missense mutation in the amyloid precursor protein gene with familial Alzheimer's disease. Nature 349(6311): 704–706
8. Scheuner D et al (1996) Secreted amyloid [beta]-protein similar to that in the senile plaques of Alzheimer's disease is increased in vivo by the presenilin 1 and 2 and APP mutations linked to familial Alzheimer's disease. Nat Med 2(8): 864–870
9. Kumar-Singh S et al (2006) Mean age-of-onset of familial Alzheimer disease caused by presenilin mutations correlates with both increased Aβ42 and decreased Aβ40. Hum Mutat 27(7): 686–695
10. Bentahir M et al (2006) Presenilin clinical mutations can affect γ-secretase activity by different mechanisms. J Neurochem 96(3):732–742
11. De Strooper B et al (1998) Deficiency of presenilin-1 inhibits the normal cleavage of amyloid precursor protein. Nature 391(6665):387–390
12. Braak H, Braak E (2000) Pathoanatomy of Parkinson's disease. J Neurol 247:II3–II10
13. Kowal SL, Dall TM, Chakrabarti R, Storm MV, Jain A (2013) The current and projected economic burden of Parkinson's disease in the United States. Mov Disord 28:311–318. doi:10.1002/mds.25292
14. Fearnley JM, Lees AJ (1991) Ageing and Parkinson's disease: substantia nigra regional selectivity. Brain 114:2283–2301. doi:10.1093/brain/114.5.2283
15. Campion D et al (1999) Early-onset autosomal dominant Alzheimer disease: prevalence, genetic heterogeneity, and mutation spectrum. Am J Hum Genet 65(3):664–670
16. Genin E et al (2011) APOE and Alzheimer disease: a major gene with semi-dominant inheritance. Mol Psychiatry 16(9):903–907
17. Farrer LA et al (1997) Effects of age, sex, and ethnicity on the association between apolipoprotein e genotype and Alzheimer disease: a meta-analysis. JAMA 278(16):1349–1356
18. Mounsey RB, Teismann P (2010) Mitochondrial dysfunction in Parkinson's disease: pathogenesis and neuroprotection. Parkinson's Dis 2011(Article ID 617472): 18 pages
19. Klein C, Westenberger A (2012) Genetics of Parkinson's disease. Cold Spring Harb Perspect Med 2(1)
20. Kuusisto E, Parkkinen L, Alafuzoff I (2003) Morphogenesis of Lewy bodies: dissimilar incorporation of [alpha]-synuclein, ubiquitin, and p62. J Neuropathol Exp Neurol 62:1241–1253
21. Wakabayashi K, Tanji K, Mori F, Takahashi H (2007) The Lewy body in Parkinson's disease: molecules implicated in the formation and degradation of α-synuclein aggregates. Neuropathology 27:494–506
22. Maries E et al (2003) The role of [alpha]-synuclein in Parkinson's disease: insights from animal models. Nat Rev Neurosci 4(9):727–738
23. Singleton AB et al (2003) {alpha}-Synuclein locus triplication causes Parkinson's disease. Science 302(5646):841
24. Ibáñez P et al (2004) Causal relation between A-synuclein locus duplication as a cause of familial Parkinson's disease. The Lancet 364(9440): 1169–1171
25. Miyake Y et al (2012) SNCA polymorphisms, smoking, and sporadic Parkinson's disease in Japanese. Parkinsonism Relat Disord 18(5): 557–561
26. Pan F et al (2012) SNP rs356219 of the α-synuclein (SNCA) gene is associated with Parkinson's disease in a Chinese Han population. Parkinsonism Relat Disord 18(5):632–634
27. Westerlund M, Hoffer B, Olson L (2010) Parkinson's disease: exit toxins, enter genetics. Prog Neurobiol 90(2):146–156
28. MacDonald ME et al (1993) A novel gene containing a trinucleotide repeat that is expanded and unstable on Huntington's disease chromosomes. Cell 72:971–983. doi:http://dx.doi.org/10.1016/0092-8674(93)90585-E
29. Estrada Sánchez AM, Mejía-Toiber J, Massieu L (2008) Excitotoxic neuronal death and the pathogenesis of Huntington's disease. Arch Med Res 39:265–276. doi:http://dx.doi.org/10.1016/j.arcmed.2007.11.011
30. Reiner A et al (1988) Differential loss of striatal projection neurons in Huntington disease. Proc Natl Acad Sci 85:5733–5737
31. Lobsiger CS, Cleveland DW (2007) Glial cells as intrinsic components of non-cell-autonomous neurodegenerative disease. Nat Neurosci 10(11): 1355–1360
32. Langbehn DR et al (2004) A new model for prediction of the age of onset and penetrance for Huntington's disease based on CAG length. Clin Genet 65:267–277. doi:10.1111/j.1399-0004.2004.00241.x
33. Rosenblatt A et al (2006) The association of CAG repeat length with clinical progression in Huntington disease. Neurology 66:1016–1020
34. Walker FO (2007) Huntington's disease. Lancet 369:218–228. doi:http://dx.doi.org/10.1016/S0140-6736(07)60111-1
35. Rice CM, Kemp K, Wilkins A, Scolding NJ (2013) Cell therapy for multiple sclerosis: an evolving concept with implications for other neurodegenerative diseases. Lancet 382:1204–1213. doi:http://dx.doi.org/10.1016/S0140-6736(13)61810-3
36. Compston A, Coles A (2008) Multiple sclerosis. Lancet 372:1502–1517. doi:http://dx.doi.org/10.1016/S0140-6736(08)61620-7
37. Olsson T et al (1990) Autoreactive T lymphocytes in multiple sclerosis determined by antigen-induced secretion of interferon-gamma. J Clin Invest 86:981–985. doi:10.1172/jci114800

38. Disanto G et al (2013) Epstein–Barr virus, latitude and multiple sclerosis. Mult Scler J 19:362–365. doi:10.1177/1352458512451942
39. Simpson S, Blizzard L, Otahal P, Van der Mei I, Taylor B (2011) Latitude is significantly associated with the prevalence of multiple sclerosis: a meta-analysis. J Neurol Neurosurg Psychiatry 82:1132–1141. doi:10.1136/jnnp.2011.240432
40. Orton S-M et al (2006) Sex ratio of multiple sclerosis in Canada: a longitudinal study. Lancet Neurol 5:932–936. doi:http://dx.doi.org/10.1016/S1474-4422(06)70581-6
41. Stüve O, Oksenberg J (2010) In: Pagon RA, Adam MP, Bird TD (eds) GeneReviews
42. Barcellos LF et al (2006) Heterogeneity at the HLA-DRB1 locus and risk for multiple sclerosis. Hum Mol Genet 15:2813–2824. doi:10.1093/hmg/ddl223
43. Alcina A et al (2012) Multiple sclerosis risk variant *HLA-DRB1*1501* associates with high expression of *DRB1* gene in different human populations. PLoS One 7:e29819. doi:10.1371/journal.pone.0029819
44. Pesiridis GS, Lee VM-Y, Trojanowski JQ (2009) Mutations in TDP-43 link glycine-rich domain functions to amyotrophic lateral sclerosis. Hum Mol Genet 18:R156–R162. doi:10.1093/hmg/ddp303
45. Kiernan MC et al (2011) Amyotrophic lateral sclerosis. Lancet 377:942–955. doi:http://dx.doi.org/10.1016/S0140-6736(10)61156-7
46. Orenstein SJ et al (2013) Interplay of LRRK2 with chaperone-mediated autophagy. Nat Neurosci 16(4):394–406
47. Friedman LG et al (2012) Disrupted autophagy leads to dopaminergic axon and dendrite degeneration and promotes presynaptic accumulation of α-synuclein and LRRK2 in the brain. J Neurosci 32(22):7585–7593
48. Badiola N et al (2011) Tau enhances A-synuclein aggregation and toxicity in cellular models of synucleinopathy. PLoS One 6(10):e26609
49. Clinton LK et al (2010) Synergistic interactions between ABeta, Tau, and A-Synuclein: acceleration of neuropathology and cognitive decline. J Neurosci 30(21):7281–7289
50. Takahashi K, Yamanaka S (2006) Induction of pluripotent stem cells from mouse embryonic and adult fibroblast cultures by defined factors. Cell 126:663–676
51. Gurdon JB (1962) The developmental capacity of nuclei taken from intestinal epithelium cells of feeding tadpoles. J Embryol Exp Morphol 10:622–640
52. Takahashi K et al (2007) Induction of pluripotent stem cells from adult human fibroblasts by defined factors. Cell 131:861–872
53. Yu J et al (2007) Induced pluripotent stem cell lines derived from human somatic cells. Science 318:1917–1920. doi:10.1126/science.1151526
54. Niknejad H, Deihim T, Ahmadiani A, Jorjani M, Peirovi H (2012) Permanent expression of midbrain dopaminergic neurons traits in differentiated amniotic epithelial cells. Neurosci Lett 506(1):22–27
55. Yan X et al (2009) iPS cells reprogrammed from human mesenchymal-like stem/progenitor cells of dental tissue origin. Stem Cells Dev 19:469–480
56. Narsinh KH et al (2010) Generation of adult human induced pluripotent stem cells using nonviral minicircle DNA vectors. Nat Protoc 6:78–88
57. Ye Z et al (2009) Human-induced pluripotent stem cells from blood cells of healthy donors and patients with acquired blood disorders. Blood 114:5473–5480. doi:10.1182/blood-2009-04-217406
58. Yuin-Han L et al (2009) Generation of induced pluripotent stem cells from human blood. Blood 113:5476–5479
59. Brown ME et al (2010) Derivation of induced pluripotent stem cells from human peripheral blood T lymphocytes. PLoS One 5:e11373
60. Zhou T et al (2012) Generation of human induced pluripotent stem cells from urine samples. Nat Protocols 7:2080–2089. doi:http://www.nature.com/nprot/journal/v7/n12/abs/nprot.2012.115.html#supplementary-information
61. Okita K, Ichisaka T, Yamanaka S (2007) Generation of germline-competent induced pluripotent stem cells. Nature 448:313–317
62. Yu J et al (2009) Human induced pluripotent stem cells free of vector and transgene sequences. Science 324:797–801. doi:10.1126/science.1172482
63. Zhou W, Freed CR (2009) Adenoviral gene delivery can reprogram human fibroblasts to induced pluripotent stem cells. Stem Cells 27:2667–2674. doi:10.1002/stem.201
64. Fusaki N, Ban H, Nishiyama A, Saeki K, Hasegawa M (2009) Efficient induction of transgene-free human pluripotent stem cells using a vector based on Sendai virus, an RNA virus that does not integrate into the host genome. Proc Jpn Acad Ser B 85:348–362
65. Jia F et al (2010) A nonviral minicircle vector for deriving human iPS cells. Nat Methods 7:197–199. doi:http://www.nature.com/nmeth/journal/v7/n3/suppinfo/nmeth.1426_S1.html
66. Okita K et al (2011) A more efficient method to generate integration-free human iPS cells. Nat Methods 8:409–412. doi:http://www.nature.com/nmeth/journal/v8/n5/abs/nmeth.1591.html#supplementary-information
67. Park HY et al (2012) Efficient generation of virus-free iPS cells using liposomal magnetofection. PLoS One 7:e45812. doi:10.1371/journal.pone.0045812
68. Rais Y et al (2013) Deterministic direct reprogramming of somatic cells to pluripotency. Nature 502: 65–70. doi:10.1038/nature12587, http://www.nature.com/nature/journal/v502/n7469/abs/nature12587.html#supplementary-information
69. Kaji K et al (2009) Virus-free induction of pluripotency and subsequent excision of reprogramming factors. Nature 458:771–775. doi:http://www.nature.com/nature/journal/v458/n7239/suppinfo/nature07864_S1.html

70. Woltjen K et al (2009) piggyBac transposition reprograms fibroblasts to induced pluripotent stem cells. Nature 458:766–770. doi:http://www.nature.com/nature/journal/v458/n7239/suppinfo/nature07863_S1.html
71. Sommer CA et al (2010) Excision of reprogramming transgenes improves the differentiation potential of iPS cells generated with a single excisable vector. Stem Cells 28:64–74. doi:10.1002/stem.255
72. Kim D et al (2009) Generation of human induced pluripotent stem cells by direct delivery of reprogramming proteins. Cell Stem Cell 4:472–476. doi:http://dx.doi.org/10.1016/j.stem.2009.05.005
73. Warren L et al (2010) Highly efficient reprogramming to pluripotency and directed differentiation of human cells with synthetic modified mRNA. Cell Stem Cell 7:618–630. doi:http://dx.doi.org/10.1016/j.stem.2010.08.012
74. Miyoshi N et al (2011) Reprogramming of mouse and human cells to pluripotency using mature microRNAs. Cell Stem Cell 8:633–638
75. Hou P et al (2013) Pluripotent stem cells induced from mouse somatic cells by small-molecule compounds. Science 341:651–654. doi:10.1126/science.1239278
76. Martins-Taylor K, Xu R-H (2012) Concise review: genomic stability of human induced pluripotent stem cells. Stem Cells 30:22–27. doi:10.1002/stem.705
77. Baker DEC et al (2007) Adaptation to culture of human embryonic stem cells and oncogenesis in vivo. Nat Biotechnol 25:207–215. doi:http://www.nature.com/nbt/journal/v25/n2/suppinfo/nbt1285_S1.html
78. Ben-David U, Benvenisty N (2011) The tumorigenicity of human embryonic and induced pluripotent stem cells. Nat Rev Cancer 11:268–277
79. Lanford RE et al (2011) Therapeutic silencing of microRNA-122 in primates with chronic hepatitis C virus infection. Science 327:198–201. doi:10.1126/science.1178178
80. Li Y, Masaki T, Yamane D, McGivern DR, Lemon SM (2013) Competing and noncompeting activities of miR-122 and the 5′ exonuclease Xrn1 in regulation of hepatitis C virus replication. Proc Natl Acad Sci 110:1881–1886. doi:10.1073/pnas.1213515110
81. Grinnemo K-H, Sylvén C, Hovatta O, Dellgren G, Corbascio M (2008) Immunogenicity of human embryonic stem cells. Cell Tissue Res 331:67–78
82. Palmisano I et al (2012) Amino acid starvation induces reactivation of silenced transgenes and latent HIV-1 provirus via down-regulation of histone deacetylase 4 (HDAC4). Proc Natl Acad Sci 109:E2284–E2293. doi:10.1073/pnas.1202174109
83. Toivonen S et al (2013) Comparative analysis of targeted differentiation of human induced pluripotent stem cells (hiPSCs) and human embryonic stem cells reveals variability associated with incomplete transgene silencing in retrovirally derived hiPSC lines. Stem Cells Transl Med 2:83–93. doi:10.5966/sctm.2012-0047
84. Hussein SM et al (2011) Copy number variation and selection during reprogramming to pluripotency. Nature 471:58–62. doi:http://www.nature.com/nature/journal/v471/n7336/abs/10.1038-nature09871-unlocked.html#supplementary-information
85. Gore A et al (2011) Somatic coding mutations in human induced pluripotent stem cells. Nature 471:63–67. doi:http://www.nature.com/nature/journal/v471/n7336/abs/10.1038-nature09805-unlocked.html#supplementary-information
86. Kim K et al (2010) Epigenetic memory in induced pluripotent stem cells. Nature 467:285–290. doi:http://www.nature.com/nature/journal/v467/n7313/abs/nature09342.html#supplementary-information
87. Sommer CA et al (2012) Residual expression of reprogramming factors affects the transcriptional program and epigenetic signatures of induced pluripotent stem cells. PLoS One 7:e51711. doi:10.1371/journal.pone.0051711
88. Byers B et al (2011) SNCA triplication Parkinson's patient's iPSC-derived DA neurons accumulate a-synuclein and are susceptible to oxidative stress. PLoS One 6:e26159
89. Devine MJ et al (2011) Parkinson's disease induced pluripotent stem cells with triplication of the A-synuclein locus. Nat Commun 2:440
90. Soldner F et al (2011) Generation of isogenic pluripotent stem cells differing exclusively at two early onset Parkinson point mutations. Cell 146:318–331
91. Reinhardt P et al (2013) Genetic correction of a LRRK2 mutation in human iPSCs links Parkinsonian neurodegeneration to ERK-dependent changes in gene expression. Cell Stem Cell 12:354–367. doi:http://dx.doi.org/10.1016/j.stem.2013.01.008
92. Liu G-H et al (2012) Progressive degeneration of human neural stem cells caused by pathogenic LRRK2. Nature 491:603–607. doi:http://www.nature.com/nature/journal/v491/n7425/abs/nature11557.html#supplementary-information
93. An MC et al (2012) Genetic correction of Huntington's disease phenotypes in induced pluripotent stem cells. Cell Stem Cell 11:253–263. doi:http://dx.doi.org/10.1016/j.stem.2012.04.026
94. Liu G-H et al (2011) Recapitulation of premature ageing with iPSCs from Hutchinson-Gilford progeria syndrome. Nature 472:221–225. doi:http://www.nature.com/nature/journal/v472/n7342/abs/10.1038-nature09879-unlocked.html#supplementary-information
95. Liu G-H et al (2011) Targeted gene correction of laminopathy-associated LMNA mutations in patient-specific iPSCs. Cell Stem Cell 8:688–694. doi:http://dx.doi.org/10.1016/j.stem.2011.04.019
96. Howden SE et al (2011) Genetic correction and analysis of induced pluripotent stem cells from a patient with gyrate atrophy. Proc Natl Acad Sci 108:6537–6542. doi:10.1073/pnas.1103388108

97. Hockemeyer D et al (2009) Efficient targeting of expressed and silent genes in human ESCs and iPSCs using zinc-finger nucleases. Nat Biotechnol 27: 851–857. doi:10.1038/nbt.1562, http://www.nature.com/nbt/journal/v27/n9/abs/nbt.1562.html#supplementary-information
98. Song H, Chung S-K, Xu Y (2010) Modeling disease in human ESCs using an efficient BAC-based homologous recombination system. Cell stem cell 6:80–89. doi:http://dx.doi.org/10.1016/j.stem.2009.11.016
99. Hockemeyer D et al (2011) Genetic engineering of human pluripotent cells using TALE nucleases. Nat Biotechnol 29:731–734. doi:http://www.nature.com/nbt/journal/v29/n8/abs/nbt.1927.html#supplementary-information
100. Cong L et al (2013) Multiplex genome engineering using CRISPR/Cas systems. Science 339:819–823. doi:10.1126/science.1231143
101. Gruenert DC, Sargent RG (2012) Virus-mediated genetic surgery: homologous recombination with a little [ldquo]helper[rdquo] from my friends. Mol Ther Nucleic Acids 1:e2. doi:10.1038/mtna.2011.7
102. Ramirez CL et al (2008) Unexpected failure rates for modular assembly of engineered zinc fingers. Nat Methods 5:374–375. doi:http://www.nature.com/nmeth/journal/v5/n5/suppinfo/nmeth0508-374_S1.html
103. Ma N et al (2013) TALEN-mediated gene correction in integration-free β-thalassemia iPSCs. J Biol Chem. doi:10.1074/jbc.M113.496174
104. Fu Y et al (2013) High-frequency off-target mutagenesis induced by CRISPR-Cas nucleases in human cells. Nat Biotechnol 31: 822–826. doi:10.1038/nbt.2623, http://www.nature.com/nbt/journal/v31/n9/abs/nbt.2623.html#supplementary-information
105. Ran FA, HSU et al (2013) Double-nicking by RNA-guided CRISPR CAS9 for enhanced genome editing specificity. Cell 154(6):1380–1389
106. Chung CY et al (2013) Identification and rescue of α-synuclein toxicity in Parkinson patient–derived neurons. Science. doi:10.1126/science.1245296
107. Ring KL et al (2012) Direct reprogramming of mouse and human fibroblasts into multipotent neural stem cells with a single factor. Cell Stem Cell 11:100–109. doi:http://dx.doi.org/10.1016/j.stem.2012.05.018
108. Vierbuchen T et al (2010) Direct conversion of fibroblasts to functional neurons by defined factors. Nature 463:1035–1041
109. Jakovcevski I, Filipovic R, Mo Z, Rakic S, Zecevic N (2009) Oligodendrocyte development and the onset of myelination in the human fetal brain. Frontiers Neuroanat 3. doi:10.3389/neuro.05.005.2009
110. Pang ZP et al (2011) Induction of human neuronal cells by defined transcription factors. Nature 476:220–223. doi:http://www.nature.com/nature/journal/v476/n7359/abs/nature10202.html#supplementary-information
111. Ambasudhan R et al (2011) Direct reprogramming of adult human fibroblasts to functional neurons under defined conditions. Cell Stem Cell 9:113–118. doi:http://dx.doi.org/10.1016/j.stem.2011.07.002
112. Yoo AS et al (2011) MicroRNA-mediated conversion of human fibroblasts to neurons. Nature 476:228–231. doi:http://www.nature.com/nature/journal/v476/n7359/abs/nature10323.html#supplementary-information
113. Qiang L et al (2011) Directed conversion of Alzheimer's disease patient skin fibroblasts into functional neurons. Cell 146:359–371. doi:http://dx.doi.org/10.1016/j.cell.2011.07.007
114. Caiazzo M et al (2011) Direct generation of functional dopaminergic neurons from mouse and human fibroblasts. Nature 476:224–227. doi:http://www.nature.com/nature/journal/v476/n7359/abs/nature10284.html#supplementary-information
115. Pfisterer U et al (2011) Direct conversion of human fibroblasts to dopaminergic neurons. Proc Natl Acad Sci U S A 108:10343–10348. doi:10.1073/pnas.1105135108
116. Addis RC et al (2011) Efficient conversion of astrocytes to functional midbrain dopaminergic neurons using a single polycistronic vector. PLoS One 6:e28719. doi:10.1371/journal.pone.0028719
117. Kim J et al (2011) Functional integration of dopaminergic neurons directly converted from mouse fibroblasts. Cell Stem Cell 9:413–419. doi:http://dx.doi.org/10.1016/j.stem.2011.09.011
118. Liu X et al (2012) Direct reprogramming of human fibroblasts into dopaminergic neuron-like cells. Cell Res 22:321–332. doi:http://www.nature.com/cr/journal/v22/n2/suppinfo/cr2011181s1.html
119. Meng F, Wang X, Gu P, Wang Z, Guo W (2013) Induction of retinal ganglion-like cells from fibroblasts by adenoviral gene delivery. Neuroscience 250:381–393. doi:http://dx.doi.org/10.1016/j.neuroscience.2013.07.001
120. Son EY et al (2011) Conversion of mouse and human fibroblasts into functional spinal motor neurons. Cell Stem Cell 9:205–218. doi:http://dx.doi.org/10.1016/j.stem.2011.07.014
121. Kim J et al (2011) Direct reprogramming of mouse fibroblasts to neural progenitors. Proc Natl Acad Sci 108:7838–7843. doi:10.1073/pnas.1103113108
122. Thier M et al (2012) Direct conversion of fibroblasts into stably expandable neural stem cells. Cell Stem Cell 10:473–479. doi:http://dx.doi.org/10.1016/j.stem.2012.03.003
123. Han DW et al (2012) Direct reprogramming of fibroblasts into neural stem cells by defined factors. Cell Stem Cell 10:465–472. doi:http://dx.doi.org/10.1016/j.stem.2012.02.021
124. Lujan E, Chanda S, Ahlenius H, Südhof TC, Wernig M (2012) Direct conversion of mouse fibroblasts to self-renewing, tripotent neural precursor cells. Proc Natl Acad Sci 109:2527–2532. doi:10.1073/pnas.1121003109

125. Sheng C et al (2012) Direct reprogramming of Sertoli cells into multipotent neural stem cells by defined factors. Cell Res 22:208–218. doi:http://www.nature.com/cr/journal/v22/n1/suppinfo/cr2011175s1.html
126. He X et al (1989) Expression of a large family of POU-domain regulatory genes in mammalian brain development. Nature 340:35–42
127. Crompton LA et al (2013) Stepwise, non-adherent differentiation of human pluripotent stem cells to generate basal forebrain cholinergic neurons via hedgehog signaling. Stem Cell Res 11:1206–1221. doi:http://dx.doi.org/10.1016/j.scr.2013.08.002
128. Delli Carri A et al (2013) Human pluripotent stem cell differentiation into authentic striatal projection neurons. Stem Cell Rev Rep 9:461–474. doi:10.1007/s12015-013-9441-8
129. Delli Carri A et al (2013) Developmentally coordinated extrinsic signals drive human pluripotent stem cell differentiation toward authentic DARPP-32+ medium-sized spiny neurons. Development 140:301–312. doi:10.1242/dev.084608
130. Kriks S et al (2011) Dopamine neurons derived from human ES cells efficiently engraft in animal models of Parkinson's disease. Nature 480(7378):547–551
131. Amoroso MW et al (2013) Accelerated high-yield generation of limb-innervating motor neurons from human stem cells. J Neurosci 33:574–586. doi:10.1523/jneurosci.0906-12.2013
132. Sundberg M et al (2011) A xeno-free culturing protocol for pluripotent stem cell-derived oligodendrocyte precursor cell production. Regen Med 6:449–460. doi:10.2217/rme.11.36
133. Kirkeby A et al (2012) Generation of regionally specified neural progenitors and functional neurons from human embryonic stem cells under defined conditions. Cell Rep 1:703–714
134. Alexander T, Nolte C, Krumlauf R (2009) Hox genes and segmentation of the hindbrain and axial skeleton. Ann Rev Cell Dev Biol 25:431–456. doi:10.1146/annurev.cellbio.042308.113423
135. Millet S et al (1999) A role for Gbx2 in repression of Otx2 and positioning the mid/hindbrain organizer. Nature 401:161–164
136. Brodski C et al (2003) Location and size of dopaminergic and serotonergic cell populations are controlled by the position of the midbrain–hindbrain organizer. J Neurosci 23:4199–4207
137. Mishima Y, Lindgren AG, Chizhikov VV, Johnson RL, Millen KJ (2009) Overlapping function of Lmx1a and Lmx1b in anterior hindbrain roof plate formation and cerebellar growth. J Neurosci 29:11377–11384. doi:10.1523/jneurosci.0969-09.2009
138. Liang X et al (2011) Isl1 Is required for multiple aspects of motor neuron development. Mol Cell Neurosci 47:215–222. doi:http://dx.doi.org/10.1016/j.mcn.2011.04.007
139. Dasen JS, De Camilli A, Wang B, Tucker PW, Jessell TM (2008) Hox repertoires for motor neuron diversity and connectivity gated by a single accessory factor, FoxP1. Cell 134:304–316. doi:http://dx.doi.org/10.1016/j.cell.2008.06.019
140. Rousso DL, Gaber ZB, Wellik D, Morrisey EE, Novitch BG (2008) Coordinated actions of the forkhead protein Foxp1 and Hox proteins in the columnar organization of spinal motor neurons. Neuron 59:226–240
141. Binamé F, Sakry D, Dimou L, Jolivel V, Trotter J (2013) NG2 regulates directional migration of oligodendrocyte precursor cells via Rho GTPases and polarity complex proteins. J Neurosci 33:10858–10874. doi:10.1523/jneurosci.5010-12.2013
142. Mabie PC et al (1997) Bone morphogenetic proteins induce astroglial differentiation of oligodendroglial–astroglial progenitor cells. J Neurosci 17:4112–4120
143. Thomson JA et al (1998) Embryonic stem cell lines derived from human blastocysts. Science 282:1145–1147. doi:10.1126/science.282.5391.1145
144. Becker AJ, McCulloch EA, Till JE (1963) Cytological demonstration of the clonal nature of spleen colonies derived from transplanted mouse marrow cells. Nature 197:452–454
145. Moss FP, Leblond CP (1971) Satellite cells as the source of nuclei in muscles of growing rats. Anat Rec 170:421–435. doi:10.1002/ar.1091700405
146. Mauro A (1961) Satellite cell of skeletal muscle fibers. J Biophys Biochem Cytol 9:493–495. doi:10.1083/jcb.9.2.493
147. Friedenstein AJ, Petrakova KV, Kurolesova AI, Frolova GP (1968) Heterotopic of bone marrow. Analysis of precursor cells for osteogenic and hematopoietic tissues. Transplantation 6:230–247
148. Broxmeyer HE et al (1990) Human umbilical cord blood: a clinically useful source of transplantable hematopoietic stem/progenitor cells. Int J Cell Cloning 8:76–91. doi:10.1002/stem.5530080708
149. Emerson SG et al (1985) Purification of fetal hematopoietic progenitors and demonstration of recombinant multipotential colony-stimulating activity. J Clin Invest 76:1286–1290. doi:10.1172/jci112087
150. Reynolds B, Weiss S (1992) Generation of neurons and astrocytes from isolated cells of the adult mammalian central nervous system. Science 255:1707–1710
151. Zuk PA et al (2002) Human adipose tissue is a source of multipotent stem cells. Mol Biol Cell 13:4279–4295. doi:10.1091/mbc.E02-02-0105
152. Zuk PA et al (2001) Multilineage cells from human adipose tissue: implications for cell-based therapies. Tissue Eng 7:211–228
153. De Coppi P et al (2007) Isolation of amniotic stem cell lines with potential for therapy. Nat Biotechnol 25:100–106. doi:http://www.nature.com/nbt/journal/v25/n1/suppinfo/nbt1274_S1.html
154. Zhao Y et al (2009) Human cord blood stem cell-modulated regulatory T lymphocytes reverse

the autoimmune-caused type 1 diabetes in nonobese diabetic (NOD) mice. PLoS One 4:e4226. doi:10.1371/journal.pone.0004226
155. Zhao Y et al (2012) Reversal of type 1 diabetes via islet beta cell regeneration following immune modulation by cord blood-derived multipotent stem cells. BMC Med 10:3
156. Neirinckx V, Marquet A, Coste C, Rogister B, Wislet-Gendebien S (2013) Adult bone marrow neural crest stem cells and mesenchymal stem cells are not able to replace lost neurons in acute MPTP-lesioned mice. PLoS One 8:e64723. doi:10.1371/journal.pone.0064723
157. Khoo MLM, Tao H, Meedeniya ACB, Mackay-Sim A, Ma DDF (2011) Transplantation of neuronal-primed human bone marrow mesenchymal stem cells in hemiparkinsonian rodents. PLoS One 6:e19025. doi:10.1371/journal.pone.0019025
158. Wislet-Gendebien S et al (2012) *In vivo* tumorigenesis was observed after injection of *in vitro* expanded neural crest stem cells isolated from adult bone marrow. PLoS One 7:e46425. doi:10.1371/journal.pone.0046425
159. Xiong N et al (2010) Long-term efficacy and safety of human umbilical cord mesenchymal stromal cells in rotenone-induced hemiparkinsonian rats. Biol Blood Marrow Transplant 16:1519–1529
160. Somoza R, Juri C, Baes M, Wyneken U, Rubio FJ (2010) Intranigral transplantation of epigenetically induced BDNF-secreting human mesenchymal stem cells: implications for cell-based therapies in Parkinson's disease. Biol Blood Marrow Transplant 16:1530–1540
161. Glavaski-Joksimovic A et al (2010) Glial cell line-derived neurotrophic factor–secreting genetically modified human bone marrow-derived mesenchymal stem cells promote recovery in a rat model of Parkinson's disease. J Neurosci Res 88:2669–2681. doi:10.1002/jnr.22435
162. Whone AL, Kemp K, Sun M, Wilkins A, Scolding NJ (2012) Human bone marrow mesenchymal stem cells protect catecholaminergic and serotonergic neuronal perikarya and transporter function from oxidative stress by the secretion of glial-derived neurotrophic factor. Brain Res 1431:86–96. doi:http://dx.doi.org/10.1016/j.brainres.2011.10.038
163. Kingham P et al (2013) Stimulating the neurotrophic and angiogenic properties of human adipose derived stem cells enhances nerve repair. Stem Cells 23(7):741–754
164. Dunnett SB, Björklund A, Stenevi U, Iversen SD (1981) Grafts of embryonic substantia nigra reinnervating the ventrolateral striatum ameliorate sensorimotor impairments and akinesia in rats with 6-OHDA lesions of the nigrostriatal pathway. Brain Res 229:209–217. doi:http://dx.doi.org/10.1016/0006-8993(81)90759-9
165. Dunnett SB, Isacson O, Sirinathsinghji DJS, Clarke DJ, Björklund A (1988) Striatal grafts in rats with unilateral neostriatal lesions—III. Recovery from dopamine-dependent motor asymmetry and deficits in skilled paw reaching. Neuroscience 24:813–820. doi:http://dx.doi.org/10.1016/0306-4522(88)90069-3
166. Zetterström T et al (1986) In vivo measurement of spontaneous release and metabolism of dopamine from intrastriatal nigral grafts using intracerebral dialysis. Brain Res 362:344–349. doi:http://dx.doi.org/10.1016/0006-8993(86)90460-9
167. Björklund A, Stenevi U, Dunnett SB, Iversen SD (1981) Functional reactivation of the deafferented neostriatum by nigral transplants. Nature 289:497–499
168. Björklund A, Dunnett SB, Stenevi U, Lewis ME, Iversen SD (1980) Reinnervation of the denervated striatum by substantia nigra transplants: Functional consequences as revealed by pharmacological and sensorimotor testing. Brain Res 199:307–333. doi:http://dx.doi.org/10.1016/0006-8993(80)90692-7
169. Lindvall O, Rehncrona S, Brundin P et al (1989) Human fetal dopamine neurons grafted into the striatum in two patients with severe Parkinson's disease: a detailed account of methodology and a 6-month follow-up. Arch Neurol 46:615–631. doi:10.1001/archneur.1989.00520420033021
170. Lindvall O et al (1990) Grafts of fetal dopamine neurons survive and improve motor function in Parkinson's disease. Science 247:574–577. doi:10.1126/science.2105529
171. Freed CR et al (1992) Survival of implanted fetal dopamine cells and neurologic improvement 12 to 46 months after transplantation for Parkinson's disease. N Engl J Med 327:1549–1555. doi:10.1056/NEJM199211263272202
172. Widner H et al (1992) Bilateral fetal mesencephalic grafting in two patients with Parkinsonism induced by 1-methyl-4-phenyl-L,2,3,6-tetrahydropyridine (MPTP). N Engl J Med 327:1556–1563. doi:10.1056/NEJM199211263272203
173. Spencer DD et al (1992) Unilateral transplantation of human fetal mesencephalic tissue into the caudate nucleus of patients with Parkinson's disease. N Engl J Med 327:1541–1548. doi:10.1056/NEJM199211263272201
174. Kordower JH et al (1995) Neuropathological evidence of graft survival and striatal reinnervation after the transplantation of fetal mesencephalic tissue in a patient with Parkinson's disease. N Engl J Med 332:1118–1124. doi:10.1056/NEJM199504273321702
175. Peschanski M et al (1994) Bilateral motor improvement and alteration of L-dopa effect in two patients with Parkinson's disease following intrastriatal transplantation of foetal ventral mesencephalon. Brain 117:487–499. doi:10.1093/brain/117.3.487
176. Henderson BH, Clough CG, Hughes RC, Hitchcock ER, Kenny BG (1991) IMplantation of human fetal ventral mesencephalon to the right caudate nucleus in advanced Parkinson's disease. Arch Neurol 48:822–827. doi:10.1001/archneur.1991.00530200062020

177. Freed CR et al (2001) Transplantation of embryonic dopamine neurons for severe Parkinson's disease. N Engl J Med 344:710–719. doi:10.1056/NEJM200103083441002
178. Olanow CW et al (2003) A double-blind controlled trial of bilateral fetal nigral transplantation in Parkinson's disease. Ann Neurol 54:403–414. doi:10.1002/ana.10720
179. Barker RA, Barrett J, Mason SL, Björklund A (2013) Fetal dopaminergic transplantation trials and the future of neural grafting in Parkinson's disease. Lancet Neurol 12:84–91. doi:http://dx.doi.org/10.1016/S1474-4422(12)70295-8
180. Kordower JH, Chu Y, Hauser RA, Freeman TB, Olanow CW (2008) Lewy body-like pathology in long-term embryonic nigral transplants in Parkinson's disease. Nat Med 14:504–506
181. Li J-Y et al (2008) Lewy bodies in grafted neurons in subjects with Parkinson's disease suggest host-to-graft disease propagation. Nat Med 14:501–503
182. Mendez I et al (2008) Dopamine neurons implanted into people with Parkinson's disease survive without pathology for 14 years. Nat Med 14:507–509
183. Politis M et al (2010) Serotonergic neurons mediate dyskinesia side effects in Parkinson's patients with neural transplants. Sci Transl Med 2:38ra46. doi:10.1126/scitranslmed.3000976
184. Carta M, Carlsson T, Kirik D, Bjorklund A (2007) Dopamine released from 5-HT terminals is the cause of L-DOPA-induced dyskinesia in parkinsonian rats. Brain 130:1819–1833. Epub 2007 Apr 1823 (2007)
185. Evans JR, Mason SL, Barker RA (2012) In: Dunnett Stephen B, Anders B (eds) Progress in brain research, vol 200. Elsevier, pp 169–198
186. Winkler C, Kirik D, Björklund A (2005) Cell transplantation in Parkinson's disease: how can we make it work? Trends Neurosci 28:86–92. doi:http://dx.doi.org/10.1016/j.tins.2004.12.006
187. Rosser AE et al (2002) Unilateral transplantation of human primary fetal tissue in four patients with Huntington's disease: NEST-UK safety report ISRCTN no 36485475. J Neurol Neurosurg Psychiatry 73:678–685. doi:10.1136/jnnp.73.6.678
188. Keene CD et al (2009) A patient with Huntington's disease and long-surviving fetal neural transplants that developed mass lesions. Acta Neuropathol 117:329–338. doi:10.1007/s00401-008-0465-0
189. Bachoud-Lévi A-C et al (2006) Effect of fetal neural transplants in patients with Huntington's disease 6 years after surgery: a long-term follow-up study. Lancet Neurol 5:303–309. doi:http://dx.doi.org/10.1016/S1474-4422(06)70381-7
190. Dawson TM, Ko HS, Dawson VL (2010) Genetic animal models of Parkinson's disease. Neuron 66:646–661
191. Nguyen HN et al (2011) LRRK2 mutant iPSC-derived DA neurons demonstrate increased susceptibility to oxidative stress. Cell Stem Cell 8: 267–280
192. Polo JM et al (2010) Cell type of origin influences the molecular and functional properties of mouse induced pluripotent stem cells. Nat Biotechnol 28:848–855. doi:http://www.nature.com/nbt/journal/v28/n8/abs/nbt.1667.html#supplementary-information
193. Lashuel HA et al (2002) α-Synuclein, especially the Parkinson's disease-associated mutants, forms pore-like annular and tubular protofibrils. J Mol Biol 322(5):1089–1102
194. Osafune K et al (2008) Marked differences in differentiation propensity among human embryonic stem cell lines. Nat Biotechnol 26:313–315. doi:http://www.nature.com/nbt/journal/v26/n3/suppinfo/nbt1383_S1.html
195. Lund RJ, Narva E, Lahesmaa R (2012) Genetic and epigenetic stability of human pluripotent stem cells. Nat Rev Genet 13:732–744. doi:http://www.nature.com/nrg/journal/v13/n10/suppinfo/nrg3271_S1.html
196. Heiskanen A et al (2007) N-glycolylneuraminic acid xenoantigen contamination of human embryonic and mesenchymal stem cells is substantially reversible. Stem Cells 25:197–202. doi:10.1634/stemcells.2006-0444
197. Martin MJ, Muotri A, Gage F, Varki A (2005) Human embryonic stem cells express an immunogenic nonhuman sialic acid. Nat Med 11:228–232
198. Draper JS et al (2004) Recurrent gain of chromosomes 17q and 12 in cultured human embryonic stem cells. Nat Biotechnol 22:53–54
199. Catalina P, et al (2008) Human ESCs predisposition to karyotypic instability: Is a matter of culture adaptation or differential vulnerability among hESC lines due to inherent properties? 7:76
200. Takahashi K, Narita M, Yokura M, Ichisaka T, Yamanaka S (2009) Human induced pluripotent stem cells on autologous feeders. PLoS One 4:e8067
201. Gurjala AN, Liu WR, Mogford JE, Procaccini PSA, Mustoe TA (2005) Age-dependent response of primary human dermal fibroblasts to oxidative stress: cell survival, pro-survival kinases, and entrance into cellular senescence. Wound Repair Regen 13:565–575
202. Awe J et al (2013) Generation and characterization of transgene-free human induced pluripotent stem cells and conversion to putative clinical-grade status. Stem Cell Res Ther 4:87
203. Karumbayaram S et al (2012) From skin biopsy to neurons through a pluripotent intermediate under good manufacturing practice protocols. Stem Cells Transl Med 1:36–43. doi:10.5966/sctm.2011-0001
204. Yagi T et al (2011) Modeling familial Alzheimer's disease with induced pluripotent stem cells. Hum Mol Genet 20:4530–4539. doi:10.1093/hmg/ddr394
205. Israel MA et al (2012) Probing sporadic and familial Alzheimer's disease using induced pluripotent stem cells. Nature 482:216–220. doi:http://www.nature.com/nature/journal/v482/n7384/abs/nature10821.html#supplementary-information

206. Kondo T et al (2013) Modeling Alzheimer's disease with iPSCs reveals stress phenotypes associated with intracellular Aβ and differential drug responsiveness. Cell Stem Cell 12:487–496. doi:http://dx.doi.org/10.1016/j.stem.2013.01.009
207. Cole GM, Frautschy SA (2006) Docosahexaenoic acid protects from amyloid and dendritic pathology in an Alzheimer's disease mouse model. Nutr Health 18:249–259. doi:10.1177/026010600601800307
208. Cooper O et al (2012) Pharmacological rescue of mitochondrial deficits in iPSC-derived neural cells from patients with familial Parkinson's disease. Sci Transl Med 4:141ra190. doi:10.1126/scitranslmed.3003985
209. Sánchez-Danés A et al (2012) Disease-specific phenotypes in dopamine neurons from human iPS-based models of genetic and sporadic Parkinson's disease. EMBO Mol Med 4:380–395. doi:10.1002/emmm.201200215
210. Sanders LH et al (2013) LRRK2 mutations cause mitochondrial DNA damage in iPSC-derived neural cells from Parkinson's disease patients: Reversal by gene correction. Neurobiol Dis. doi:http://dx.doi.org/10.1016/j.nbd.2013.10.013
211. Gloeckner CJ et al (2006) The Parkinson disease causing LRRK2 mutation I2020T is associated with increased kinase activity. Hum Mol Genet 15:223–232. doi:10.1093/hmg/ddi439
212. West AB et al (2005) Parkinson's disease-associated mutations in leucine-rich repeat kinase 2 augment kinase activity. Proc Natl Acad Sci U S A 102:16842–16847. doi:10.1073/pnas.0507360102
213. Camnasio S et al (2012) The first reported generation of several induced pluripotent stem cell lines from homozygous and heterozygous Huntington's disease patients demonstrates mutation related enhanced lysosomal activity. Neurobiol Dis 46:41–51. doi:http://dx.doi.org/10.1016/j.nbd.2011.12.042
214. Pearson CE, Edamura KN, Cleary JD (2005) Repeat instability: mechanisms of dynamic mutations. Nat Rev Genet 6:729–742. doi:http://www.nature.com/nrg/journal/v6/n10/suppinfo/nrg1689_S1.html
215. Cannella M et al (2009) DNA instability in replicating Huntington's disease lymphoblasts. BMC Med Genet 10:11
216. Dragileva E et al (2009) Intergenerational and striatal CAG repeat instability in Huntington's disease knock-in mice involve different DNA repair genes. Neurobiol Dis 33:37–47. doi:http://dx.doi.org/10.1016/j.nbd.2008.09.014
217. Kennedy L et al (2003) Dramatic tissue-specific mutation length increases are an early molecular event in Huntington disease pathogenesis. Hum Mol Genet 12:3359–3367. doi:10.1093/hmg/ddg352
218. Jeon I et al (2012) Neuronal properties, in vivo effects, and pathology of a Huntington's disease patient-derived induced pluripotent stem cells. Stem Cells 30:2054–2062. doi:10.1002/stem.1135
219. The HD iPSc Consortium (2012) Induced pluripotent stem cells from patients with Huntington's disease show CAG-repeat-expansion-associated phenotypes. Cell Stem Cell 11:264–278. doi:http://dx.doi.org/10.1016/j.stem.2012.04.027
220. Egawa N et al (2012) Drug screening for ALS using patient-specific induced pluripotent stem cells. Sci Transl Med 4:145ra104. doi:10.1126/scitranslmed.3004052
221. Tollervey JR et al (2011) Characterizing the RNA targets and position-dependent splicing regulation by TDP-43. Nat Neurosci 14:452–458. doi:http://www.nature.com/neuro/journal/v14/n4/abs/nn.2778.html#supplementary-information
222. Polymenidou M et al (2011) Long pre-mRNA depletion and RNA missplicing contribute to neuronal vulnerability from loss of TDP-43. Nat Neurosci 14:459–468. doi:http://www.nature.com/neuro/journal/v14/n4/abs/nn.2779.html#supplementary-information
223. Hagell P et al (2002) Dyskinesias following neural transplantation in Parkinson's disease. Nat Neurosci 5:627–628
224. Bohnen NI et al (2013) Gait speed in Parkinson disease correlates with cholinergic degeneration. Neurology 81:1611–1616
225. Fox SH (2013) Non-dopaminergic treatments for motor control in Parkinson's disease. Drugs 73:1405–1415. doi:10.1007/s40265-013-0105-4
226. Emre M (2003) Dementia associated with Parkinson's disease. Lancet Neurol 2:229–237. doi:http://dx.doi.org/10.1016/S1474-4422(03)00351-X
227. Luk KC et al (2012) Pathological α-synuclein transmission initiates Parkinson-like neurodegeneration in nontransgenic mice. Science 338:949–953. doi:10.1126/science.1227157
228. Sandberg MK, Al-Doujaily H, Sharps B, Clarke AR, Collinge J (2011) Prion propagation and toxicity in vivo occur in two distinct mechanistic phases. Nature 470:540–542
229. Frost B, Jacks RL, Diamond MI (2009) Propagation of Tau misfolding from the outside to the inside of a cell. J Biol Chem 284:12845–12852. doi:10.1074/jbc.M808759200
230. Desplats P et al (2009) Inclusion formation and neuronal cell death through neuron-to-neuron transmission of α-synuclein. Proc Natl Acad Sci 106:13010–13015. doi:10.1073/pnas.0903691106
231. de Calignon A et al (2012) Propagation of tau pathology in a model of early Alzheimer's disease. Neuron 73:685–697. doi:http://dx.doi.org/10.1016/j.neuron.2011.11.033
232. Masuda-Suzukake M et al (2013) Prion-like spreading of pathological α-synuclein in brain. Brain 136:1128–1138. doi:10.1093/brain/awt037
233. Clavaguera F et al (2009) Transmission and spreading of tauopathy in transgenic mouse brain. Nat Cell Biol 11:909–913. doi:http://www.nature.com/ncb/journal/v11/n7/suppinfo/ncb1901_S1.html

234. Moreno JA et al (2013) Oral treatment targeting the unfolded protein response prevents neurodegeneration and clinical disease in prion-infected mice. Sci Transl Med 5:206ra138. doi:10.1126/scitranslmed.3006767
235. Juopperi T et al (2012) Astrocytes generated from patient induced pluripotent stem cells recapitulate features of Huntington's disease patient cells. Mol Brain 5:17
236. Han J et al (2013) ER-stress-induced transcriptional regulation increases protein synthesis leading to cell death. Nat Cell Biol 15:481–490. doi:10.1038/ncb2738, http://www.nature.com/ncb/journal/v15/n5/abs/ncb2738.html#supplementary-information
237. Kon T et al (2013) ALS-associated protein FIG4 is localized in Pick and Lewy bodies, and also neuronal nuclear inclusions, in polyglutamine and intranuclear inclusion body diseases. Neuropathology 34(1):19–26. doi:10.1111/neup.12056
238. Wang IF, Wu L-S, Shen CKJ (2008) TDP-43: an emerging new player in neurodegenerative diseases. Trends Mol Med 14:479–485. doi:http://dx.doi.org/10.1016/j.molmed.2008.09.001
239. Xu Z-S (2012) Does a loss of TDP-43 function cause neurodegeneration? Mol Neurodegeneration 7:27
240. Nonaka T et al (2013) Prion-like properties of pathological TDP-43 aggregates from diseased brains. Cell Rep 4(1):124–134
241. Wolswijk G (2002) Oligodendrocyte precursor cells in the demyelinated multiple sclerosis spinal cord. Brain 125:338–349. doi:10.1093/brain/awf031
242. Shen S et al (2008) Age-dependent epigenetic control of differentiation inhibitors is critical for remyelination efficiency. Nat Neurosci 11:1024–1034. doi:http://www.nature.com/neuro/journal/v11/n9/suppinfo/nn.2172_S1.html
243. Payne NL et al (2013) Distinct immunomodulatory and migratory mechanisms underpin the therapeutic potential of human mesenchymal stem cells in autoimmune demyelination. Cell Transplant 22:1409–1425. doi:10.3727/096368912x657620
244. Bai L et al (2009) Human bone marrow-derived mesenchymal stem cells induce Th2-polarized immune response and promote endogenous repair in animal models of multiple sclerosis. Glia 57:1192–1203. doi:10.1002/glia.20841
245. Connick P et al (2012) Autologous mesenchymal stem cells for the treatment of secondary progressive multiple sclerosis: an open-label phase 2a proof-of-concept study. Lancet Neurol 11:150–156. doi:http://dx.doi.org/10.1016/S1474-4422(11)70305-2
246. Rice CM et al (2010) Safety and feasibility of autologous bone marrow cellular therapy in relapsing-progressive multiple sclerosis. Clin Pharmacol Ther 87:679–685. doi:http://www.nature.com/clpt/journal/v87/n6/suppinfo/clpt201044s1.html

Stem Cells and Neuronal Differentiation

Indrani Datta, Debanjana Majumdar, Kavina Ganapathy, and Ramesh R. Bhonde

Stem cells are being increasingly considered alternative and viable sources of treatment for debilitating nervous system disorders and neurodegenerative diseases. Stem cells specific to nervous tissue, i.e., neural stem cells (NSCs), exist in two neurogenic regions of the adult brain – subventricular zone (SVZ) in the lateral ventricle and the subgranular zone (SGZ) in hippocampal dentate gyrus [1–4]. The inaccessibility and unavailability of NSCs deep in the brain makes it a difficult proposition to use them in clinical applications. Different stem cells are thus being tested for their neuronal differentiation capability, as a cell source for generation of functional mature neurons and glial cells. The "gold standard" of stem cells are embryonic stem cells (ESCs) as they not only retain long-term self-renewal capacity but also exhibit pluripotency to all three germ lineages. Recent advances in technology have brought the advent of another pluripotent stem cells called "inducible pluripotent stem cells" (iPSc), derived through "reprogramming" of terminally differentiated cells by the addition of a select set of genes [5–7]. However, several limitations still exist for the use of iPScs in therapeutic applications, such as the use of viral vectors for transfer of genes, inclusion of oncogenes, and teratoma formation [5, 6, 8–10]. Stem cells may also be isolated from several tissue sources and these are termed as adult stem cells (ASCs). The first ASCs to be identified were the hematopoietic stem cells (HSCs) derived from bone marrow, but the second population of stem cells from bone marrow called mesenchymal stem cells (MSCs) gained prominence due to their unique properties [11–13]. MSCs are nontumorigenic and immunomodulatory in addition to possessing multilineage differentiation potential not only towards mesodermal lineage derivatives but also to phenotypes of other germ layer cells like neuronal, hepatocytes, and islet cells [14–17]. Although fetal and adult origin MSCs possess some common characteristics with respect to expression of mesenchymal markers and absence of hematopoietic and HLA-DR markers, their neuronal differentiation efficacy is still to be evaluated for consideration as suitable candidates for nervous system disorders and neurodegenerative diseases.

Birth of a Neuron

The primary unit of the nervous system is neurons. Neurons are specialized cells of the nervous system consisting of axons and dendrites that

Debanjana Majumdar and Kavina Ganapathy contributed equally to the work

I. Datta (✉) • D. Majumdar • K. Ganapathy
R.R. Bhonde
School of Regenerative Medicine, Manipal Institute of Regenerative Medicine, Manipal University, GKVK Post, Bellary Road, Allalasandra, Near Hotel Royal Orchid, Yelahanka, Bangalore 560065, India
e-mail: indrani.datta@manipal.edu; indrani.datta@gmail.com; diamajumdar88@gmail.com; kavina.ganapathy@manipal.edu; rr.bhonde@manipal.edu

I. Somasundaram (ed.), *Stem Cell Therapy for Organ Failure*,
DOI 10.1007/978-81-322-2110-4_5, © Springer India 2014

receive, integrate, and transduce electrochemical signals. Neurons in the nervous system are highly polarized and form an ordered communication system with both neuronal and nonneuronal cells through synapses. The initial step in development of nervous system in vertebrates involves the segregation of ectoderm into epidermal and neural primordia. Initially, the neural plate in mammalian and avian embryos forms through apicobasal cell elongation of neuroepithelial cells and convergent extension [18–22]. This is followed by bending of the neural plate at localized regions termed hinge points – a single median hinge point and paired dorsolateral hinge points [23, 24]. The fusion of the neural folds forms the neural tube. In mammalian central nervous system, neurons are generated from the neuroepithelial cells near the lumen of the neural tube termed the ventricular zone. A vertical cleavage of neuroepithelial cells during symmetric division gives rise to two identical daughter cells that resemble the precursor cell, but a horizontal cleavage during asymmetric division produces basal daughter cells that retain contact with the basal surface and an apical daughter cell that loses contact with the lumen (Fig. 1). This apical daughter cell migrates away, and the time of this horizontal division is termed as the birthday of the neuron (apical daughter cell) and the basal cell remains in the proliferative zone. During this asymmetric division, there is a switch in the mitotic state generating one daughter cell (basal cell) mitotically active as a stem cell, while the other apical cell remains in the cell cycle for a number of divisions and is committed to generate neurons [25].

Differentiation and Specification of Neuroepithelial Cells

The nervous system consists of a diverse neural cell type population, derived from these multipotent neuroepithelial/precursor cells. The differentiation of the neural precursor cells from the neural tube across its rostrocaudal axis gives rise to the neurons and glia of the central nervous system encompassing the brain and spinal cord. The differentiation of the neural tube begins with the formation of three primary brain vesicles from which the brain develops – prosencephalon, mesencephalon, and rhombencephalon. The dorsal fusion of the neural plate occurs by the third week to form the neural tube, which by the end of the fourth week further extends across the rostrocaudal axis. This creates spatiotemporal differences within neuroepithelial cells at the

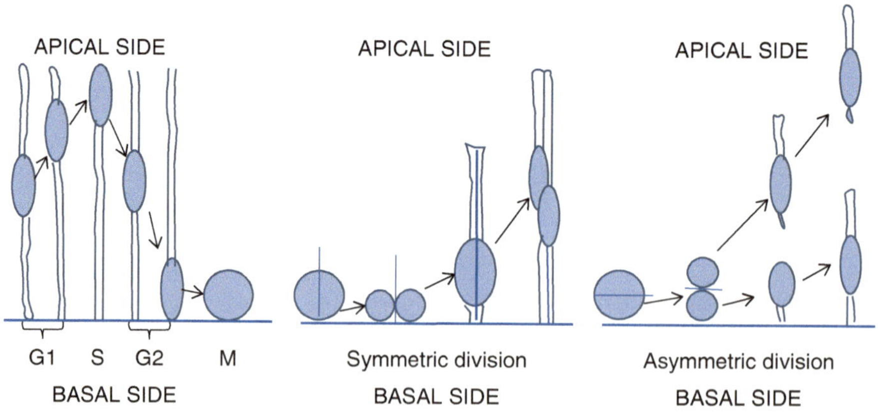

Fig. 1 Ventricular neuroepithelial cell nuclei undergo intracellular migration during cell cycle. A vertical cleavage (perpendicular to ventricular surface) gives rise to two daughter cells that sit side by side, both retaining apical connections. Both the daughter cells reenter the cell cycle. A horizontal (parallel) cleavage produces a basal daughter cell that retains contact with the basal surface and an apical daughter that loses contact with the lumen. The basal daughter stays in the epithelium, while the apical daughter migrates away (Adopted from Chenn and McConnell [25])

time of neural tube formation. The diversity in the cell fate specification of the neural precursors across the rostrocaudal axis is primarily dictated by the nature of environmental cues over time during development. The complex interplay of extrinsic microenvironment in the form of extracellular matrix and morphogens regulates intrinsic specifiers, often the master key transcription factors involved in patterning [26]. In addition, the developmental commitment with respect to the time of birthday for neural precursor cells is supported by the fact that neurogenesis precedes gliogenesis during the differentiation of neural tube. A developmental restriction with time in differentiation potential between early and late cortical progenitors is also noted. Late cortical progenitors can only produce upper layer neurons even in much younger microenvironment indicating that they lose the potency to produce earlier generated phenotypes [27]. The positional identities of the neural progenitors along the dorsoventral axis of the neural tube are through a gradient of signaling molecules secreted at the floor plate and roof plate (Fig. 2). The generation of functional specialized neuronal and glial cells involves a stepwise process starting with the specification of neuroepithelial cells to rapidly dividing transit-amplifying cells and then to migrating neuroblasts and glioblasts which in turn get specified to functional terminally differentiated cells under the influence of the adjacent microenvironment.

Isolation of Neural Stem Cells (NSCs)

Neural stem cells are multipotent cells possessing self-renewal capacity and differentiation capability towards cells of the central nervous system. During embryogenesis the germinal neuroepithelial cells along the neural tube compromises the NSC population which in adult brain gets restrained in two primary neurogenic areas, viz., the SVZ of the lateral ventricle and the SGZ in the dentate gyrus (Fig. 3; [2–4]). The SVZ of the adult brain harbors at least three distinct cell phenotypes: A, B, and C cells, besides the ependymal cells lining the lateral ventricle. Experimental evidence [28] suggests that the SVZ astrocytes (type B cells) represent quiescent stem cells that normally proliferate at a low rate and generate the migratory neuronal precursors (type A cells), through the generation of a third, intermediate cell type, the C cell (or D cell in the hippocampus), which has the characteristics of the classic fast-proliferating, transit-amplifying progenitor cells found in many self-renewing tissues [29, 30]. Adult neurogenesis is regulated by intrinsic specifiers and extrinsic modulators. The intrinsic programs include genetic and epigenetic factors essential for controlling NSC self-renewal and multipotency. The extrinsic factors include the niche where NSCs physically reside and primarily comprise periventricular astrocytes, ependymal cells, vasculature,

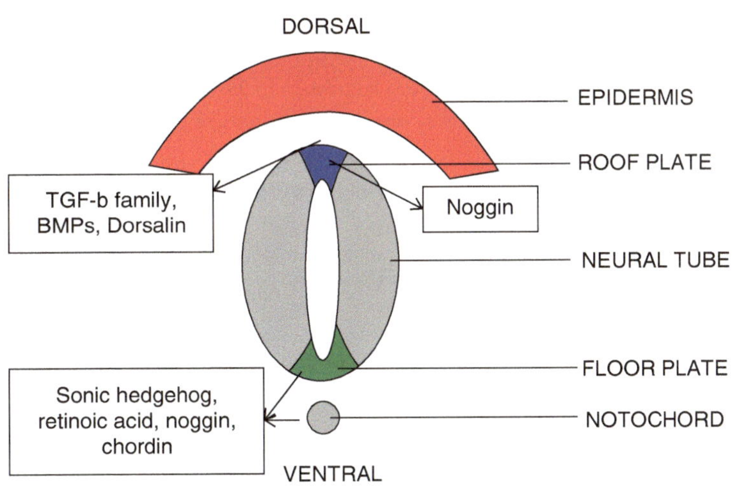

Fig. 2 Schematic representations showing the positional identities of the neural progenitors along the dorsoventral axis of the neural tube is through a gradient of signaling molecules secreted at the floor plate and roof plate

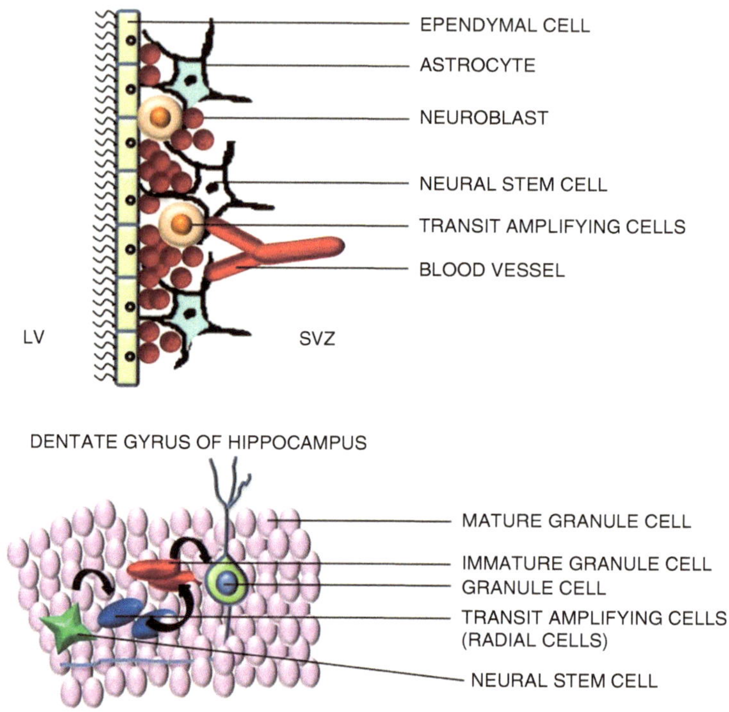

Fig. 3 Schematic representation of NSCs in the two primary neurogenic areas of the adult brain; viz., the subventricular zone (*SVZ*) of the lateral ventricle and the subgranular zone (*SGZ*) in the dentate gyrus

neurotransmitters, and the basal lamina. NSCs isolated from neurogenic areas of adult brain or embryonic CNS region are cultured in vitro by two methods – (1) two-dimensional culture in the form of monolayer on specific extracellular matrix and (2) three-dimensional suspension culture in the form of neurospheres. NSCs derived in serum- and feeder-free culture conditions as floating aggregates are known as neurospheres. Neurospheres have been extensively used for studying molecular mechanisms that regulate self-renewal and differentiation of NSCs. Although NSCs responsive to two mitogens bFGF and EGF have been reported, many labs use a combination of both [31–33]. In the presence of these mitogens, they undergo continuous symmetrical cell division (self-renewal) while retaining the differentiation potential to neurons, astrocytes, and oligodendrocytes. For mouse NSCs, several passages in the presence of these mitogens resulted in homogeneous morphology in culture that uniformly expressed nestin and Sox2. These cells are stem cells as they are clonogenic and maintain indefinitely the capacity to generate both neurons and astrocytes. It was further established that NSCs cultured in the absence of EGF and in the presence of exogenous Jagged1 showed enhanced neurogenic potential when put for differentiation [34], in turn suggesting that activation of Notch receptors on NSCs are pivotal for maintenance of undifferentiated state and differentiation potential. NSCs have also been isolated from brain tissue by fluorescent labeled sorting of cells through FACS for cell surface markers like CD133, CD24, or GFP expression driven by NSC-specific promoters such as nestin, Sox2, Sox1, and FGF1 [35, 36]. These NSCs once sorted were cultured in the presence of growth factors and in suspension culture form.

Differentiation of NSCs to Neuronal Cells

The derivation of NSCs from human fetal brain is characterized by classical bipolar morphology and other cellular morphologies. It also displays interkinetic nuclear migration along the cell process, a characteristic feature depicted by neuroepithelial cells in vivo. There are reports of human fetal NSCs having been maintained in culture up to 35 passages retaining normal karyotype and differentiation ability. NSCs are immunopositive for set of neural precursor/radial glial markers such as nestin, vimentin, Sox2, brain lipid-binding protein (BLBP), RC2, GLAST, and 3CB2 [36–38]. Besides, the molecular markers include the *Pax* and *Hes* gene families, members of Notch and Wnt signaling pathways, RNA-binding proteins (musashi12), and cell surface markers CD24 and CD133 [39, 40]. Human NSCs are reported to express moderate levels of GFAP too. NSCs upon plating onto poly-ornithine/laminin substrate with removal of the mitogens show spontaneous differentiation to a mixed culture of differentiated mature neuronal (Tuj1+ and Map2ab+) and astroglial (GFAP+) cells. Neuronal maturation can be further achieved by exposing them to Neurobasal media with B27 along with neurotrophic factors BDNF and NGF [41]. Functional electrophysiological experiments suggest that the matured neurons indeed exhibit voltage-gated Na+, K+, and Ca2+ channels, similar to those observed in primary neurons [42]. Exposure of NSCs to BMP4 agonists has exhibited differentiation towards astrocytes [43, 44]. BMP2 in combination with CNTF facilitated generation of GFAP-positive astrocytes [45]. In addition, Glaser et al. [46] have reported that sequential exposure to bFGF, PDGF, and forskolin followed by thyroid hormone (T3) and ascorbic acid can yield differentiated oligodendrocytes (20 %) in culture expressing O4, CNPase, and myelin proteolipid protein. The differentiation potential of NSCs in vivo is generally obtained from transplantation studies into embryonic or neonatal brain or the subventricular zone of adult rodent brain. In such a neurogenic environment, neural progenitors have exhibited "site-specific" differentiation to neuronal cells. On the other hand, transplantation of multipotent NSCs into the injured brain has indicated that the host microenvironment has strong gliogenic signals imparted majorly by the proinflammatory cytokines (TFNα, IL6, and IFNγ) present in the vicinity [47–49]. Other cytokines like PDGF, SDF1, MCP1, and HGF also play a role in chemoattraction and migration of NSCs [50–52]. The number of neurons generated by transplanted NSCs is lower in the lesion area than in intact host striatum. They are localized mainly in partially injured or intact regions and do not repopulate neuron-depleted areas. NSCs transplanted in nonneurogenic areas during spinal cord injury have shown differentiation to astrocytes and oligodendrocytes and not neurons [53]. However, NSCs transplanted in Parkinson's disease in vivo model have shown increase in tyrosine hydroxylase-positive neurons in the midbrain area [54]. NSCs from neonatal rat upon transplantation in the lesion area of the brain of adult PD rat model have shown distinct neuronal differentiation indicating sufficient availability of intrinsic cues for dopaminergic traits in the lesioned brain [55]. Thus, transplantation of NSCs has indeed shown diverse response in acquisition of specific phenotypes depending on the recipient environment, i.e., whether in the control or injured brain and neurogenic or nonneurogenic region. Due to limitations in analyzing differentiation of NSCs in vivo, many groups of scientists have used the advantage of neurosphere assays to address the role of intrinsic gene function for self-renewal and differentiation of NSCs. Gene targeting and knockouts by homologous recombination in NSCs are explored to understand loss/gain of function of genes with respect to self-renewal and differentiation. Insertion of master key genes is also performed on NSCs for achieving terminally differentiated functional neurons.

Differentiation of Human Embryonic Stem Cells to Neuronal Cells

Due to the inaccessibility and serious ethical concerns surrounding the use of human neural stem cells, other pluripotent stem cells are tapped into for generation of human origin neural progenitors. Human embryonic stem cells (hESCs) and induced pluripotent stem cells (iPSCs) from human skin fibroblast are presently considered the best resource for generation of human origin neural stem cells. In the last few years, there have been several reports on differentiation of hESCs and iPSCs to neural progenitors and specialized neuronal and glial cells too. The utility of pluripotent stem cells as banks for neural progenitors depends on the availability of standardized robust defined protocols of neuronal differentiation from these cells. Finally, the quality check for in vitro generated differentiated neural progenitors will be determined by their functional integration after transplantation in vivo.

ESCs are known to recapitulate embryonic developmental stages and so protocols of neuronal differentiation usually involve a multistep process. Indeed, the functional genomic screen of human stem cell differentiation revealed pathways involved in [56]. As per Spemann and Mangold [57], the Spemann organizer dictates the fate commitment of the neuroectodermal primordial cells through a set of crucial signaling factors like fibroblast growth factor (FGF), Wnt, Sonic Hedgehog (SHH), retinoic acid (RA), and bone morphogenetic proteins (BMP) inhibitors [58, 59]. Generation of neural progenitors from hESCs are initiated through two culture techniques – (1) embryoid body formation in serum-free condition and (2) presence of RA monolayer culture of ESCs on feeder-free specialized ECM-coated Petri dishes or coculture with stromal cell cultures. EBs usually mimic gastrulation processes and express markers from the three germ lineages [60, 61]. Embryoid body at specific day points are usually plated on suitable ECM with NSC media for the neuroectodermal outgrowth to occur. After eliminating the remnant of the EB, the outgrowth of neural progenitors can be propagated in the NSC media and defined ECM. This procedure is explained in detail in our published work that resulted in a good yield of neural progenitors from day 4 EBs under defined culture conditions [62] with more than 95 % population of cells expressing early neural markers like nestin, Sox2, musashi12, and negative for pluripotent marker Oct4. Initially, these cells organize into rosettes and express cell adhesion molecule N-cadherin [63]. A similar EB method of differentiation of hESCs to an enriched yield of NSCs was obtained in cGMP conditions by Swistowski et al. [64]. These NSCs could be maintained in xeno-free defined media for a prolonged period of time while retaining their ability to differentiate in vitro into functional dopaminergic neurons and upon in vivo transplantation could survive and give rise to differentiated dopaminergic neurons [65–67]. Transplantation of hESC-derived DA neurons has been found to attenuate locomotion deficits in PD rat model. Further analysis has also established that the increase in TH+ve cells in the midbrain is largely generated from grafted neural progenitors. Daadi et al. [68] have shown engraftment of hESC-derived midbrain dopaminergic neurons in a monkey model of Parkinson's disease. After transplantation the TH-expressing cells did not co-localize with GAD and maintained their DA-induced phenotype, extended neurite outgrowths, and expressed synaptic markers.

Besides dopaminergic neuronal differentiation, studies have also shown generation of different subtype neurons and glial cells of the CNS [69]. Treatment of these neuroepithelial cells by RA represses expression of anterior genes such as Otx2, Foxg1, and Pax6 and induces posterior Hox genes [70]. In later neuroepithelial cells (>15 days), this effect of RA is not seen. Neuroepithelial cells generated in absence of morphogens shows a dorsal fate expressing Pax6, Emx1, Ngn2, and Tbr1 but not ventral transcription factor Nkx2.1 indicating a predominant dorsal telencephalic fate [71]. This dorsal fate is determined by high levels of Wnt and low levels of SHH during hESC differentiation to telencephalic progenitors. The presence of SHH, one of the crucial morphogens for patterning midbrain DA neurons, represses the dorsal characteristics and induces ventral forebrain fate. Vazin et al. have depicted that SHH also generated GABAergic neurons [72]. Taking cues from development, it is also observed during in vitro differentiation of neural progenitors from hESCs that manipulating the timing of SHH exposure can give rise to distinct neuronal populations with specific transcriptional profiles and neurotransmitter phenotypes [72–74]. hESC-derived ventral forebrain progenitors have been reported by few groups [75, 76] with highly enriched populations of NKX2.1:GFP-positive progenitors, including cells with telencephalic identity. Liu et al. [77] have shown that hESCs can be differentiated to NKX2.1(+)medial ganglionic eminence (MGE)-like progenitor cells which, after transplantation into the hippocampus of mice with mu P75-saporin depleted basal forebrain cholinergic neurons (BFCNs) and GABA neurons in the medial septum, produced BFCNs that synaptically connected with endogenous neurons and generated GABAergic neurons too. In the presence of low gradients of SHH and Wnts, the neuroepithelial cells exhibit phenotypes of LGE cells which express Gsx2 and low levels of Pax6 but not Nkx2.1 [78]. On removal of SHH, these LGE progenitor cells can differentiate to projection GABAergic neurons expressing GAD 65/67, DARPP32, Meis2, and Ctip2 [78]. Generation of cholinergic neurons from hESCs has been recently been reported by BMP9 and NGF treatment [79, 80]. Treatment of neuroepithelia with RA (0.1 μM) in a chemically defined media for 10–17 days suppresses anterior transcription factors Otx2 and Foxg1 and activates posterior transcription factors Hoxb4, Hoxc5, and Hoxc8. Further ventralization of the caudal neuroepithelia is brought about by SHH from day 14–21 resulting in Olig2-expressing motor neuron progenitors. These progenitors on removal of RA and SHH express Mnx1, Lhx3, and Isl1/2, markers for postmitotic motor neurons [81–83]. Transplantation of hESC-derived motor neurons into chick embryonic spinal cord and mouse spinal cord expressed Nkx6.1 and Mnx1, and the grafted cells survived after transplantation [83]. Neural progenitors derived from EBs under IGF-1/insulin signaling gave rise to retinal pigment epithelial (RPE) cells and upon transplantation into a rat model of retinal degeneration resulted in the formation of a donor-derived RPE monolayer that rescues photoreceptor cells [84]. Generation of neural progenitors from hESCs is also achieved by culturing the cells in a monolayer subjected to synergistic inhibition of glycogen synthase kinase3 (GSK3), transforming growth factor β (TGF-β), and Notch signaling pathways by small molecules [85]. The self-renewal of these neuroepithelial cells can be maintained in the presence of leukemia inhibitory factor, GSK3 inhibitor (CHIR99021), and TGF-β receptor inhibitor (SB431542). Further, they retain neurogenic potential and responsiveness to instructive neural patterning cues towards midbrain and hindbrain neuronal subtypes and exhibit in vivo integration. Unlike the neuronal differentiation reported, astroglial differentiation from hESCs is quite rare. This is primarily due to the lack of knowledge of astroglial fate commitment during embryonic development. Systemic analysis has indicated that astroglial progenitor markers have been detected after long-term culturing of the neural progenitors derived from hESCs specifically maintained in EGF-containing media. The glial progenitors express the markers NF1A, S100b, CD44, and GFAP [86]. Neural progenitors generated from hESCs cultured in presence of RA and SHH agonist purmorphamine (Pur) under defined culture conditions have shown differentiation to oligodendrocytes by expressing Olig2, Nkx2.2, and Sox10 [87, 88]. In vivo transplantation of these glial derivatives has not yet been tested.

Neuronal differentiation is achieved not only by regulation of external microenvironment but also through intrinsic regulators. MicroRNA expression profiling of NPs and neuronal progenies shows gain- and loss-of-function of miR-153, miR-324-5p/3p, and miR-181a/a that contribute to the shift of NPs from self-renewal to neuronal differentiation. Stappert et al. [89] have shown that miR-125b and miR-181a specifically promote the generation of neurons of dopaminergic fate, whereas miR-181a inhibits the development of this neurotransmitter subtype. Studies have also shown conditional induction of master key transcription factors along with extrinsic cues for faster generation of specialized neuronal cells.

Differentiation of Human-Induced Pluripotent Stem Cells (iPSCs) to Neuronal Cells

It was long believed that once programmed to commit to a specialized differentiated cell type, cells rarely undergo dramatic fate changes in vivo as a result of an irreversible loss of developmental potency. However, with the advent of nuclear transfer technology, the cloning of an animal from the nucleus of a terminally differentiated cell explicitly proved that epigenetic modifications to the genome acquired during development are reversible and that nuclei from even the most functionally specialized cells maintained the potential to generate an adult organism. Further, pathbreaking work by Yamanaka's research group showed that a combination of four transcription factors (Oct4, Sox2, Klf4, and cMyc) was sufficient to reprogram diverse somatic cell types in vitro to a pluripotent state [90]. These newly reprogrammed cells are known as induced pluripotent stem cells (iPSCs). Reprogramming to pluripotency showed a return to the developmental "ground state" mirroring the features of ESCs. Differentiation of these cells to a neuronal lineage too follows a similar combination of induction factors that are required to differentiate hESCs. The process of neuronal differentiation for iPSCs begins with the initiation of primitive neuroectoderm which is manifested by rosette formation. The developmental clock of the rosettes show onset of early neural markers OTX2, PAX6, Sox1, Nestin, NR2F1, NR2F2, and IRX2 followed by glial-like cells at the later day points [91]. Moreover, the cells that emerged from the rosettes during spontaneous differentiation were capable of differentiating into dopaminergic neurons in vitro and into mature-appearing pyramidal and serotonergic neurons weeks after being injected into the motor cortex of NOD-SCID mice. For some human iPSCs that do not differentiate efficiently to neural progenitors, inhibition of BMP by Noggin and SB31542 has been used to increase yield [92, 93]. hiPSCs were also differentiated by coculturing them with rat primary neuronal and glial cells and on matrigel-coated tissue culture dish with differentiation medium [94]. Distinct maturation properties were attained depending on the protocol used, and functional maturation was achieved the best when cultured along with primary neuronal culture. hiPSCs have been efficiently differentiated to region and transmitter-specific neuronal cells including glutamatergic, GABAergic, cholinergic, dopaminergic, and motor neurons as well as astrocytes and oligodendrocytes [95, 96]. The neural progenitors derived from hiPSCs upon transplantation into the fetal mouse brain migrated into various brain regions and showed in vivo differentiation into glutamatergic, GABAergic, and dopaminergic subtypes. Even differentiated Parkinson patient-derived iPSCs grew in the adult rodent brain and reduced motor asymmetry in Parkinsonian rats [97, 98]. A recent detailed gene expression microarray study indicated that expression of ion channels such as voltage-gated Ca^{2+}, Na^{+}, and K^{+} channels, ionotropic neurotransmitter receptors, and ionotropic purinergic receptors is distinctly upregulated in the differentiated progeny of iPSCs in comparison to the starting cell type [99]. Furthermore, electrophysiological recordings and morphological analysis showed that the grafted cells had attained neuronal integration and synaptic activity. Also, neuroepithelial cells derived from hiPSCs after grafting in stroke-damaged brain have shown improvement in recovery [100]. Transplantation of neuroepithelial cells obtained from hiPSCs has also shown improvement of neurological function in rats with experimental intracerebral hemorrhage [101], and differentiation of neural progenitors derived from hiPSCs in a transgenic rat model of ALS carrying a human mutated SOD1 (G93A) was reported by Popescu et al. [102]. In a recent study, successful differentiation of hESCs and hiPSCs to retinal ganglion cells in the presence of Notch inhibitor N-[N-(3,5-difluorophenacetyl)-l-alanyl]-S-phenylglycine t-butyl ester (DAPT) has been reported [103]. Neural crest progenitors were also derived from iPSCs and compared with those from hESCs, followed by differentiation to functional

Schwann cells [104]. More than the therapeutic use of neural progenitors derived from hiPSCs, this in vitro cell-based technology is in the spotlight to reproduce cellular models of poorly understood diseases – such as Down syndrome, Friedreich's ataxia, Gaucher disease [105–107], amyotrophic lateral sclerosis [108], spinal muscular atrophy [109], Parkinson's disease [110, 111], schizophrenia [112], Huntington's disease [110], and Alzheimer's disease [113]. Furthermore, correction of genetic mutations in disease-specific iPS cells can rescue phenotypes in cultured cells [111, 114] or in mouse models of human diseases, such as sickle cell anemia [115]. The advent of iPSCs brought about an opportunity to study for the first time the cell biology and genetics of neurons derived from any individual. Furthermore, by recapitulating in vitro developmental steps for neuronal cells, it can provide indication for factors responsible for typical and atypical development. Fibroblasts of patients suffering from these diseases can be efficiently converted into iPSCs that are then differentiated into neurons to study the pathogenesis of these diseases (reviewed in [116–118]. Dimos et al. [108] derived iPSCs from an 82-year-old ALS patient, and these patient-specific iPSCs were efficiently differentiated even to motor neurons, the cell type destroyed in ALS. Cellular models are also made on neural progenitors generated from hiPSCs derived from fibroblasts of patients with central nervous system neuropathies [119, 120]. The studies reporting neuronal differentiation from pluripotent stem cells are listed in Table 1.

Direct Reprogramming of Fibroblasts to Induced NSCs (iNSCs) or Neural Progenitors (iNPs)

The primary limitation related to derivation of functional neuronal cells from hESCs and hiPSCs is the involvement of multiple steps, variability, and slow procedures. Generating neurons by differentiation of hESCs or iPSCs requires months of tissue culture procedures and renders large-scale studies difficult [90]. Moreover, the differentiation protocols are dependent on specific chemicals or growth factors such as pharmacological agents and bioactive proteins that may vary in consistency, thus introducing a further element of variability [120] and can form teratomas in vivo. This has led to the advent of direct lineage conversion or reprogramming to lineage-specific stem/progenitor cells of another germ layer in one step, bypassing the intermediate pluripotent stage, and these cells in the neuronal lineage are known as induced neurons (iN).

To reprogram fibroblasts to NSCs, two broad approaches have been used (Fig. 4). Thier et al. [121] used the same four factors (Oct4, Sox2, cMyc, and Klf4) applied for iPSC reprogramming but restricted Oct4 expression for the first 5 days using either protein transduction or mRNA transfection. This method tried to create a scenario in which reprogramming intermediates that have begun to acquire pluripotency are placed under the control of three factors (Sox2, cMyc, and Klf4). Sox2, which is known to strongly regulate neuroectodermal development with concomitant inhibition of mesendodermal development [122], in turn led to the acquisition of an NSC fate by the presumed pluripotent intermediates. Conceptually, this method is similar to that taken by Kim et al. to produce induced neural progenitors from fibroblasts [123] although well-controlled Oct4 expression in this study allowed the generation of tripotent induced neural stem cells (iNSCs). These iNSCs have extensive self-renewal capacity in comparison to the bipotent cells with limited passaging ability. Han et al. [124] and Lujan et al. [127] took a different approach for this direct conversion of fibroblasts. Each of these two groups started with a list of 11 candidate factors that resulted in generation of iNSCs or induced neural progenitor cells (iNPCs). The systematic elimination of factors narrowed the list down to the minimum combination of factors required. Han et al. identified four factors (Sox2, cMyc, Klf4, and Brn4/Pou3f4), whereas Lujan et al. arrived at a three-factor combination (Sox2, FoxG1, and Brn2). Both the nonuse of Oct4 and the inevitable role of Sox2 are noteworthy. In fact, Ring et al. [125] managed to generate iNSCs from mouse and human fibroblasts by

Table 1 List of neuronal differentiation studies from pluripotent stem cells

Source	Intermediate cell type	Markers	Target population	Morphogens/factors used	Markers	References
hESC	Neural progenitors	Nestin, Sox2, Musashi1, (-ve for OCT 4), Ki67	DA neuronal subtype	SHH, FGF8, polyornithin and fibronectin	TH, Nurr1, Map2ab	Datta et al. [62]
hESC	(Primitive anterior neuroepithelia)	Pax6, Sox2, N-cadherin	Neuro epithelial cells	RETINOIC ACID (RA)	Pax6, Otx2, Sox1, N-Cadherin, Hox	Pankratz et al. [63]
hESC	NSC	Nestin, Sox2, Glast, Vim and Blbp	DA neuronal subtype, Astrocytes	SHH, FGF8, cAMP, BDNF, GDNF, TGFß3, poly-l-ornithine, laminin	TH, LMX1a, VMAT, Girk2, β III Tubulin	Swistowski et al. [64]
hESC	Neuroepithelial cells	SOX1, SOX2, SOX3, Nestin, Musashi 1	Postmitotic neurons and astrocytes	FGF2, ± EGF	PSNCAM, A2B5, MAPII, GFAP, O4	Shin et al. [65]
mESC	Ventral midbrain cells		DA neuronal subtype	(FGF2, FGF 8, SHH) + Transfected with Wnt 5a	Nurr1, TH, β III Tubulin	Parish et al. [66]
hESC	Neural progenitors	Pax6, Sox1, Sox2, Nestin	neurons, astrocytes, GABAergic, dopaminergic, motor neurons and oligodendrocytes			Yan et al. [69]
hESC			GABAergic, dopaminergic cell types	Multivalent Shh conjugates		Vazin et al. [72]
hESC	Retinal progenitors	Six3, Crx, Rx, Pax6, Otx2, and Chx10	Retinal cells	SHH	S-Opsin	Amirpouret al. [73]
hPSCs	Neuroepithelial cells		Motoneuron progenitors	RA, SHH and purmorphamine	OLIG2	Hu et al. (2009)
hESC	Neural precursor_		Functional Motoneuron	FGF2, RA, SHH agonist	OLIG2, RALDH2, FOXP1	Patani et al. [74]
				FGF2, SHH agonist	OLIG2	
hESC	Neuroepithelial cells	Pax6, Sox1	Motoneuron	RA	HB9, HoxC8, ChAT and VAChT	Li et al. [81]
hESC	Neural progenitors		Glial progenitors	EGF	NF1A, S100b, CD44 and GFAP	Deneen et al. [86]

hESC	Neural progenitors		Oligodendrocytes	RA and purmorphamine	OLIG2, Nkx2.2 and Sox10	Jiang et al. [87] and Alsanie et al. [88]
hESC	Neural progenitors		DA cell type	miR-125b and miR-181a	NURR1, DAT, TH and GAD1	Stappert et al. [89]
hiPSCs/hESC's	Neural rosette	OTX2, PAX6, Sox1, Nestin, NR2F1, NR2F2, and IRX2	Dopaminergic neurons-in vitro and mature appearing pyramidal and serotonergic neurons in vivo		TH	Malchenko et al. [91]
hiPSCs	Neural progenitors	βIII tubulin, Ki67, Nestin, Pax6 and Sox2	Excitatory neurons	Cultured NP's on E18 rat primary cortical neurons	NeuN, MAP2, VGLUT and Synaptophysin	Verpelli et al. [94]
				Cultured NP's on rat primary glial cells	MAP2, synaptophysin and VGLUT	
				Cultured NP's on matrigel	βIII tubulin, MAP2	
hiPSC lines	Neuroepithelial cells	OTX2, LHX2, HOXB4, FOXG1	fore brain neurons, dopaminergic neurons, spinal motor neurons	Ornithine/laminin-coated coverslips + RA + SHH + FGF8 + BD NF + GDNF + IGF1	TBR1+, MAP2, CTIP2, VGLUT1, TH+	Zeng et al. [95]
miPSCs cell lines	Embryoid bodies to neuroepithelial like cells	Nestin, Sox2, and Brn2	Neurons, astrocytes, oligodendrocytes	Removal of FGF 2 from culture	βIII tubulin, GFAP, 04	Wernig et al. [98]
			Dopaminergic neurons	SHH, FGF8	βIII tubulin, TH, VMAT2, Enl, Pitx3, and Nurrl	
			Neurons and glia	in-vivo	GFAP, NeuN and β-III-tubulin	
			Glutamatergic neurons		EAAC1	
			GABAergic neurons		GAD67	
			Catecholaminergic neurons		TH	
hiPSCs/hESC's	Neural rosettes		Retinal ganglion cells	Notch inhibitor N-[N-(3, 5-difluorophenacetyl)-l-alanyl]-S-phenylglycine t-butyl ester (DAPT)	BRN3A, BRN3B, ATOH7/Math5, y-synuclein, Islet-1, and THY-1	Riazifar H et al. [103]

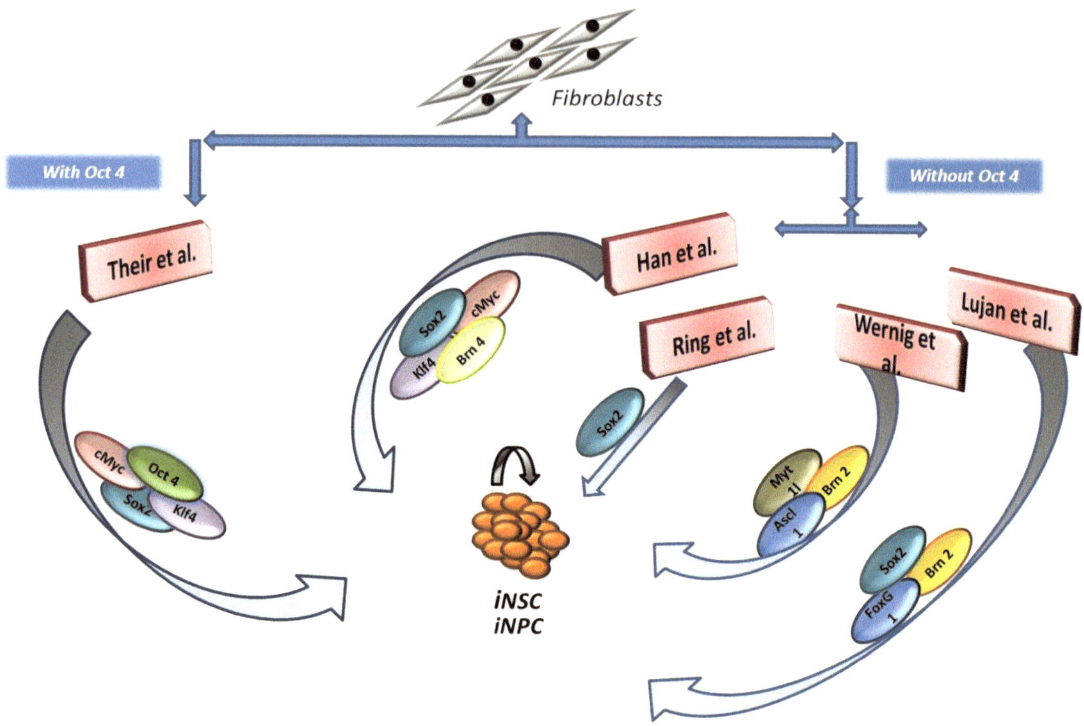

Fig. 4 Schematic representation of cell reprogramming approaches by introducing different sets of genes, adopted by different scientific research groups for deriving iNSCs or iNPCs from fibroblast cells

using the single factor Sox2 for reprogramming. In 2011, Pang et al. [126] reported a detailed method of reprogramming by screening a pool of 19 genes. Their study suggested that Ascl1 alone was sufficient to induce neuronal traits in fibroblasts, but the combination of *Ascl1*, *Brn2*, *and Mytl1* provided the best efficiency of reprogramming to iNSCs [126]. The iNSCs possess surprisingly robust self-renewal capability, with up to 130 passages reported in culture [124]. The iNSCs and iNPCs can undergo trilineage differentiation (neuron, astrocyte, and oligodendrocyte) in culture that when transplanted give rise to neuronal and glial progenies. This is consistent with their gene expression signatures that resemble endogenous NPCs. Lujan et al. reported successful myelination by oligodendrocytes differentiated from iNPCs. Further in the same year Caiazzo et al. reported the generation of differentiated functional dopaminergic neurons by direct conversion from mouse and human fibroblasts. Direct conversion in vivo for generation of iNSCs has also been reported recently [128, 129]. Transplanted human fibroblasts and human astrocytes engineered to express inducible forms of neural reprogramming genes converted into neurons when these genes were activated after transplantation [129]. Using a transgenic mouse model to specifically direct expression of reprogramming genes to parenchymal astrocytes residing in the striatum, it was shown that endogenous mouse astrocytes can be directly converted into neuronal nuclei (NeuN)-expressing neurons in situ. This provides evidence that direct neural conversion can take place in the adult rodent brain when using transplanted human cells or endogenous mouse cells as a starting point for neural conversion. Similarly, hiNPs transplanted to lateral ventricle were able to differentiate into NeuN-positive terminal neurons just by responding to induction of the surrounding environment. These differentiated terminal neurons were found in a variety of locations, displaying their migration ability in the brain [130]. Direct lineage conversion is not only restricted to NSCs or neural progenitors but also for terminally differentiated specialized neuronal cells. A recent report [131] showed that two small molecules (forskolin and dorsomorphin) enable the transcription factor Neurogenin-2 (NGN2) to convert human fetal lung fibroblasts into cholinergic neurons with high purity (>90 %) and efficiency (up to 99 % of NGN2-expressing cells). These human induced cholinergic neurons (hiCN) show mature electrophysiological properties and exhibit motor neuron-like features, including morphology, gene expression, and the formation of functional neuromuscular junctions. NSCs exist in vivo in a highly regionalized manner and produce region-specific neuronal types, and it appears that iNSCs may also share this regional identity. Han et al. reported strong expression of ventral hindbrain markers in the iNSCs generated by Sox2, cMyc, Klf4, and Brn4/Pou3f4. Although the reason for this hindbrain signature is unclear, one can imagine that region-specific factors could be deliberately added to the reprogramming mix, directly inducing region-specific iNSCs that are both expandable and able to produce defined neuronal subtypes. Such a scenario could give iNSCs a potential advantage over iPSCs.

Differentiation of Mesenchymal Stem Cells

Mesenchymal stromal cells (MSCs) are multipotent somatic stem cells shown to reside within the connective tissues of most organs. These non-hematopoietic stem cells are tissue specific and more restricted than embryonic stem cells in terms of differentiation. MSCs are regarded as strong candidates for cell replacement therapies because of their ability to self-renew, differentiate to multilineage, migrate and home in on injury sites, and for being immunomodulatory [132–134]. MSCs were first derived from bone marrow by Friedenstein et al. in 1976 [135]. These cells were characterized based on plastic adherence, marker expression, and the ability to differentiate to adipogenic, chondrogenic, and osteogenic lineages [136]. Although BM-MSCs are preferred for therapies, the procedure for procurement of bone marrow is extremely invasive and painful for patients. Apart from the bone marrow, MSCs are also located in adult and fetal tissues (Fig. 5). There are an increasing number of reports describing their presence in adipose tissue [137–139]; periodontal tissues such as dental pulp, dental ligament, follicle, and papilla [137, 140]; peripheral blood [141]; umbilical cord Wharton's jelly, cord blood, and chorionic villi of the placenta [142–145]; amniotic fluid [146]; fetal liver [147]; and lung [148]. The stromal cell population isolated from these tissues of origin need to be characterized by the set of MSC criteria such as plastic adherence; expression of CD105, CD73, and CD90; negative expression of CD45, CD34, CD14 or CD11b, CD79 alpha or CD19, and HLA-DR surface molecules; and in vitro differentiation to osteoblasts, adipocytes, and chondroblasts [149]. These cells have received extensive attention in the field of tissue engineering and regenerative medicine due to their availability and multilineage potential.

MSCs being a connective tissue derivative obviously show differentiation potential towards mesoderm lineage. However, recent studies suggest that MSCs have the capacity to transdifferentiate to cells of the other two lineages too (Fig. 5), including islet cells, myoblasts, cardiomyocytes, hepatocytes, and neuronal cells [150–156]. A microSAGE analysis of 2,353 expressed genes in a single cell-derived colony of undifferentiated human mesenchymal stem cells reveals mRNAs of multiple cell lineages [157]. Although NSCs or neural progenitors have been successfully generated from ESCs [62, 74] and iPSCs [158, 159], differentiating MSCs to functional neuronal cells is considered to be more difficult as it is a more committed cell type.

In Vitro Neuronal Differentiation of MSCs

Neuronal Differentiation Using Extrinsic Cues

In vitro neuronal differentiation of MSCs or adult stem cells (ASCs) is performed primarily to estimate the neuronal plasticity of these

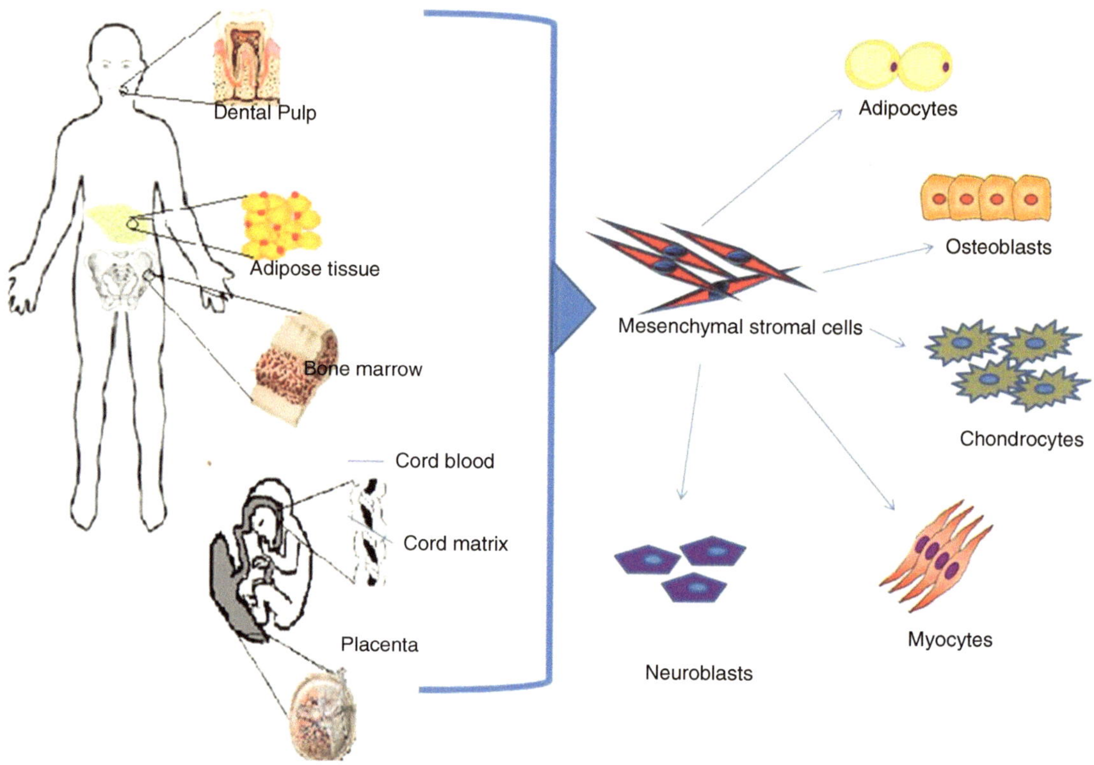

Fig. 5 Schematic representation of isolation of MSCs from adult and fetal tissues and their differentiation potential to different lineages

cells under defined regulated environment. Studies on differentiation of MSCs to neuronal phenotype have typically followed two broad approaches – modulating the microenvironment by extrinsic cues and delivery of key transcription factors along with extrinsic cues to obtain better efficiency and yield [160]. The extrinsic cues used by researchers for the past decade portray a vast diversity – ranging from chemical inducers to embryonic growth factors and conditioned media from rodent primary neuronal and glial cultures. Woodbury et al. [161] were the first group to report neuronal differentiation from BM-MSCs. They obtained a fast transition of neuronal phenotype under the influence of chemical induction such as dimethyl sulfoxide (DMSO), butylated hydroxyanisole (BHA), and β-mercaptoethanol (BME). However, studies by Lu et al. [165] in 2004 indicated that pure chemical exposure of MSCs induces neuronal-like morphology but does not yield functional neurons and additionally leads to cell toxicity and cell shrinkage. Several other neuronal differentiation protocols involving 3-isobutyl-1-methylxanthine (IBMX), bFGF, dimethyl sulfoxide (DMSO), butylated hydroxyanisole (BHA), or epigenetic reprogramming by 5-azacytidine [163] reported rapid changes in morphology with rounded cell bodies, which were eventually reversible in the absence of the induction factors and showed drastic cell death with time [163]. Although 5-azacytidine treatments showed neuronal morphology by day 4, here too the decrease in cell density with time is noteworthy [163]. The cytoskeletal changes resulting in pseudo-neuronal morphology thus indicated cell toxicity. So, the use of embryonic morphogens and neurotrophic factors for neuronal differentiation of MSCs started gaining favor. Trzaska et al. depicted that BM-MSCs exposed to midbrain cues (SHH and FGF8) were capable of differentiating them to dopaminergic neurons [164]. These ontogenically relevant cues were also used by Datta et al. (2011) to assess the neuronal plasticity of WJ-MSCs in comparison to BM-MSCs. The author's group [155, 156] in their study has shown phenotypical and functional characterization of BM-MSCs, WJ-MSCs, and DPSCs, not only in the presence of the morphogens but also when the differentiated cells were maintained in maintenance medium (absence of the morphogens). The MSCs from all the three sources spontaneously showed abundant expression of early neuronal markers such as nestin, musashi12, and A2B5 along with the mesenchymal markers. Upon induction, the upregulation of mature neuronal markers β-tubulin III and Map2ab was followed by a decrease in these early neuronal markers. Moreover, a distinct increase in dopaminergic-specific transcription factors (En1, Nurr1, Pitx3) and dopaminergic marker TH was observed. Functionally these cells could secrete dopamine constitutively and upon stimulation with ATP. The differentiated DPSCs though could secrete dopamine upon KCl stimulation too, in turn indicating the presence of purinergic receptors and potassium ion channels in the induced cells. In mature neuronal cells, the neurotransmitter is stored in vesicles within the cell and its release can be triggered by the influx of intracellular Ca^{2+}. The midbrain cues were thus capable of inducing the DPSCs to excitable cells, mimicking the physiology of neurotransmitter release of native neurons [156].

Studies have also targeted the signaling pathways for neural differentiation of MSCs by providing ligand molecules, molecular effectors, and inhibitors that are involved in upregulating transcription factors followed by gene expression. One such pathway frequently targeted for neuronal differentiation by many research groups is cyclic adenosine monophosphate (cAMP)-activated PKA (protein kinase A) pathway, which in the downstream mechanism phosphorylates CREB followed by regulation of different genes coded for cFOS, BDNF, and TH in turn aiding neural differentiation [166–168]. Based on this evidence, many studies have successfully induced neural differentiation of BM-MSCs and UCB-MSCs by using forskolin, dibutyryl-cAMP (db-cAMP), and 3-isobutyl-1-methylxanthine (IBMX). These studies determined the expression of neurofilament (NF), TH isoforms, and nuclear receptor related 1 (Nurr1) and also showed significant voltage-dependent ionic currents [169, 170]. However, the role of cAMP and the downstream effect in neural differentiation is not explicitly defined [171, 172]. Forskolin along with GDNF and embryonic midbrain morphogens were also used for neuronal induction of human exfoliated deciduous teeth (SHED) to differentiate to dopaminergic neurons [173]. Alexanian et al. [174] used the combination of small molecules that affect the regulation of chromatin structure and function and agents that favor neural differentiation to generate neural-like cells from human MSCs. The efficiency of neuronal differentiation and maturation was improved by two specific inhibitors of SMAD signaling (SMAD1/3 and SMAD3/5/8) that play an important role in neuronal differentiation of ESCs and were added to chromatin-modifying enzymes. Results demonstrated that human MSCs grown in these culture conditions exhibited higher expression of several mature neuronal genes, formed synapse-like structures, and exhibited electrophysiological properties of differentiating neural stem cells [174]. Recent studies have shown that the inhibition of histone deacetylases (HDACs) induces the differentiation of diverse cancer and stem cells which in turn suggests that HDAC inhibitors may be good candidates for neuronal induction of MSCs too. Jang et al. [175] investigated the effects of a HDAC inhibitor, valproic acid (VPA), for the neuronal differentiation of BM-MSCs. VPA-treated MSCs had significant increase in expression of neuro-progenitor marker nestin, musashi, CD133, and GFAP. VPA-pretreated MSCs upon differentiation with neuronal induction media (VPA-dMSCs) exhibited a cell body and dendritic morphology similar to neurons and neuronal-specific marker genes, including nestin, musashi, CD133, GFAP, NeuN, MAP-2, NF-M,

KCNH1, and KCNH5, but no functional study is reported. Some studies have also shown that MSCs cultured in the presence of rodent primary neuronal or glial culture-conditioned media or as coculture enhanced the expression of neuronal markers in the MSCs [177]. This could be due to the presence of potentially rich sources of neuronal differentiation promoting signals in the culture derivatives [178]. However, the use of rodent culture derivatives not only introduces undefined additives but is also xenogeneic.

Neuronal Differentiation Using a Defined Medium Composition

Researchers have also targeted enriching neuronal marker expression in MSCs by culturing them in suspension culture in NSC medium consisting of serum-free media with bFGF and EGF [176, 178]. MSCs, ADSCs, and DPSCs were capable of forming neurosphere-like structures and expressed neural progenitor markers in these conditions. Some studies have used the NSC media to prime the MSCs or ASCs towards neuronal lineage followed by ontogenic morphogens and growth factors as induction factors [179]. Besides specialized dopaminergic neuronal cells, MSCs have been differentiated to cholinergic cell types as well. The spindle-shaped or fibroblast-like WJ-MSCs changed into bulbous cells and positively expressed cholinergic neuronal markers, along with elevation of secretion of acetylcholine in the induced WJ-MSCs [180]. Transdifferentiation of BM-MSCs to cholinergic neurons was also demonstrated by Naghdi et al. [181] by the use of BME and NGF. This treatment with BME led to the generation of NF68-positive neuroblasts, which generated close to 80 % cholinergic marker-positive cells upon addition of NGF. When treated with neuronal induction medium consisting of brain-derived neurotrophic factor (BDNF), low-serum media and supplemented with hippocampal cholinergic neuro-stimulating peptide (HCNP) or rat denervated hippocampal extract (rDHE) or in combination, it enhanced the action of choline acetyltransferase (ChAT) [182]. A new multistep induction protocol has been reported by Darabi et al. [183] for the transdifferentiation of bone marrow stromal stem cells into GABAergic neuron-like cells. Rat BM-MSCs were pre-induced using β-mercaptoethanol (BME) and induced using retinoic acid (RA) and creatine. Immunostaining of neurofilament 200 kDa, neurofilament 160 kDa, nestin, fibronectin, gamma-aminobutyric acid (GABA), and glutamic acid decarboxylase (GAD) 65/67 was performed in the induced MSCs. Neuronal differentiation has been attained by adipose-derived stem cells too. Factors like bFGF, EGF, insulin, retinoic acid, and hydroxycortisone have been used for neuronal differentiation of AD-MSCs [184–186]. AD-MSCs expressed sodium current on treatment with bFGF and forskolin by increasing the intracellular cAMP levels, which was found to be useful in neural induction [171, 187]. A recent study has shown neuronal induction of ADSCs using BME, glial cell line-derived neurotrophic factor (GNDF), brain-derived neurotrophic factor (BDNF), retinoic acid (RA), 5-azacytidine, as well as their combinations [188]. NSC media consisting of EGF and bFGF along with IBMX and BDNF have also been used for ADSC neuronal induction [189]. A few studies have demonstrated the differentiation of MSCs towards glial cell type as well. A recent study has shown that norepinephrine, a neurotransmitter, when added in vitro can generate oligodendrocytes from the umbilical cord-derived multipotent progenitor cells in a three-dimensional environment [190]. Researchers have also shown that BM-MSCs and ADSCs can differentiate into Schwann-like cells (SLC) that have the potential to myelinate neuronal cells during regeneration [191].

Direct Conversion of MSCs to Neuronal Cell Types by Transduction

Apart from these studies, various other strategies have been employed to obtain neural cells types from different sources of MSCs. It has been shown that the microRNAs play an

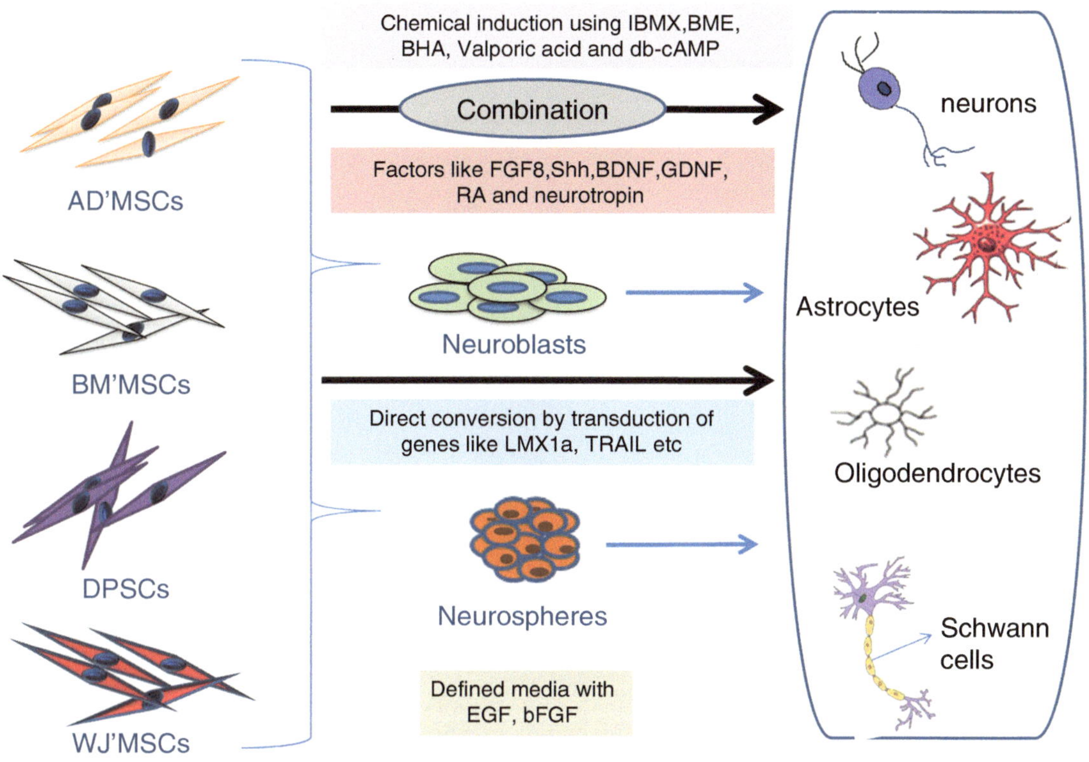

Fig. 6 Schematic representation of various in vitro neuronal induction strategies being reported in literature for obtaining differentiated neuronal and glial cells from adult and fetal tissue-derived stem cells

important role in the neuronal differentiation of WJ-MSCs [192]. These small RNAs of 18–24 nucleotides in length were involved in the regulation of gene expression and found to enhance motility and oxidative phosphorylation in neural cells derived from the WJ-MSCs. Studies aiming at the direct insertion of target genes for differentiation have also been reported. Transcription factor such as Neurogenin-1 was effective in converting MSCs into neuron-specific protein-expressing cells [193, 194]. The lentiviral delivery of transcription factor LMX1a showed enhancement of dopaminergic phenotype in differentiated human BM-MSCs [193, 194]. These neurons could synthesize higher level of the enzyme tyrosine hydroxylase (TH). It has been shown that the neuron-restrictive silence factors promote neural differentiation with enhanced electrophysiological properties [195]. AD-MSCs transfected with TRAIL (tumor necrosis factor-related apoptosis-inducing ligand) showed therapeutic efficacy against brainstem gliomas, and these MSCs were able to differentiate into neural cell types, thus reducing the tumor volume in vivo [196]. The schematic representation depicting the various in vitro neuronal induction strategies adopted for transdifferentiation of adult and fetal tissue-derived stem cells to differentiated neuronal and glial cells is provided in (Fig. 6). The in vitro neuronal differentiation studies discussed over here is summarized in Table 2.

In vitro studies on neuronal differentiation clearly demonstrate that MSCs and adult tissue-derived stem cells possess a certain extent of neuronal plasticity and can respond to ontologically relevant embryonic cues. However, for transplantation in vivo, the microenvironment cannot be controlled as the embryonic developmental clock and most of the embryonic cues and morphogens would not be available. The in vitro neuronal plasticity of MSCs was indeed replicated in

Table 2 List of in vitro neuronal differentiation studies from mesenchymal stem cells isolated from different adult and fetal tissue sources

Mesenchymal stromal cells	Defined factors	Results	References
BM'MSCs	Dimethylesulfoxide, butylatedhydroxyanisol, β-mercaptoethanol	Neuronal like cells positive for Enolase, NeuN, Tau and Neurofilament-M.	Woodbury et al. [161]
BM'MSCs	Dimethylesulfoxide, butylatedhydroxyanisol, β-mercaptoethanol, high molarity sodium chloride, detergent	Neuronal like morphology due to cell shrinkage and toxicity, positive for NeuN and NF-M but described to be Non functional neurons.	Lu et al. [165]
BM'MSCs	Dimethylesulfoxide, butylatedhydroxyanisol, β-mercaptoethanol	Non functional Neuronal like cells positive for NF-200, S100β, (βIII-tubulin, NSE and MAP2. Negative for Na(+), K(+) currents and action potentials	Barnabe et al. [162]
DPSCs	bFGF, 5-azacytidine, db-cAMP, neurotrophin-3 and supplementary components	Neuronal like cells expressing Neurogenin-2, NSE, NF-M, GFAP and positive for active voltage-dependent channels.	Kiraly et al. [163]
BM'MSCs	Midbrain cues such as SHH and FGF8	DA like neurons positive for TH, Pitx3, nurr1, DAT and VMAT2and also expressed NeuN and beta III tubulin	Trzaska et al. (2007, 2011)
BM'MSCs, WJ'MSCs and DPSCs	SHH, FGF8 and bFGF	Assessed the early neuronal markers in naive MSCs from three sources, obtained neuronal like cells expressing EN1, Nurr1, Pitx3 and TH. Studies also showed the functional DA neurons through the dopamine release and increase of intracellular calcium	Datta et al. [155, 156]
BM'MSCs & UCB'MSCs	Forskolin, dibutyryl-cAMP (db-cAMP) and (IBMX)	Cells were positive for NF-M, TH isoforms and nuclear receptor related 1 (Nurr1) and also showed significant voltage dependent ionic currents	Wang et al. [169] and Lepski et al. [170]
DPSCs(SHED)	SHH, FGF8, Forskolin and GDNF	Dopaminergic cells positive for beta III tubulin nestin, TH and MAP2	Wand et al. [173]
BM'MSCs	Trichostatin A (TSA) RG-108, 8-BrcAMP 1 μM Rolipram	DA like neurons expressing Nurr1 and TH. Also determined the secretion of neurotrophins and dopamine	Alexanian et al. [174]
BM'MSCs	Combination of histone deacetylase and valproic acid	Neural progenitors positive for nestin, Musashi, CD133, and GFAP, NeuN, Map2, NF-M, KCNHl and KCNH5	Jeong et al. (2013)
BM'MSCs	bFGF and EGF	Neurospheres which was positive for nestin and musashi were co cultured with primary neurons and obtain specific neural cell type	Fu et al. [176]
WJ'MSCs	bFGF and EGF	Neurospheres positive for neural progenitor markers nestin, Sox2 and Pax6 transcription factors. When differentiated on fibronectin coated dishes these were able to generate neuron/glial - like cells which expressed Nfl, Map2 and GFAP	Balasubramanian et al. [178]

(continued)

Table 2 (continued)

Mesenchymal stromal cells	Defined factors	Results	References
DPSCs	EGF, FGF, ITS, retinoic acid	Neural progenitors positive for Nestin, beta 3 tubulin, EGF, FGF, ITS, retinoic acid Neuronal Nestin, PSA-NCAM,	Arthur et al. [179]
BM'MSCs	FGFβ, BDNF, EGF and NGF	Cholinergic neurons positive for GAP-43, NF-H, Neu-N	Naghdi et al. [181]
WJ'MSCs	BDNF, Low serum media, hippocampal cholinergic neuro-stimulating peptide or hippocampal extract and in combination	Functional cholinergic like cells were obtained with enhanced ChAT and secretion of Ach	Zhang et al. (2012)
BM'MSCs	β mercaptoethanol (BME), retinoic acid and creatine	GABAergic like cells positive for NF-200, NF-160, nestin, fibronectin, GABA and GAD65/67	Darabi et al. [183]
AD'MSCs	Valporicacid, insulin, hydroxyanisole, hydrocortisone, EGF, FGF.	Neuronal GFAP, Neu-N, nestin, IF-M	Schaffler and Buchler [185] and Safford et al. [184]
AD'MSCs	BDNF and retinoic acid	Neuronal cells positive for MAP 2, Neu-N, nestin, GalC, S 100, GFAP, TH	Anghileri et al. [186]
AD'MSCs	bFGF and EGF	Neuronal cells positive for Nestin, Sox2, vimentin, A2B5, GFAP, tuj1.	Lim et al. (2010)
AD'MSCs	bFGF, forskolin, ciliary neurotrophic factor, GDNF.	Neuronal cells positive for Map-2ab, NF-M, GFAP, GalC, O4, TH, DAT	Kim et al. [187] and Rooney et al. [171]
AD'MSCs	BME, GDNF, BDNF, RA and 5-azacytidine and combinations	Neural cells expressed nestin, BIITub & ENO2. *In vivo* brain promoted their migration from the transplantation site to the recipient cerebral parenchyma.	Pavlova et al. [188]
AD'MSCs	1st step-EGF, bFGF with IBMX, BDNF; 2nd step-BDNF	Both induced Neural cells were positive for GFAP and TUJ1 markers but 1st step method showed higher expression then 2nd step method	Ying et al. [189]
WJ'MSCs	Defined medium containing the neurotransmitter norepinephrine (NE)	Under two-dimensional conditions, differentiated into oligodendrocyte precursors. In a three-dimensional environment, the MLPCs differentiated into committed oligodendrocytes that expressed myelin basic protein	Hedvika et al. [190]
AD'MSCs	Glial growth factors (GGF-2, bFGF, PDGF and forskolin	Spindle-like morphology similar to Schwann cells, expressed the glial markers, GFAP, S100 and p75, When co-cultured with NG108-15 motor neuron-like cells, induced neurite growth in NG108-15 cells	Mantovani et al. [191]
AD'MSCs	β-mercaptoethanol, all-trans-RA, and mixture of forskolin, bFGF, PDGF and heregulin	Schwaan like cells express S100 and GFAP. Enhance neurite outgrowth in co-culture with sensory neurons	Jiang et al. [87]

(continued)

Table 2 (continued)

Mesenchymal stromal cells	Defined factors	Results	References
WJ'MSCs	B27 supplement, RA, and bFGF and carried out miRNA analysis	Gene Ontology database showed that 136 genes were associated with cell motility, energy production, oxidative phosphorylation and actin cytoskeleton organization, indicating that miR-34a plays a critical role in cell migration	Chang et al. [192]
BM'MSCS	LMX1a transduction	LMX1a protein was concentrated in the cells' nuclei and specific dopaminergic developmental genes were upregulated, expressed higher levels of tyrosine hydroxylase, and secreted significantly higher level of dopamine	Brazilay et al. [193]
AD'MSCs	TRAIL transduction	Type of cells assessed Astrocyte, oligodendrocyte positive forTuj 1, GFAP, CNPase, adiponectin, sialoprotein. Efficacy was tested *In vivo* and *In vitro*	Choi et al. [196]

in vivo embryonic brain. Muñoz-Elias et al. [197] for the first time had demonstrated that both rat and human BM-MSCs assume neuronal functions in vivo in an embryonic CNS microenvironment. The transplanted adult MSCs not only survived and migrated in embryonic day 15.5 (E15.5) rat ventricles in utero but also differentiated in a regionally and temporally specific manner. Transplantation of MSCs during the gliogenic clock in neonatal mouse brains showed that BM-MSCs could migrate throughout the forebrain and cerebellum and differentiate into astrocytes after injection [198]. The in vivo differentiation of MSCs to neuronal cells thus gets restricted from embryonic to neonatal and then for adult CNS. The analysis of postmortem brain samples from females who had received bone marrow transplants from male donors showed that marrow cells can enter the brain and generate new neurons just as rodent cells do [199]. The underlying diseases of the patients were lymphocytic leukemia and genetic deficiency of the immune system, and they survived between 1 and 9 months after transplant. This was one of the first indicative studies suggesting the neuronal plasticity of marrow cells and probable neurogenic environment of the adult CNS. Most studies reported have targeted the transplantation of undifferentiated MSCs in the disease or injury model, and some studies have transplanted these MSCs after priming the cells with neuronal induction cues or growth factors. DPSCs upon transplantation in neurogenic area of the CNS, hippocampus, of mice underwent proliferation and maturation, forming NPCs and neurons [200]. Transplantation of these cells in a nonneurogenic area of the CNS shows survival, engraftment, and improvement of behavioral scores of the disease rodent model but rarely shows differentiation to mature neuronal cell type. An early study in 1998 [201] showed engraftment and migration of human BM-MSCs implanted in the brains of albino rats – similar to astrocyte grafts. Human SHED derived stem cells and WJ-MSCs transplanted post-neural induction with SHH, FGF8 have differentiated into dopaminergic neurons in vivo and further elevated the dopamine content [179, 176, 173]. Regeneration by neural-induced human BM-MSCs in rat models of Parkinson's disease has also been reported [202–204]. MSCs treated with stromal derived factor-1 (SDF-1) increased the release of dopamine and also helped in preserving the TH-positive cells [205] in Parkinsonian rat model. hMSC treatment had a protective effect on progressive loss of dopaminergic neurons induced by MG-132 in vivo through differentiation and trophic effect [206]. Human umbilical cord MSC-derived neuron-like cells have shown rescue of memory deficits and reduced amyloid-beta deposition in an AβPP/PS1 transgenic mouse model [207]. In addition, neuroectodermally converted BM-MSCs led to decrease of Aβ peptides by regulation of two genes F-spondin and neprilysin [208]. All these data suggest that MSCs induced by recently developed methodologies could be a potential source of cells to replace damaged neurons and glia in injured spinal cord and/or to promote cell survival and axonal growth of host tissue. Protection of dopaminergic neurons against the neurotoxic effects and motor deficits was obtained too by transplantation of undifferentiated ADSCs, WJ-MSCs, and BM-MSCs in MPTP-induced rats [209–212] indicating the paracrine neurotrophic effect of the cells rather than the graft cells differentiating in the adult CNS. Transplantation of GLP-1 transected hMSCs in the right ventricle of double transgenic mice mutant expressing APP and presenilin-1 showed a reduction in Aβ40/42 positively stained plaques, and the number of reactive astrocytes measured in the dentate gyrus of the hippocampus also decreased [213] again representing its paracrine effect. Contribution towards reduction in ischemic damage has been reported in ischemic stroke mouse model after transplantation of BM-MSCs and ADSCs [214, 215]. Bang et al. [216] have reported that autologous BM-MSC transplantation shows improvement of Barthel index and Rankin score in stroke patients. A long-term follow-up study of intravenous autologous mesenchymal stem cell transplantation in patients with ischemic stroke showed clinical improvement (Lee et al. 2010). Presently several clinical studies have established the safety of transplantation of autologous BM-MSCs. Four patients showed a significant slowing down of the linear decline of the forced vital capacity and of the ALS-FRS score [217]. The effect of a combination of autologous undifferentiated and neural-induced bone marrow mesenchymal stem cells (MSCs) on behavioral improvement in rats after inducing spinal cord injury has also been examined. In all treatment groups (differentiated, undifferentiated, and mix), there was less cavitation than lesion sites in the control group. The Basso-Beattie-Bresnahan (BBB) score was significantly higher in rats transplanted with a combination of cells and in rats transplanted with neural-induced MSCs alone than in undifferentiated and control rats [218]. Put together, the data obtained from the in vivo transplantation of MSCs in rodent models and clinical trials suggests that MSCs can promote endogenous reparative mechanisms that may prove applicable and beneficial for neurodegenerative disease treatment.

Conclusion

Regeneration of neurons in CNS has always remained as an enigma mainly because the neurogenic area that exists in the adult brain is deep-seated, and with currently available techniques, it is quite unlikely to engineer any modulation and be of further use in degenerative cases. This also limits the clinical use of hNSCs for any nervous system disorder. Important for effective differentiation of neuronal and glial subtypes is the patterning of primitive neuroepithelial cells. Taking inspiration from the ontogenic cues, directed differentiation has been performed from pluripotent stem cells both by modulating the extrinsic signals and changing the intrinsic master key transcription factors. Indeed, hESC and hiPSC neuronal differentiation quite efficiently mimics the developmental clock, and functional specialized neurons have been obtained by several scientific groups across the globe. However, immunological issues along with the risk of teratoma formation limit the therapeutic use of pluripotent stem cell-derived NSCs. MSCs score precisely on these points over pluripotent stem cells, and thus studies to determine the neurogenic plasticity of MSCs have been very much in the spotlight. A lot of experimental work on MSCs suggests that they bring about neuro-rescue in a multipronged way. While there are several research publications both in vitro and in vivo models, the exact mechanism of neuroprotection has not yet been understood. Moreover, the functional integration of these differentiated neuronal cells with the host brain neurons under both normal and disease CNS environments needs to be further elucidated to make the use of MSCs as candidates for cell replacement therapy a clinical reality.

References

1. Reynolds BA, Weiss S (1992) Generation of neurons and astrocytes from isolated cells of the adult mammalian central nervous system. Science 255: 1707–1710
2. Quinones-Hinojosa A, Sanai N, Soriano-Navarro M, Gonzalez-Perez O, Mirzadeh Z, Gil-Perotin S, Romero Rodriguez R, Berger MS, Garcia-Verdugo JM, Alvarez-Buylla A (2006) Cellular composition and cytoarchitecture of the adult human subventricular zone: a niche of neural stem cells. J Comp Neurol 494(3):415–434
3. Hsu YC, Lee DC, Chiu IM (2007) Neural stem cells, neural progenitors, and neurotrophic factors. Cell Transplant 16(2):133–150
4. Ma DK, Bonaguidi MA, Ming GL, Song H (2009) Adult neural stem cells in the mammalian central nervous system. Cell Res 19(6):672–682
5. Takahashi K, Yamanaka S (2006) Induction of pluripotent stem cells from mouse embryonic and adult fibroblast cultures by defined factors. Cell 126:663–676
6. Okita K, Ichisaka T, Yamanaka S (2007) Generation of germline-competent induced pluripotent stem cells. Nature 448:313–317
7. Wernig M, Meissner A, Foreman R, Brambrink T, Ku M, Hochedlinger K, Bernstein BE, Jaenisch R (2007) In-vitro reprogramming of fibroblasts into a pluripotent ES-cell-like state. Nature 448: 318–324
8. Mikkelsen TS, Hanna J, Zhang X, Ku M, Wernig M, Schorderet P, Bernstein BE, Jaenisch R, Lander ES, Meissner A (2008) Dissecting direct reprogramming through integrative genomic analysis. Nature 454:49–55
9. Sridharan R, Tchieu J, Mason MJ, Yachechko R, Kuoy E, Horvath S, Zhou Q, Plath K (2009) Role of the murine reprogramming factors in the induction of pluripotency. Cell 136:364–377
10. Sommer CA, Sommer AG, Longmire TA, Christodoulou C, Thomas DD, Gostissa M, Alt FW, Murphy GJ, Kotton DN, Mostoslavsky G (2010) Excision of reprogramming transgenes improves the differentiation potential of iPS cells generated with a single excisable vector. Stem Cells 28:64–74
11. Dezawa M, Kanno H, Hoshino M, Cho H, Matsumoto N, Itokazu Y, Tajima N, Yamada H, Sawada H, Ishikawa H, Mimura T, Kitada M, Suzuki Y, Ide C (2004) Specific induction of neuronal cells from bone marrow stromal cells and application for autologous transplantation. J Clin Invest 113(12): 1701–1710
12. Syková E, Homola A, Mazanec R, Lachmann H, Konrádová SL, Kobylka P, Pádr R, Neuwirth J, Komrska V, Vávra V, Stulík J, Bojar M (2006) Autologous bone marrow transplantation in patients with subacute and chronic spinal cord injury. Cell Transplant 15(8–9):675–687
13. Giordano A, Galderisi U, Marino IR (2007) From the laboratory bench to the patient's bedside: an update on clinical trials with mesenchymal stem cells. J Cell Physiol 211(1):27–35
14. Keyser KA, Beagles KE, Kiem HP (2007) Comparison of mesenchymal stem cells from different tissues to suppress T-cell activation. Cell Transplant 16(5):555–562
15. Le Blanc K (2003) Immunomodulatory effects of fetal and adult mesenchymal stem cells. Cytotherapy 5:485–489
16. Tse WT, Pendleton JD, Beyer WM, Egalka MC, Guinan EC (2003) Suppression of allogeneic T-cell proliferation by human marrow stromal cells: implications in transplantation. Transplantation 75: 389–397
17. Pittenger MF, Mackay AM, Beck SC, Jaiswal RK, Douglas R, Mosca JD, Moorman MA, Simonetti DW, Craig S, Marshak DR (1999) Multilineage potential of adult human mesenchymal stem cells. Science 284:143–147
18. Schoenwolf GC, Alvarez IS (1989) Roles of neuroepithelial cell rearrangement and division in shaping of the avian neural plate. Development 106(3):427–439
19. Keller R, Shih J, Sater AK, Moreno C (1992) Planar induction of convergence and extension of the neural plate by the organizer of Xenopus. Dev Dyn 193(3):218–234
20. Keller R, Davidson L, Edlund A, Elul T, Ezin M, Shook D, Skoglund P (2000) Mechanisms of convergence and extension by cell intercalation. Trans R Soc Lond B Biol Sci 355(1399): 897–922
21. Wallingford JB, Harland RM (2001) Xenopus Dishevelled signaling regulates both neural and mesodermal convergent extension: parallel forces elongating the body axis. Development 128(13): 2581–2592
22. Wallingford JB, Harland RM (2002) Neural tube closure requires Dishevelled-dependent convergent extension of the midline. Development 129(24):5815–5825
23. Smith JL, Schoenwolf GC (1987) Cell cycle and neuroepithelial cell shape during bending of the chick neural plate. Anat Rec 218(2):196–206
24. Smith JL, Schoenwolf GC (1988) Role of cell-cycle in regulating neuroepithelial cell shape during bending of the chick neural plate. Cell Tissue Res 252(3):491–500

25. Chenn A, McConnell SK (1995) Cleavage orientation and the asymmetric inheritance of notch1 immunoreactivity in mammalian neurogenesis. Cell 82(4):631–641
26. Maxwell SL, Li M (2005) Midbrain dopaminergic development in vivo and in vitro from embryonic stem cells. J Anat 207:209–218
27. Frantz GD, McConnell SK (1996) Restriction of late cerebral cortical progenitors to an upper-layer fate. Neuron 17(1):55–61
28. Galli R, Gritti A, Bonfanti L, Vescovi AL (2003) Neural stem cells: an overview. Circ Res 92(6): 598–608
29. Doetsch F, Caillé I, Lim DA, García-Verdugo JM, Alvarez-Buylla A (1999) Subventricular zone astrocytes are neural stem cells in the adult mammalian brain. Cell 97(6):703–716
30. Alvarez-Buylla A, García-Verdugo JM, Tramontin AD (2001) A unified hypothesis on the lineage of neural stem cells. Nat Rev Neurosci 2(4):287–293
31. Reynolds BA, Weiss S (1996) Clonal and population analyses demonstrate that an EGF-responsive mammalian embryonic CNS precursor is a stem cell. Dev Biol 175:1–13
32. Seaberg RM, van der Kooy D (2002) Adult rodent neurogenic regions: the ventricular subependyma contains neural stem cells, but the dentate gyrus contains restricted progenitors. J Neurosci 22(5): 1784–1793
33. Bull ND, Bartlett PF (2005) The adult mouse hippocampal progenitor is neurogenic but not a stem cell. J Neurosci 25(47):10815–10821
34. Nyfeler Y, Kirch RD, Mantei N, Leone DP, Radtke F, Suter U, Taylor V (2005) Jagged1 signals in the postnatal subventricular zone are required for neural stem cell self-renewal. EMBO J 24(19):3504–3515. Epub 2005 Sep 15
35. Roy NS, Benraiss A, Wang S, Fraser RA, Goodman R, Couldwell WT, Nedergaard M, Kawaguchi A, Okano H, Goldman SA (2000) Promoter-targeted selection and isolation of neural progenitor cells from the adult human ventricular zone. J Neurosci Res 59(3):321–331
36. Keyoung HM, Roy NS, Benraiss A, Louissaint A Jr, Suzuki A, Hashimoto M, Rashbaum WK, Okano H, Goldman SA (2001) High-yield selection and extraction of two promoter-defined phenotypes of neural stem cells from the fetal human brain. Nat Biotechnol 19(9):843–850
37. Götz M (2003) Glial cells generate neurons–master control within CNS regions: developmental perspectives on neural stem cells. Neuroscientist 9(5): 379–397
38. Rakic S, Zecevic N (2003) Early oligodendrocyte progenitor cells in the human fetal telencephalon. Glia 41(2):117–127
39. Johansson CB, Momma S, Clarke DL, Risling M, Lendahl U, Frisén J (1999) Identification of a neural stem cell in the adult mammalian central nervous system. Cell 96(1):25–34
40. Kageyama R, Ohtsuka T, Kobayashi T (2008) Roles of Hes genes in neural development. Dev Growth Differ 50(Suppl 1):S97–S103. doi:10.1111/j.1440-169X.2008.00993.x. Epub 2008 Apr 22
41. Frederiksen K, McKay RD (1988) Proliferation and differentiation of rat neuro-epithelial precursor cells in vivo. J Neurosci 8(4):1144–1151
42. Lendahl U, Zimmerman LB, McKay RD (1990) CNS stem cells express a new class of intermediate filament protein. Cell 60(4):585–595
43. Ciceroni C, Mosillo P, Mastrantoni E, Sale P, Ricci-Vitiani L, Biagioni F, Stocchi F, Nicoletti F, Melchiorri D (2010) mGLU3 metabotropic glutamate receptors modulate the differentiation of SVZ-derived neural stem cells towards the astrocytic lineage. Glia 58(7):813–822. doi:10.1002/glia.20965
44. Andersson T, Duckworth JK, Fritz N, Lewicka M, Södersten E, Uhlén P, Hermanson O (2011) Noggin and Wnt3a enable BMP4-dependent differentiation of telencephalic stem cells into GluR-agonist responsive neurons. Mol Cell Neurosci 47(1):10–18. doi:10.1016/j.mcn.2011.01.006. Epub 2011 Jan 14
45. Okano-Uchida T, Naruse M, Ikezawa T, Shibasaki K, Ishizaki Y (2013) Cerebellar neural stem cells differentiate into two distinct types of astrocytes in response to CNTF and BMP2. Neurosci Lett 552:15–20. doi:10.1016/j.neulet.2013.07.021. Epub 2013 July 26
46. Glasser T, Pollard SM, Smith A, Brustle O (2007) Tripotential differentiation of adherently expandable neural stem (NS) cells. PLoS One 2(3):e298. doi:10.1371/journal.pone.0000298
47. Bonni A, Sun Y, Nadal-Vicens M, Bhatt A, Frank DA, Rozovsky I, Stahl N, Yancopoulos GD, Greenberg ME (1997) Regulation of gliogenesis in the central nervous system by the JAK-STAT signaling pathway. Science 278(5337):477–483
48. Vallières L, Campbell IL, Gage FH, Sawchenko PE (2002) Reduced hippocampal neurogenesis in adult transgenic mice with chronic astrocytic production of interleukin-6. J Neurosci 22(2):486–492
49. Ben-Hur T, Einstein O, Mizrachi-Kol R, Ben-Menachem O, Reinhartz E, Karussis D, Abramsky O (2003) Transplanted multipotential neural precursor cells migrate into the inflamed white matter in response to experimental autoimmune encephalomyelitis. Glia 41(1):73–80
50. Imitola J, Raddassi K, Park KI, Mueller FJ, Nieto M, Teng YD, Frenkel D, Li J, Sidman RL, Walsh CA, Snyder EY, Khoury SJ (2004) Directed migration of neural stem cells to sites of CNS injury by the stromal cell-derived factor 1alpha/CXC chemokine receptor 4 pathway. Proc Natl Acad Sci U S A 101(52):18117–18122. Epub 2004 Dec 17
51. Belmadani A, Tran PB, Ren D, Miller RJ (2006) Chemokines regulate the migration of neural progenitors to sites of neuroinflammation. J Neurosci 26(12):3182–3191
52. Lalive PH, Paglinawan R, Biollaz G, Kappos EA, Leone DP, Malipiero U, Relvas JB, Moransard

M, Suter T, Fontana A (2005) TGF-beta-treated microglia induce oligodendrocyte precursor cell chemotaxis through the HGF-c-Met pathway. Eur J Immunol 35(3):727–737

53. Vroemen M, Aigner L, Winkler J, Weidner N (2003) Adult neural progenitor cell grafts survive after acute spinal cord injury and integrate along axonal pathways. Eur J Neurosci 18(4):743–751
54. Anton R, Kordower JH, Maidment NT, Manaster JS, Kane DJ, Rabizadeh S, Schueller SB, Yang J, Rabizadeh S, Edwards RH et al (1994) Neural-targeted gene therapy for rodent and primate hemi-parkinsonism. Exp Neurol 127(2):207–218
55. Lundberg C, Horellou P, Mallet J, Björklund A (1996) Generation of DOPA-producing astrocytes by retroviral transduction of the human tyrosine hydroxylase gene: in vitro characterization and in vivo effects in the rat Parkinson model. Exp Neurol 139(1):39–53
56. Zhang Y, Pak C, Han Y, Ahlenius H, Zhang Z, Chanda S, Marro S, Patzke C, Acuna C, Covy J, Xu W, Yang N, Danko T, Chen L, Wernig M, Südhof TC (2013) Rapid single-step induction of functional neurons from human pluripotent stem cells. Neuron 78(5):785–798
57. Spemann H, Mangold H (1924) Induction of embryonic primordia by implantation of organizers from a different species. Roux's Arch Entw Mech 100: 599–638
58. De Robertis EM, Wessely O, Oelgeschläger M, Brizuela B, Pera E, Larraín J, Abreu J, Bachiller D (2001) Molecular mechanisms of cell-cell signaling by the Spemann-Mangold organizer. Int J Dev Biol 45(1):189–197
59. Niehrs C (2004) Regionally specific induction by the Spemann-Mangold organizer. Nat Rev Genet 5(6):425–434
60. Pal R, Totey S, Mamidi MK, Bhat VS, Totey S (2009) Propensity of human embryonic stem cell lines during early stage of lineage specification controls their terminal differentiation into mature cell types. Exp Biol Med (Maywood) 234(10): 1230–1243
61. Player A, Wang Y, Rao M, Kawasaki E (2007) Gene expression analysis of RNA purified from embryonic stem cells and embryoid body-derived cells using a high-throughput microarray platform. Curr Protoc Stem Cell Biol Chapter 1: Unit 1B.2. doi:10.1002/9780470151808.sc01b02s2
62. Datta I, Ganapathy K, Tattikota SM, Bhonde R (2013) Directed differentiation of human embryonic stem cell-line HUES9 to dopaminergic neurons in a serum-free defined culture niche. Cell Biol Int 37(1):54–64
63. Pankratz MT, Li XJ, Lavaute TM, Lyons EA, Chen X, Zhang SC (2007) Directed neural differentiation of human embryonic stem cells via an obligated primitive anterior stage. Stem Cells 25:1511–1520
64. Swistowski A, Peng J, Han Y, Swistowska AM, Rao MS, Zeng X (2009) Xeno-free defined conditions for culture of human embryonic stem cells, neural stem cells and dopaminergic neurons derived from them. PLoS One 4(7):e6233
65. Shin S, Mitalipova M, Noggle S, Tibbitts D, Venable A, Rao R, Stice SL (2006) Long-term proliferation of human embryonic stem cell-derived neuroepithelial cells using defined adherent culture conditions. Stem Cells 24(1):125–138
66. Parish CL, Castelo-Branco G, Rawal N, Tonnesen J, Sorensen AT, Salto C, Kokaia M, Lindvall O, Arenas E (2008) Wnt5a-treated midbrain neural stem cells improve dopamine cell replacement therapy in parkinsonian mice. J Clin Invest 118(1):149–160
67. Ferrari D, Sanchez-Pernaute R, Lee H, Studer L, Isacson O (2006) Transplanted dopamine neurons derived from primate ES cells preferentially innervate DARPP-32 striatal progenitors within the graft. Eur J Neurosci 24(7):1885–1896
68. Daadi MM, Grueter BA, Malenka RC, Redmond DE Jr, Steinberg GK (2012) Dopaminergic neurons from midbrain-specified human embryonic stem cell-derived neural stem cells engrafted in a monkey model of Parkinson's disease. PLoS One 7(7):e41120
69. Yan Y, Shin S, Jha BS, Liu Q, Sheng J, Li F, Zhan M, Davis J, Bharti K, Zeng X, Rao M, Malik N, Vemuri MC (2013) Efficient and rapid derivation of primitive neural stem cells and generation of brain subtype neurons from human pluripotent stem cells. Stem Cells Transl Med 2(11):862–870
70. Li T, Zheng J, Xie Y, Wang S, Zhang X, Li J, Jin L, Ma Y, Wolf DP, Zhou Q, Ji W (2005) Transplantable neural progenitor populations derived from rhesus monkey embryonic stem cells. Stem Cells 23(9):1295–1303. Epub 2005 July 28
71. Eiraku M, Watanabe K, Matsuo-Takasaki M, Kawada M, Yonemura S, Matsumura M, Wataya T, Nishiyama A, Muguruma K, Sasai Y (2008) Self-organized formation of polarized cortical tissues from ESCs and its active manipulation by extrinsic signals. Cell Stem Cell 3:519–532
72. Vazin T, Ashton RS, Conway A, Rode NA, Lee SM, Bravo V, Healy KE, Kane RS, Schaffer DV (2014) The effect of multivalent Sonic hedgehog on differentiation of human embryonic stem cells into dopaminergic and GABAergic neurons. Biomaterials 35(3):941–948
73. Amirpour N, Karamali F, Rabiee F, Rezaei L, Esfandiari E, Razavi S, Dehghani A, Razmju H, Nasr-Esfahani MH, Baharvand H (2012) Differentiation of human embryonic stem cell-derived retinal progenitors into retinal cells by Sonic hedgehog and/or retinal pigmented epithelium and transplantation into the subretinal space of sodium iodate-injected rabbits. Stem Cells Dev 21(1):42–53. doi:10.1089/scd.2011.0073. Epub 2011 June 1
74. Patani R, Hollins AJ, Wishart TM, Puddifoot CA, Alvarez S, de Lera AR, Wyllie DJ, Compston DA, Pedersen RA, Gillingwater TH, Hardingham GE, Allen ND, Chandran S (2011) Retinoid-independent

motor neurogenesis from human embryonic stem cells reveals a medial columnar ground state. Nat Commun 2:214. doi:10.1038/ncomms1216

75. Germain ND, Banda EC, Becker S, Naegele JR, Grabel LB (2013) Derivation and isolation of NKX2.1-positive basal forebrain progenitors from human embryonic stem cells. Stem Cells Dev 22(10):1477–1489
76. Nicholas CR, Chen J, Tang Y, Southwell DG, Chalmers N, Vogt D, Arnold CM, Chen YJ, Stanley EG, Elefanty AG, Sasai Y, Alvarez-Buylla A, Rubenstein JL, Kriegstein AR (2013) Functional maturation of hPSC-derived forebrain interneurons requires an extended timeline and mimics human neural development. Cell Stem Cell 12(5): 573–586
77. Liu Y, Weick JP, Liu H, Krencik R, Zhang X, Ma L, Zhou GM, Ayala M, Zhang SC (2013) Medial ganglionic eminence-like cells derived from human embryonic stem cells correct learning and memory deficits. Nat Biotechnol 31(5):440–447. doi:10.1038/nbt.2565. Epub 2013 Apr 21
78. Ma LX, Hu B, Liu Y, Liu H, Vermilyea SC, Liu H, Zhang X, Sun Y, Gao L, Li J, Ayala M, Zhang SC (2010) Specification of functional striatal GABAergic projection neurons from human stem cells. Soc Neurosci Abstr 331.7
79. Bissonnette CJ, Lyass L, Bhattacharyya BJ, Belmadani A, Miller RJ, Kessler JA (2011) The controlled generation of functional basal forebrain cholinergic neurons from human embryonic stem cells. Stem Cells 29(5):802–811
80. Nilbratt M, Porras O, Marutle A, Hovatta O, Nordberg A (2010) Neurotrophic factors promote cholinergic differentiation in human embryonic stem cell-derived neurons. J Cell Mol Med 14(6B):1476–1484. doi:10.1111/j.1582-4934.2009.00916.x. Epub 2009 Oct 3
81. Li XJ, Du ZW, Zarnowska ED, Pankratz M, Hansen LO, Pearce RA, Zhang SC (2005) Specification of motoneurons from human embryonic stem cells. Nat Biotechnol 23:215–221
82. Singh Roy N, Nakano T, Xuing L, Kang J, Nedergaard M, Goldman SA (2005) Enhancer-specified GFP based FACS purification of human spinal motor neurons from embryonic stem cells. Exp Neurol 196:224–234. [PubMed: 16198339]
83. Lee H, Shamy GA, Elkabetz Y, Schofield CM, Harrsion NL, Panagiotakos G, Socci ND, Tabar V, Studer L (2007) Directed differentiation and transplantation of human embryonic stem cell-derived motoneurons. Stem Cells 25:1931–1939
84. Diniz B, Thomas P, Thomas B, Ribeiro R, Hu Y, Brant R, Ahuja A, Zhu D, Liu L, Koss M, Maia M, Chader G, Hinton DR, Humayun MS (2013) Subretinal implantation of retinal pigment epithelial cells derived from human embryonic stem cells: improved survival when implanted as a monolayer. Invest Ophthalmol Vis Sci 54(7):5087–5096
85. Li W, Sun W, Zhang Y, Wei W, Ambasudhan R, Xia P, Talantova M, Lin T, Kim J, Wang X, Kim WR, Lipton SA, Zhang K, Ding S (2011) Rapid induction and long-term self-renewal of primitive neural precursors from human embryonic stem cells by small molecule inhibitors. Proc Natl Acad Sci U S A 108(20):8299–8304
86. Deneen B, Ho R, Lukaszewicz A, Hochstim CJ, Gronostajski RM, Anderson DJ (2006) The transcription factor NFIA controls the onset of gliogenesis in the developing spinal cord. Neuron 52: 953–968
87. Jiang P, Selvaraj V, Deng W (2010) Differentiation of embryonic stem cells into oligodendrocyte precursors. J Vis Exp 39
88. Alsanie WF, Niclis JC, Petratos S (2013) Human embryonic stem cell-derived oligodendrocytes: protocols and perspectives. Stem Cells Dev 22(18):2459–2476
89. Stappert L, Borghese L, Roese-Koerner B, Weinhold S, Koch P, Terstegge S, Uhrberg M, Wernet P, Brüstle O (2013) MicroRNA-based promotion of human neuronal differentiation and subtype specification. PLoS One 8(3):e59011
90. Takahashi K, Tanabe K, Ohnuki M, Narita M, Ichisaka T, Tomoda K, Yamanaka S (2007) Induction of pluripotent stem cells from adult human fibroblasts by defined factors. Cell 131(5): 861–872
91. Malchenko S, Xie J, de Fatima Bonaldo M, Vanin EF, Bhattacharyya BJ, Belmadani A, Xi G, Galat V, Goossens W, Seftor RE, Tomita T, Crispino J, Miller RJ, Bohn MC, Hendrix MJ, Soares MB (2014) Onset of rosette formation during spontaneous neural differentiation of hESC and hiPSC colonies. Gene 534(2):400–407
92. Salewski RP, Buttigieg J, Mitchell RA, van der Kooy D, Nagy A, Fehlings MG (2013) The generation of definitive neural stem cells from PiggyBac transposon-induced pluripotent stem cells can be enhanced by induction of the NOTCH signaling pathway. Stem Cells Dev 22(3):383–396
93. Waschek JA (2012) Noggin on heaven's door: a factor that promotes the selective production of serotonergic neurons from murine embryonic stem cells and induced pluripotent stem cells. J Neurochem 122(1):1–3
94. Verpelli C, Carlessi L, Bechi G, FusarPoli E, Orellana D, Heise C, Franceschetti S, Mantegazza R, Mantegazza M, Delia D, Sala C (2013) Comparative neuronal differentiation of self-renewing neural progenitor cell lines obtained from human induced pluripotent stem cells. Front Cell Neurosci 7:175
95. Zeng H, Guo M, Martins-Taylor K, Wang X, Zhang Z, Park JW, Zhan S, Kronenberg MS, Lichtler A, Liu HX, Chen FP, Yue L, Li XJ, Xu RH (2010) Specification of region-specific neurons including forebrain glutamatergic neurons from human induced pluripotent stem cells. PLoS One 5:e11853

96. Yan Y, Shin S, Jha BS, Liu Q, Sheng J, Li F, Zhan M, Davis J, Bharti K, Zeng X, Rao M, Malik N, Vemuri MC (2013) Efficient and rapid derivation of primitive neural stem cells and generation of brain subtype neurons from human pluripotent stem cells. Stem Cells Transl Med 2(11):862–870. doi:10.5966/sctm.2013-0080. Epub 2013 Oct 10
97. Hargus G, Cooper O, Deleidi M, Levy A, Lee K, Marlow E, Yow A, Soldner F, Hockemeyer D, Hallett PJ, Osborn T, Jaenisch R, Isacson O (2010) Differentiated Parkinson patient-derived induced pluripotent stem cells grow in the adult rodent brain and reduce motor asymmetry in Parkinsonian rats. Proc Natl Acad Sci U S A 107(36):15921–15926. doi:10.1073/pnas.1010209107. Epub 2010 Aug 23
98. Wernig M, Zhao JP, Pruszak J, Hedlund E, Fu D, Soldner F, Broccoli V, Constantine-Paton M, Isacson O, Jaenisch R (2008) Neurons derived from reprogrammed fibroblasts functionally integrate into the fetal brain and improve symptoms of rats with Parkinson's disease. Proc Natl Acad Sci U S A 105(15):5856–5861
99. Linta L, Stockmann M, Lin Q, Lechel A, Proepper C, Boeckers TM, Kleger A, Liebau S (2013) Microarray-based comparisons of ion channel expression patterns: human keratinocytes to reprogrammed hiPSCs to differentiated neuronal and cardiac progeny. Stem Cells Int 2013:784629
100. Oki K, Tatarishvili J, Wood J, Koch P, Wattananit S, Mine Y, Monni E, Tornero D, Ahlenius H, Ladewig J, Brüstle O, Lindvall O, Kokaia Z (2012) Human-induced pluripotent stem cells form functional neurons and improve recovery after grafting in stroke-damaged brain. Stem Cells 30(6):1120–1133. doi:10.1002/stem.1104
101. Qin J, Song B, Zhang H, Wang Y, Wang N, Ji Y, Qi J, Chandra A, Yang B, Zhang Y, Gong G, Xu Y (2013) Transplantation of human neuro-epithelial-like stem cells derived from induced pluripotent stem cells improves neurological function in rats with experimental intracerebral hemorrhage. Neurosci Lett 548:95–100. doi:10.1016/j.neulet.2013.05.007. Epub 2013 May 13
102. Popescu IR, Nicaise C, Liu S, Bisch G, Knippenberg S, Daubie V, Bohl D, Pochet R (2013) Neural progenitors derived from human induced pluripotent stem cells survive and differentiate upon transplantation into a rat model of amyotrophic lateral sclerosis. Stem Cells Transl Med 2(3):167–174. doi:10.5966/sctm.2012-0042. Epub 2013 Feb 14
103. Riazifar H, Jia Y, Chen J, Lynch G, Huang T (2014) Chemically induced specification of retinal ganglion cells from human embryonic and induced pluripotent stem cells. Stem Cell Transl Med 3(4):424–432
104. Liu Q, Spusta SC, Mi R, Lassiter RN, Stark MR, Höke A, Rao MS, Zeng X (2012) Human neural crest stem cells derived from human ESCs and induced pluripotent stem cells: induction, maintenance, and differentiation into functional schwann cells. Stem Cells Transl Med 1(4):266–278
105. Lu HF, Chai C, Lim TC, Leong MF, Lim JK, Gao S, Lim KL, Wan AC (2014) A defined xeno-free and feeder-free culture system for the derivation, expansion and direct differentiation of transgene-free patient-specific induced pluripotent stem cells. Biomaterials 35(9):2816–2826
106. Hick A, Wattenhofer-Donzé M, Chintawar S, Tropel P, Simard JP, Vaucamps N, Gall D, Lambot L, André C, Reutenauer L, Rai M, Teletin M, Messaddeq N, Schiffmann SN, Viville S, Pearson CE, Pandolfo M, Puccio H (2013) Neurons and cardiomyocytes derived from induced pluripotent stem cells as a model for mitochondrial defects in Friedreich's ataxia. Dis Model Mech 6(3):608–621
107. Panicker LM, Miller D, Park TS, Patel B, Azevedo JL, Awad O, Masood MA, Veenstra TD, Goldin E, Stubblefield BK, Tayebi N, Polumuri SK, Vogel SN, Sidransky E, Zambidis ET, Feldman RA (2012) Induced pluripotent stem cell model recapitulates pathologic hallmarks of Gaucher disease. Proc Natl Acad Sci U S A 109(44):18054–18059. doi:10.1073/pnas.1207889109. Epub 2012 Oct 15
108. Dimos JT, Rodolfa KT, Niakan KK, Weisenthal LM, Mitsumoto H, Chung W, Croft GF, Saphier G, Leibel R, Goland R, Wichterle H, Henderson CE, Eggan K (2008) Induced pluripotent stem cells generated from patients with ALS can be differentiated into motor neurons. Science 321(5893):1218–1221
109. Ebert AD, Yu J, Rose FF Jr, Mattis VB, Lorson CL, Thomson JA, Svendsen CN (2009) Induced pluripotent stem cells from a spinal muscular atrophy patient. Nature 457(7227):277–280
110. Park IH, Arora N, Huo H, Maherali N, Ahfeldt T, Shimamura A, Lensch MW, Cowan C, Hochedlinger K, Daley GQ (2008) Disease-specific induced pluripotent stem cells. Cell 134(5):877–886. doi:10.1016/j.cell.2008.07.041. Epub 2008 Aug 7
111. Soldner F, Laganière J, Cheng AW, Hockemeyer D, Gao Q, Alagappan R, Khurana V, Golbe LI, Myers RH, Lindquist S, Zhang L, Guschin D, Fong LK, Vu BJ, Meng X, Urnov FD, Rebar EJ, Gregory PD, Zhang HS, Jaenisch R (2011) Generation of isogenic pluripotent stem cells differing exclusively at two early onset Parkinson point mutations. Cell 146(2):318–331. doi:10.1016/j.cell.2011.06.019
112. Brennand KJ, Landek-Salgado MA, Sawa A (2013) Modeling Heterogeneous Patients with a Clinical Diagnosis of Schizophrenia with Induced Pluripotent Stem Cells. Biol Psychiatry pii:S0006-3223(13)01000-7
113. Israel MA, Yuan SH, Bardy C, Reyna SM, Mu Y, Herrera C, Hefferan MP, Van Gorp S, Nazor KL, Boscolo FS, Carson CT, Laurent LC, Marsala

M, Gage FH, Remes AM, Koo EH, Goldstein LS (2012) Probing sporadic and familial Alzheimer's disease using induced pluripotent stem cells. Nature 482(7384):216–220

114. Yusa K, Zhou L, Li MA, Bradley A, Craig NL (2011) A hyperactive piggyBac transposase for mammalian applications. Proc Natl Acad Sci U S A 108(4):1531–1536
115. Hanna J, Wernig M, Markoulaki S, Sun CW, Meissner A, Cassady JP, Beard C, Brambrink T, Wu LC, Townes TM, Jaenisch R (2007) Treatment of sickle cell anemia mouse model with iPS cells generated from autologous skin. Science 318(5858):1920–1923. Epub 2007 Dec 6
116. Juopperi TA, Kim WR, Chiang CH, Yu H, Margolis RL, Ross CA, Ming GL, Song H (2012) Astrocytes generated from patient induced pluripotent stem cells recapitulate features of Huntington's disease patient cells. Mol Brain 5:17. doi:10.1186/1756-6606-5-17
117. Brennand KJ, Gage FH (2012) Modeling psychiatric disorders through reprogramming. Dis Model Mech 5(1):26–32
118. Marchetto MC, Gage FH (2012) Modeling brain disease in a dish: really? Cell Stem Cell 10(6): 642–645
119. Xia G, Santostefano KE, Goodwin M, Liu J, Subramony SH, Swanson MS, Terada N, Ashizawa T (2013) Generation of neural cells from DM1 induced pluripotent stem cells as cellular model for the study of central nervous system neuropathogenesis. Cell Reprogram 15(2):166–177
120. Soldner F, Jaenisch R (2012) Medicine iPSC disease modeling science. Science 338(6111):1155–1156. doi:10.1126/science.1227682
121. Thier M, Wörsdörfer P, Lakes YB, Gorris R, Herms S, Opitz T, Seiferling D, Quandel T, Hoffmann P, Nöthen MM, Brüstle O, Edenhofer F (2012) Direct conversion of fibroblasts into stably expandable neural stem cells. Cell Stem Cell 10(4):473–479. doi:10.1016/j.stem.2012.03.003. Epub 2012 Mar 22
122. Thomson M, Liu SJ, Zou LN, Smith Z, Meissner A, Ramanathan S (2011) Cell 145:875–889
123. Kim J, Efe JA, Zhu S, Talantova M, Yuan X, Wang S, Lipton SA, Zhang K, Ding S (2011) Proc Natl Acad Sci U S A 108:7838–7843
124. Han DW, Tapia N, Hermann A, Hemmer K, Hö̈ ing S, Arau'zo- Bravo M, Zahres H, Wu G, Frank S, Moritz S et al (2012) Cell Stem Cell 10(this issue):465–472
125. Ring KL, Tong LM, Balestra ME, Javier R, Andrews-Zwilling Y, Li G, Walker D, Zhang WR, Kreitzer AC, Huang Y (2012) Direct reprogramming of mouse and human fibroblasts into multipotent neural stem cells with a single factor. Cell Stem Cell 11(1):100–109. doi:10.1016/j.stem.2012.05.018. Epub 2012 June 7
126. Pang ZP, Yang N, Vierbuchen T, Ostermeier A, Fuentes DR, Yang TQ, Citri A, Sebastiano V, Marro S, Südhof TC, Wernig M (2011) Induction of human neuronal cells by defined transcription factors. Nature 476(7359):220–223. doi:10.1038/nature10202
127. Lujan E, Chanda S, Ahlenius H, Sü dhof TC, Wernig M (2012) Proc Natl Acad Sci USA 109: 2527–2532
128. Caiazzo M, Dell'Anno MT, Dvoretskova E, Lazarevic D, Taverna S, Leo D, Sotnikova TD, Menegon A, Roncaglia P, Colciago G, Russo G, Carninci P, Pezzoli G, Gainetdinov RR, Gustincich S, Dityatev A, Broccoli V (2011) Direct generation of functional dopaminergic neurons from mouse and human fibroblasts. Nature 476(7359):224–227. doi:10.1038/nature10284
129. Torper O, Pfisterer U, Wolf DA, Pereira M, Lau S, Jakobsson J, Björklund A, Grealish S, Parmar M (2013) Generation of induced neurons via direct conversion in vivo. Proc Natl Acad Sci U S A 110(17):7038–7043. doi:10.1073/pnas.1303829110. Epub 2013 Mar 25
130. Zou Q, Yan Q, Zhong J, Wang K, Sun H, Yi X, Lai L (2014) Direct conversion of human fibroblasts into neuronal restricted progenitors. J Biol Chem 289(8):5250–5260
131. Liu ML, Zang T, Zou Y, Chang JC, Gibson JR, Huber KM, Zhang CL (2013) Small molecules enable neurogenin 2 to efficiently convert human fibroblasts into cholinergic neurons. Nat Commun 4:2183. doi:10.1038/ncomms3183
132. Reyes M, Lund T, Lenvik T, Aguiar D, Koodie L, Verfaillie CM (2001) Purification and ex vivo expansion of postnatal, human marrow mesodermal progenitor cells. Blood 98:2615–2625
133. Fibbe WE, Noort WA (2003) Mesenchymal stem cells and hematopoietic stem cells transplantation. Ann NY Acad Sci 996:235–244
134. Egusa H, Schweizer FE, Wang CC, Matsuka Y, Nishimura I (2005) Neuronal differentiation of bone marrow-derived stromal stem cells involves suppression of discordant phenotypes through gene silencing. J Biol Chem 280:23691–23697
135. Friedenstein AJ, Gorskaja U, Kalugina NN (1976) Fibroblast precursors in normal and irradiated mouse hematopoietic organs. Exp Hematol 4: 267–274
136. Dennis JE, Charbord P (2002) Origin and differentiation of human and murine stroma. Stem Cells 20(3):205–214
137. Gronthos S, Brahim J, Li W, Fisher LW, Cherman N, Boyde A, DenBesten P, Robey PG, Shi S (2002) Stem cell properties of human dental pulp stem cells. J Dent Res 81(8):531–535
138. Zuk PA, Zhu M, Ashjian P, De Ugarte DA, Huang JI, Mizuno H, Alfonso ZC, Fraser JK, Benhaim P, Hedrick MH (2002) Human adipose tissue is a source of multipotent stem cells. Mol Biol Cell 13(12):4279–4295
139. Gimble J, Guilak F (2003) Adipose-derived adult stem cells: isolation, characterization, and differentiation potential. Cytotherapy 5(5):362–369

140. Miura M, Gronthos S, Zhao M, Lu B, Fisher LW, Robey PG, Shi S (2003) SHED: stem cells from human exfoliated deciduous teeth. Proc Natl Acad Sci U S A 100:5807–5812
141. Zvaifler NJ, Marinova-Mutafchieva L, Adams G, Edwards CJ, Moss J, Burger JA, Maini RN (2000) Mesenchymal precursor cells in the blood of normal individuals. Arthritis Res 2:477–488
142. Mitchell KE, Weiss ML, Mitchell BM, Martin P, Davis D, Morales L, Helwig B, Beerenstrauch M, Abou-Easa K, Hildreth T, Troyer D, Medicetty S (2003) Matrix cells from Wharton's jelly form neurons and glia. Stem Cells 21(1):50–60
143. Igura K, Zhang X, Takahashi K, Mitsuru A, Yamaguchi S, Takashi TA (2004) Isolation and characterization of mesenchymal progenitor cells from chorionic villi of human placenta. Cytotherapy 6: 543–553
144. Wang HS, Hung SC, Peng ST, Huang CC, Wei HM, Guo YJ, Fu YS, Lai MC, Chen CC (2004) Mesenchymal stem cells in the Wharton's jelly of the human umbilical cord. Stem Cells 22(7): 1330–1337
145. Fu YS, Shih YT, Cheng YC, Min MY (2004) Transformation of human umbilical mesenchymal cells into neurons in vitro. J Biomed Sci 11(5):652–660
146. Tsai MS, Lee JL, Chang YJ, Hwang SM (2004) Isolation of human multipotent mesenchymal stem cells from second trimester amniotic fluid using a novel two-stage culture protocol. Hum Reprod 19:1450–1456
147. Campagnoli C, Roberts IA, Kumar S, Bennet PR, Bellantuono I, Fisk NM (2001) Identification of mesenchymal stem/progenitor cells in human first-trimester fetal blood, liver and bone marrow. Blood 98:2396–2402
148. Gong X, Sun Z, Cui D, Xu X, Zhu H, Wang L, Qian W, Han X (2012) Isolation and characterization of lung resident mesenchymal stem cells capable of differentiating into alveolar epithelial type II cells. Cell Biol Int [Epub ahead of print]
149. Dominici M, Le Blanc K, Mueller S-C, Marini FC, Krause DS, Deans RJ, Keating A, Prockop DJ, Horwitz EM (2006) Minimal criteria for defining multipotent mesenchymal stromal cells. The International Society for Cellular Therapy position statement. Cytotherapy 8(4):315–317
150. Makino S, Fukuda K, Miyoshi S, Konishi F, Kodama H, Pan J, Sano M, Takahashi T, Hori S, Abe H, Hata J, Umezawa A, Ogawa S (1999) Cardiomyocytes can be generated from marrow stromal cells in vitro. J Clin Invest 103:697–705
151. Pittenger MF, Mackay AM, Beck SC, Jaiswal RK, Douglas R, Mosca JD, Moorman MA, Simoneti DW, Craig S, Marshak DR (1999) Multilineage potential of adult human mesenchymal stem cells. Science 284:143–147
152. Jiang Y, Jahagirdar BN, Reinhardt RL, Shwartz RE, Keene CD, Oritz-Gonzalez XR, Reyes M, Lenvik T, Lund T, Blackstad M, Du J, Aldrich S, Lisberg A, Low W, Largaespada DA, Verfaillie CM (2002) Pluripotency of mesenchymal stem cells derived from adult marrow. Nature 418:41–49
153. Xu W, Zhang X, Qian H, Zhu W, Sun X, Hu J, Zhou H, Chen Y (2004) Mesenchymal stem cells from adult human bone marrow differentiate into a cardiomyocyte phenotype in vitro. Exp Biol Med (Maywood) 229:623–631
154. Egusa H, Schweizer FE, Wang CC, Matsuka Y, Nishimura I (2005) Neuronal differentiation of bone marrow-derived stromal stem cells involves suppression of discordant phenotypes through gene silencing. J Biol Chem 280:23691–23697
155. Datta I, Mishra S, Mohanty L, Pulikkot S, Joshi PG (2011) Neuronal plasticity of human Wharton's jelly mesenchymal stromal cells to the dopaminergic cell type compared with human bone marrow mesenchymal stromal cells. Cytotherapy 13: 918–932
156. Kanafi M, Majumdar D, Bhonde R, Datta I (2014 Jan 29) Midbrain cues dictate differentiation of human dental pulp stem cells towards functional dopaminergic neurons. J Cell Physiol. doi:10.1002/jcp.24570
157. Tremain N, Korkko J, Ibberson D, Kopen GC, DiGirolamo C, Phinney DG (2001) MicroSAGE analysis of 2,353 expressed genes in a single cell-derived colony of undifferentiated human mesenchymal stem cells reveals mRNAs of multiple cell lineages. Stem Cells 19(5):408–418
158. Swistowski A, Peng J, Liu Q, Mali P, Rao MS, Cheng L, Zeng X (2010) Efficient generation of functional dopaminergic neurons from human induced pluripotent stem cells under defined conditions. Stem Cells 28(10):1893–1904
159. Malgrange B, Borgs L, Grobarczyk B, Purnelle A, Ernst P, Moonen G, Nguyen L (2011) Using human pluripotent stem cells to untangle neurodegenerative disease mechanisms. Cell Mol Life Sci 68(4): 635–649
160. Krick K, Tammia M, Martin R, Höke A, Mao HQ (2011) Signaling cue presentation and cell delivery to promote nerve regeneration. Curr Opin Biotechnol 22(5):741–746. doi:10.1016/j.copbio.2011.04.002. Epub 2011 Apr 29
161. Woodbury D, Schwarz EJ, Prockop DJ, Black IB (2000) Adult rat and human bone marrow stromal cells differentiate into neurons. J Neurosci Res 61:364–370
162. Barnabé GF, Schwindt TT, Calcagnotto ME, Motta FL, Martinez G, de Oliveira Jr AC, Keim LMN, D'Almeida V, Mendez-Otero R, Mello LE (2009) Chemically-induced rat mesenchymal stem cells adopt molecular properties of neuronal like cells but do not have basic neuronal properties. PLoSONE 44e5222. doi:10.1371/journal.pone.0005222
163. Király M, Porcsalmy B, Pataki A, Kádár K, Jelitai M, Molnár B, Hermann P, Gera I, Grimm WD, Ganss B, Zsembery A, Varga G (2009) Simultaneous PKC and cAMP activation induces differenti-

ation of human dental pulp stem cells into functionally active neurons. Neurochem Int 55(5):323–332. doi:10.1016/j.neuint.2009.03.017. Epub 2009 Apr 5
164. Trazaska KA, Reddy BY, Munoz JL, Li KY, Ye JH, Rameshwar P (2008) Loss of RE-1 silencing factor in mesenchymal stem cell derived dopamine progenitors induces functional maturity. Mol Cell Neurosci 39:285–290
165. Lu P, Blesch A, Tuszynski MH (2004) Induction of bone marrow stromal cells to neurons: differentiation, transdifferentiation, or artifact? J Neurosci Res 77:174–191
166. Bourtchuladze R, Frenguelli B, Blendy J, Cioffi D, Schutz G, Silva AJ (1994) Deficient long-term memory in mice with a targeted mutation of the cAMP-responsive element-binding protein. Cell 79(1):59–68
167. Lim J, Yang C, Hong SJ, Kim KS (2000) Regulation of tyrosine hydroxylase gene transcription by the cAMP-signaling pathway: involvement of multiple transcription factors. Mol Cell Biochem 212(1–2):51–60
168. Dworkin S, Heath JK, deJong-Curtain TA, Hogan BM, Lieschke GJ, Malaterre J, Ramsay RG, Mantamadiotis T (2007) CREB activity modulates neural cell proliferation, midbrain-hindbrain organization and patterning in zebrafish. Dev Biol 307(1): 127–141
169. Wang TT, Tio M, Lee W, Beerheide W, Udolph G (2007) Neural differentiation of mesenchymal-like stem cells from cord blood is mediated by PKA. Biochem Biophys Res Commun 357(4):1021–1027
170. Lepski G, Jannes CE, Maciaczyk J, Papazoglou A, Mehlhorn AT, Kaiser S, Teixeira MJ, Marie SK, Bischofberger J, Nikkhah G (2010) Limited Ca2+ and PKA-pathway dependent neurogenic differentiation of human adult mesenchymal stem cells as compared to fetal neuronal stem cells. Exp Cell Res 316(2):216–231
171. Rooney GE, Howard L, O'Brien T, Windebank AJ, Barry FP (2009) Elevation of cAMP in mesenchymal stem cells transiently upregulates neural markers rather than inducing neural differentiation. Stem Cells Dev 18(3):387–398
172. Zhang L, Tan X, Dong C, Zou L, Zhao H, Zhang X, Tian M, Jin G (2012) In vitro differentiation of human umbilical cord mesenchymal stem cells (hUCMSCs), derived from Wharton's jelly, into choline acetyltransferase (ChAT)-positive cells. Int J Dev Neurosci 30(6):471–477
173. Wand J, Wang X, Sun Z, Wang X, Yang H, Shi S, Wang S (2010) Stem cells from human-exfoliated deciduous teeth can differentiate into dopaminergic neuron-like cells. Stem Cells Dev 19:1375–1383
174. Alexanian AR, Liu QS, Zhang Z (2013) Enhancing the efficiency of direct reprogramming of human mesenchymal stem cells into mature neuronal-like cells with the combination of small molecule modulators of chromatin modifying enzymes, SMAD signaling and cyclic adenosine monophosphate levels. Int J Biochem Cell Biol 45(8):1633–1638. doi:10.1016/j.biocel.2013.04.022. Epub 2013 May 9
175. Jang S, Cho HH, Cho YB, Park JS, Jeong HS (2010) Functional neural differentiation of human adipose tissue-derived stem cells using bFGF and forskolin. BMC Cell Biol 11:25. doi:10.1186/1471-2121-11-25
176. Fu L, Zhu L, Huang Y, Lee TD, Forman SJ, Shih CC (2008) Derivation of neural stem cells from mesenchymal stem cells: evidence for a bipotential stem cell population. Stem Cells Dev 17: 1109–1121
177. Han YG, Spassky N, Romaguera-Ros M, Garcia-Verdugo JM, Aguilar A, Schneider- Maunoury S, Alvarez-Buylla A (2008) Hedgehog signalling and primary cilia are required for the formation of adult neural stem cells. Nat Neurosci 11:277–284
178. Balasubramanian S, Thej C, Venugopal P, Priya N, Zakaria Z, SundarRaj S, Majumdar AS (2013) Higher propensity of Wharton's jelly derived mesenchymal stromal cells towards neuronal lineage in comparison to those derived from adipose and bone marrow. Cell Biol Int 37(5):507–515; ISSN 1065-6995. doi:10.1002/cbin.10056
179. Arthur A, Rychkov G, Shi S, Koblar SA, Gronthos S (2008) Adult human dental pulp stem cells differentiate toward functionally active neurons under appropriate environmental cues. Stem Cells 26: 1787–1795
180. Liang J, Wu S, Zhao H, Li SL, Liu ZX, Wu J, Zhou L (2013) Human umbilical cord mesenchymal stem cells derived from Wharton's jelly differentiate into cholinergic-like neurons in vitro. Neurosci Lett 532:59–63. doi:10.1016/j.neulet.2012.11.014. Epub 2012 Nov 21
181. Naghdi M, Tiraihi T, Namin SA, Arabkheradmand J (2009) Transdifferentiation of bone marrow stromal cells into cholinergic neuronal phenotype: a potential source for cell therapy in spinal cord injury. Cytotherapy 11:137–152
182. Zhang W, Zeng YS, Wang JM, Ding Y, Li Y, Wu W (2009) Neurotrophin-3 improves retinoic acid-induced neural differentiation of skin-derived precursors through a p75NTR-dependent signaling pathway. Neurosci Res 64(2):170–176
183. Darabi S, Tiraihi T, Delshad A, Sadeghizadeh M (2013) A new multistep induction protocol for the transdifferentiation of bone marrow stromal stem cells into GABAergic neuron-like cells. Iran Biomed J 17(1):8–14
184. Safford KM, Hicok KC, Safford SD, Halvorsen YD, Wilkison WO, Gimble JM, Rice HE (2002) Neurogenic differentiation of murine and human adipose – derived stromal cells. Biochem Biophys Res Commun 294:371–379
185. Schaffler A, Buchler C (2007) Concise review: adipose tissue-derived stromal cells – basic and clinical implications for novel cell-based therapies. Stem Cells 25:818–827
186. Anghileri E, Marconi S, Pignatelli A, Cifelli P, Galie M, Sbarbati A, Krampera M, Belluzzi O

et al (2008) Neuronal differentiation potential of human adipose-derived mesenchymal stem cells. Stem Cells Dev 17:909–916
187. Kim SS, Choi JM, Kim JW, Ham DS, Ghil SH, Kim MK, Kim-Kwon Y, Hong SY, Ahn SC, Kim SU, Lee YD, Suh-Kim H (2005) cAMP induces neuronal differentiation of mesenchymal stem cells via activation of extracellular signal-regulated kinase/MAPK. Neuroreport 16(12):1357–1361
188. Pavlova G, Lopatina T, Kalinina N, Rybalkina E, Parfyonova Y, Tkachuk V, Revishchin A (2012) In vitro neuronal induction of adipose-derived stem cells and their fate after transplantation into injured mouse brain. Curr Med Chem 19:5170–5177
189. Ying C, Hu W, Cheng B, Zheng X, Li S (2012) Neural differentiation of rat adipose-derived stem cells in vitro. Cell Mol Neurobiol 32(8):1255–1263
190. Hedvika D, Xiufang G, Stephen L, Maria S, James JH (2011) Small molecule induction of human umbilical stem cells into myelin basic protein positive oligodendrocytes in a defined three-dimensional environment. ACS Chem Neurosci 3:31–39
191. Mantovani C, Mahay D, Kingham M, Terenghi G, Shawcross SG, Wiberg M (2010) Bone marrow- and adipose-derived stem cells show expression of myelin mRNAs and proteins. Regen Med 5:403–410
192. Chang SJ, Shun-Long W, Jui-Yu H, Tao-Yeuan W, Margaret DTC, Hsei-Wei W (2011) MicroRNA – 34a modulates genes involved in cellular motility and oxidative phosphorylation in neural precursors derived from human umbilical cord mesenchymal stem cells. BMC Med Genomics 4:65
193. Barzilay R, Ben-Zur T, Bulvik S, Melamed E, Offen D (2009) Lentiviral delivery of LMX1a enhances dopaminergic phenotype in differentiated human bone marrow mesenchymal stem cells. Stem Cells Dev 18:591–601
194. Barzilay R, Melamed E, Offen D (2009) Introducing transcription factors to multipotent mesenchymal stem cells: making transdifferentiation possible. Stem Cells 27:2509–2515
195. Yang Y, Li Y, Lv Y, Zhang S, Chen L, Bai C, Nan X, Yue W, Pei X (2008) NRSF silencing induces neuronal differentiation of human mesenchymal stem cells. Exp Cell Res 314(11–12):2257–2265. doi:10.1016/j.yexcr.2008.04.008. Epub 2008 May 1
196. Choi SA, Hwang SK, Wang KC, Cho BK, Phi JH, Lee JY, Jung HW, Lee DH, Kim SK (2011) Therapeutic efficacy and safety of TRAIL-producing human adipose tissue-derived mesenchymal stem cells against experimental brainstem glioma. Neuro Oncol 13:61–69
197. Muñoz-Elias G, Marcus AJ, Coyne TM, Woodbury D, Black IB (2004) Adult bone marrow stromal cells in the embryonic brain: engraftment, migration, differentiation, and long-term survival. J Neurosci 24(19):4585–4595
198. Kopen GC, Prockop DJ, Phinney DG (1999) Marrow stromal cells migrate throughout forebrain and cerebellum, and they differentiate into astrocytes after injection into neonatal mouse brains. Proc Natl Acad Sci U S A 96(19):10711–10716
199. Mezey E, Key S, Vogelsang G, Szalayova I, Lange GD, Crain B (2003) Transplanted bone marrow generates new neurons in human brains. Proc Natl Acad Sci U S A 100(3):1364–1369. Epub 2003 Jan 21
200. Huang AH, Snyder BR, Cheng PH, Chan AW (2008) Putative dental pulp-derived stem/stromal cells promote proliferation and differentiation of endogenous neural cells in the hippocampus of mice. Stem Cells 26:2654–2663
201. Azizi SA, Stokes D, Augelli BJ, DiGirolamo C, Prockop DJ (1998) Engraftment and migration of human bone marrow stromal cells implanted in the brains of albino rats–similarities to astrocyte grafts. Proc Natl Acad Sci U S A 95(7): 3908–3913
202. Dezawa M, Hayase M, Kitada M, Wakao S, Itokazu Y, Nozaki K, Hashimoto N, Takagi Y (2009) Committed neural progenitor cells derived from genetically modified bone marrow stromal cells ameliorate deficits in a rat model of stroke. J Cereb Blood Flow Metab 29(8):1409–1420
203. Khoo ML, Tao H, Meedeniya AC, Mackay-Sim A, Ma DD (2011) Transplantation of neuronal-primed human bone marrow mesenchymal stem cells in hemiparkinsonian rodents. PLoS One 6(5):e19025. doi:10.1371/journal.pone.0019025
204. Levy YS, Bahat-Stroomza M, Barzilay R, Burshtein A, Bulvik S, Barhum Y, Panet H, Melamed E, Offen D (2008) Regenerative effect of neural-induced human mesenchymal stromal cells in rat models of Parkinson's disease. Cytotherapy 10:340–352
205. Wang F, Yasuhara T, Shingo T, Kameda M, Tajiri N, Yuan WJ, Kondo A, Kadota T et al (2010) Intravenous administration of mesenchymal stem cells exerts therapeutic effects on parkinsonian model of rats: focusing on neuroprotective effects of stromal cell-derived factor-1α. BMC Neurosci 11:52
206. Park HJ, Lee PH, Bang OY, Lee G, Ahn YH (2008) Mesenchymal stem cells therapy exerts neuroprotection in a progressive animal model of Parkinson's disease. J Neurochem 107(1):141–51. doi:10.1111/j.1471-4159.2008.05589.x. Epub 2008 July 28
207. Yang H, Xie Z, Wei L, Yang H, Yang S, Zhu Z, Wang P, Zhao C, Bi J (2013) Human umbilical cord mesenchymal stem cell-derived neuron-like cells rescue memory deficits and reduce amyloid-beta deposition in an AβPP/PS1 transgenic mouse model. Stem Cell Res Ther 4(4):76
208. Habisch HJ, Schmid B, Arnim CA, Ludolph AC, Brenner R, Storch A (2010) Efficient processing of Alzheimer's Disease amyloid-beta peptides by neuroectodermally converted mesenchymal stem cells. Stem Cells Dev 19:629–633
209. McCoy MK, Martinez TN, Ruhn KA, Wrage PC, Keefer EW, Botterman BR, Tansey KE, Tansey MG (2008) Autologous transplants of Adipose-Derived Adult Stromal (ADAS) cells afford dopaminergic

neuroprotection in a model of Parkinson's disease. Exp Neurol 210(1):14–29. Epub 2007 Nov 1
210. Weiss ML, Medicetty S, Bledsoe AR, Rachakatla RS, Choi M, Merchav S, Luo Y, Rao MS, Velagaleti G, Troyer D (2006) Human umbilical cord matrix stem cells: preliminary characterization and effect of transplantation in a rodent model of Parkinson's disease. Stem Cells 24(3):781–92. Epub 2005 Oct 13
211. Pereira MC, Secco M, Suzuki DE, Janjoppi L, Rodini CO, Torres LB, Araujo BH, Cavalheiro EA et al (2011) Contamination of mesenchymal stem cells with fibroblasts accelerates neurodegeneration in an experimental model of Parkinson's disease. Stem Cell Rev 7:1006–1017
212. Maria D, Hiroshi K, Mikio H, Hirotomi C, Naoya M, Yutaka I, Nobuyoshi T, Hitoshi Y et al (2004) Specific induction of neuronal cells from bone marrow stromal cells and application for autologous transplantation. J Clin Invest 113: 1701–1710
213. Harmening K, Heile A, Miller M, Johanson CE, Wallrapp C, Brinker T, Silverberg GD, Klinge PM (2009) Microglial downregulation in a double transgenic mouse model associated with early-onset Alzheimer's disease (AD) after intraventricular implantation of alginate encapsulated glucagon-like-peptide-1 (GLP-1) producing human mesenchymal stem cells. Cerebrospinal Fluid Res 6:S15
214. Ikegame Y, Yamashita K, Hayashi S, Mizuno H, Tawada M, You F, Yamada K, Tanaka Y (2011) Comparison of mesenchymal stem cells from adipose tissue and bone marrow for ischemic stroke therapy. Cytotherapy 13:675–685
215. Yasuhara T, Matsukawa N, Hara K, Maki M, Ali MM, Yu SJ, Bae E, Yu G, Xu L, McGrogan M, Bankiewicz K, Case C, Borlongan CV (2009) Notch-induced rat and human bone marrow stromal cell grafts reduce ischemic cell loss and ameliorate behavioral deficits in chronic stroke animals. Stem Cells Dev 18(10):1501–1514
216. Bang OY, Lee JS, Lee PH, Lee G (2005) Autologous mesenchymal stem cell transplantation in stroke patients. Ann Neurol 57(6):874–882
217. Mazzini L, Mareschi K, Ferrero I, Vassallo E, Oliveri G, Nasuelli N, Oggioni GD, Testa L, Fagioli F (2008) Stem cell treatment in amyotrophic lateral sclerosis. J Neurol Sci 265(1–2):78–83. Epub 2007 June 19
218. Pedram MS, Dehghan MM, Soleimani M, Sharifi D, Marjanmehr SH, Nasiri Z (2010) Transplantation of a combination of autologous neural differentiated and undifferentiated mesenchymal stem cells into injured spinal cord of rats. Spinal Cord 48(6): 457–463
219. Hu BY, Zhang SC (2009) Differentiation of spinal motor neurons from pluripotent human stem cells. Nat Protoc 4(9):1295–1304

Part III

Stem Cells and Organ Systems: Cardiovascular System

Stem Cell Therapy for Cardiac Tissue Regeneration Post-myocardial Infarction

Uksha Saini and Konstantinos Dean Boudoulas

Myocardial infarction typically is the result of atherosclerotic plaque rupture in a coronary artery followed by platelet aggregation and thrombus formation resulting in partial or total vessel occlusion with diminished blood flow to the myocardial tissue downstream [1]. Within minutes to hours, cardiomyocytes in the ischemic region undergo apoptosis resulting in cell death [2, 3]; this results in damaged myocardium, left ventricular dysfunction, and heart failure [4, 5]. Myocardial infarction results in substantial mortality and morbidity worldwide [6]. Currently, there are limited therapies to consistently and effectively reverse the course of this process, and the possibility to replace or restore damaged heart tissue using cell therapy is an exciting concept. Stem cells are undifferentiated and unspecialized cells found in the body that have the potential to develop into all cell types [1]. The present chapter focuses on defining various stem cells and their use for cardiac tissue regeneration post-myocardial infarction. In addition, the major challenges and drawbacks associated with their use will be discussed.

U. Saini, PhD
Division of Cardiovascular Medicine, Department of Internal Medicine, Davis Heart and Lung Research Institute, The Ohio State University, Columbus, OH 43210, USA
e-mail: Uksha.Saini@osumc.edu

K.D. Boudoulas, MD (✉)
Department of Medicine, Department of Cardiovascular Medicine, The Ohio State University, Columbus, OH, USA
e-mail: konstantinos.boudoulas@osumc.edu

Types of Stem Cells

Embryonic Stem Cell (ESC)

ESCs are derived from the embryo just after fertilization, specifically from the blastocyst. They are generated by transferring cells from the embryo into a laboratory culture dish donated for research purposes with informed consent obtained from the donors. ESCs have self-renewal properties and the potential to develop into all cell types of the human body, and thus they are pluripotent. Glycolipids SSEA3 and SSEA4, and the keratan sulfate antigens Tra-1-60 and Tra-1-81, are the common cell-surface antigen markers employed to identify human ESCs. Suppression of genes that lead to differentiation and maintenance of pluripotency is critical for these cells, and the transcription factors Oct-4, Nanog, and Sox2 play a pivotal role in regulating ESC pluripotency. Mouse myocardial infarction models have shown that transplantation of mouse ESCs can regenerate cardiomyocytes, vascular smooth muscle cells, and endothelial cells resulting in the improvement of left ventricular structure and function [7–10].

Although ESCs have ideal properties for cell-based therapy, they have the potential to form tumors including teratomas. ESCs transplanted into a donor can also result in rejection due to

I. Somasundaram (ed.), *Stem Cell Therapy for Organ Failure*,
DOI 10.1007/978-81-322-2110-4_6, © Springer India 2014

an immune response. In addition, ethical issues regarding their use have limited their utilization in clinical practice.

Induced Pluripotent Stem Cell (iPSC)

Pluripotent stem cells artificially formed by genetic reprogramming of non-pluripotent adult cells, such as fibroblasts, in order to behave like ESCs are referred to as iPSCs; this allows iPSCs to express genes and factors necessary to function like an ESC. Their phenotypes can be modulated by chemical compounds or small bioactive molecules enhancing proliferation or directing differentiation. Takahasi and Yamanaka in 2006 [11] were the first to develop mouse iPSCs closely followed by human iPSCs in 2007. iPSCs can differentiate into all three germ layers in vitro; the level of expression of pluripotency genes like Oct-4, Nanog, and Sox2 in the reprogrammed cells is comparable to ESCs [12]. Animal models have shown the ability of iPSCs to improve left ventricular structure and function post-myocardial infarction and to improve cardiac function through paracrine activation [13, 14].

Even though iPSCs have been shown to have remarkable similarity to ESCs, differences have been reported including in DNA methylation [15, 16]. In addition, iPSCs have been reported to possess somatic coding mutations and immunogenicity, thereby limiting their clinical use [17–19]. Further, iPSCs have the potential to develop tumors including teratomas.

Hematopoietic Stem Cell (HSC)

HSCs are multipotent stem cells with the ability to give rise to all cells of blood lineage; they can be isolated from the blood or bone marrow. HSCs have a higher potential than other immature blood cells to pass through the bone marrow barrier and travel in the blood stream. HSCs have also been shown to differentiate into cardiomyocytes [20]. The cell-surface markers that define human HSCs in their undifferentiated state are $CD34^{+}$, $CD59^{+}$, $Thy1^{+}$, $CD38^{low/-}$, $C\text{-}kit^{-/low}$, and lin^{-} [21]. HSCs have been shown to have the ability to transform into cardiomyocytes and endothelial cells forming functional tissue after transplantation into the infracted myocardium of mice [22]; however, other studies failed to demonstrate the cardiac regenerative capabilities of HSCs [20].

Endothelial Progenitor Cell (EPC)

EPCs are found in the bone marrow and are a subset of hematopoietic cells. EPCs can differentiate into endothelial cells and are involved in angiogenesis [23]. Studies also suggest that EPCs can provide tissue repair through their paracrine effects [24]. EPCs can be identified by their cell-surface markers including CD34, VEGFR2, c-Kit, AC133, and CXCR4 [24]. Studies have shown that EPCs have the ability to home to the myocardial infarct border zone [25]. Further, studies have shown that transplantation of EPCs can improve cardiac function post-myocardial infarction [26, 27].

Mesenchymal Stem Cell (MSC)

MSCs are found in bone marrow stroma and adipose tissue. MSCs are multipotent and can differentiate into cells of mesoderm lineage including osteocytes, chondrocytes, and adipocytes, but have also been shown to have the potential to differentiation into cells of endoderm and ectoderm lineage [28]. Three criteria have been used to characterize MSCs: cells are plastic adherent under standard culture conditions; are CD105, CD73, and CD90 positive and lack expression of CD45, CD34, CD14 or CD11b, CD79α or CD19, and HLA-DR surface markers; and must differentiate into osteoblasts, adipocytes, and chondroblasts in vitro [29]. Several favorable properties make MSCs attractive for use in cell-based therapy. MSCs have low immunogenicity making them easier to apply, specifically in allogeneic recipients. They can readily be obtained from patients through a minimally invasive bone marrow biopsy and easily expanded in vitro. MSCs also have the ability to home to sites of tissue injury and differentiate, and have a paracrine effect on the surrounding environment via secretion of bioactive molecules capable of stimulating recovery of injured cells and inhibiting inflammation and immunomodulation [30]. For these reasons, MSCs have been extensively studied and used to treat diseases ranging from diabetes mellitus to heart disease. Several animal models and clinic trials demonstrate the ability of transplanted MSCs to improve left ventricular function post-myocardial infarction [31, 32]; however, even though cardiac improvement is observed, the cardiomyocyte differentiation is limited [33].

Endogenous Cardiac Stem Cell (CSC)

Endogenous CSCs are tissue-specific stem cells confined to the heart. The heart had been considered to be a terminally differentiated post-mitotic organ without regenerative capacity; however, in the recent past, undifferentiated cells and multipotent cells have been discovered. Two distinct subpopulations of endogenous CSCs have been identified including myogenic CSCs that regenerate cardiomyocytes and vasculogenic CSCs that regenerate blood vessels [34]. These cells test negative for blood lineage markers CD34, CD45, CD20, CD45RO, and CD8 (Lin-) and positive for stem cell and endothelial progenitor cell-specific surface markers Kit, Sca-1, and Isl-1. They also posses side population properties [35, 36]. There are low numbers of CSCs that are found within the adult heart; however, these cells can be isolated through endomyocardial biopsies and expanded in vitro [36]. Upon transplantation into the ischemic heart, these cells are capable of regenerating cardiomyocytes and blood vessels improving cardiac function; CSCs have been shown to have the ability to home to injured myocardium when supplied intravenously [37, 38].

Methods of Stem Cell Delivery

Stem cells can be delivered into the myocardium through various methods including intravenous infusion, intracoronary infusion via a percutaneous approach, or directly injected into the myocardium by a percutaneous approach through the endocardium or during open-heart surgery through the epicardium [1, 39]. Infusion of stem cells into the coronary arterial circulation is relatively easily preformed using catheter-based techniques via a percutaneous approach. The cells are delivered into the coronary artery circulation filled with nutrients and oxygen allowing the cells to be delivered in a more favorable environment; however, delivering the cells to the exact area of ischemia may be difficult, and washout may occur in the coronary circulation [40]. Despite this limitation, transplanted cells have been shown to home to ischemic sites in tissue [41, 42]. Stem cells can also be injected directly into the myocardium using a needle via a transepicardial approach during an open-heart surgery or via a transendocardial approach using percutaneous catheters. These approaches allow for direct delivery of cells into the region of interest; however, the ischemic myocardium lacks adequate oxygen and nutrients making it a less than ideal environment for cell survival. In addition, cells can be injected into the myocardium via the coronary sinus using a percutaneous catheter approach [43].

Strategies for Improving Survival and Engraftment of Transplanted Cells

Even though advancements have been made in the utilization of stem cells for treating cardiovascular disease, improvement is still needed to enhance cell survival and engraftment once transplanted into the injured myocardium [44, 45]. Various approaches that have been undertaken include:

Genetic Engineering of Stem Cells

As a majority of transplanted stem cells die within a few hours of injection [46], there is an imminent need to devise methods to improve cell survival in order to enhance their efficiency. Genetic engineering of stem cells to overexpress Akt, Bcl2, fibroblast growth factor, Sfrp2, and Notch is one of the many approaches used to improve cell survival once transplanted. Mangi et al. [47], demonstrated that MSCs genetically engineered to overexpress the pro-survival gene Akt1 decreased myocardial infarct size. One of the therapeutic effects of Akt is due to the release of paracrine signals promoting cell survival and angiogenesis [48]. Studies have also shown that genetically engineered stem cells that continuously inhibit apoptosis increase cell survival after transplantation [48, 49]. Bcl2 is an antiapoptotic protein, and MSCs overexpressing Bcl2 had superior survival when transplanted into infarcted myocardium resulting in decrease infarct size and improvement in cardiac function as compared to only MSC therapy [49]; however, the long-term effects of these transplanted cells, which are continuously inhibiting apoptosis, may be harmful including unwanted cell growth and development of malignancies [50, 51].

Stem Cell Homing

The process of organogenesis during development, homeostasis, and repair is critically governed by regulated migration of stem cells; this targeted journey of stem cells is critical for effective tissue regeneration. Stem cell homing to the injured myocardium has been attributed to several chemokines and growth factors that are secreted by damaged cells in order to recruit immune and stem cells for repair; these include stromal cell-derived factor-1 (SDF-1), monocyte chemoattractant protein-3 (MCP-3), growth-regulated oncogene-1 (GRO-1), fibroblast growth factor-2 (FGF-2), hepatocyte growth factor (HGF), and insulin growth factor-1 (IGF-1) [51]. SDF1 is involved in the homing of CSCs, HSCs, and progenitor cells. MCP-3 is involved in the homing of MSCs. GRO-1 is involved in the homing of bone marrow-derived EPCs, while FGF-2, HGF, and IGF-1 activate CSCs. SDF1 and its chemokine receptor, CXCR, are upregulated during myocardial infarction [52], and engineered stem cells overexpressing SDF-1 receptor CXCR4 were reported to home more efficiently leading to improved left ventricular function compared to control non-engineered MSCs [53]. Cell culture conditions like the passage number also affect homing, where freshly isolated MSCs display enhanced homing ability compared to their culture-expanded counterparts [54].

Antiapoptotic Approaches

Hypoxic tissue that results from a myocardial infarction is a potent stimulant of apoptosis resulting in cell death of transplanted stem cells. Attempts have been made to curtail apoptosis of transplanted stem cells in hope to increase their survival. MSCs overexpressing Bcl2 result in improved cell survival after transplantation into infarcted myocardium by decreasing apoptosis [49]. In addition, MSCs modified with a hypoxia-regulated heme oxygenase-1 (HO-1) plasmid where found to have decrease apoptosis and improvement in cell survival during ischemia-reperfusion injury in vitro [55]. Adenovirus E1A genes are also known inducers of apoptosis and can be inhibited by cellular repressor E1A-stimulated genes (CREG); Deng et al. [15], demonstrated that CREG overexpression inhibits MSC apoptosis and downregulation of p53 expression by activating Akt.

An alternative to genetic modification of stem cells for apoptotic attenuation is to precondition the cells prior to transplantation, thus alleviating the need to supply transplanted stem cells with continued apoptotic inhibition in which the long-term effects may be harmful. Rat MSCs pretreated with stromal-derived factor 1α, an activator of the Akt pathway, enhanced cell survival by 20 % and reduced the number of apoptotic cells in the peri-infarct region by 33 % [56]. Moreover, MSCs preconditioned with 100 % oxygen

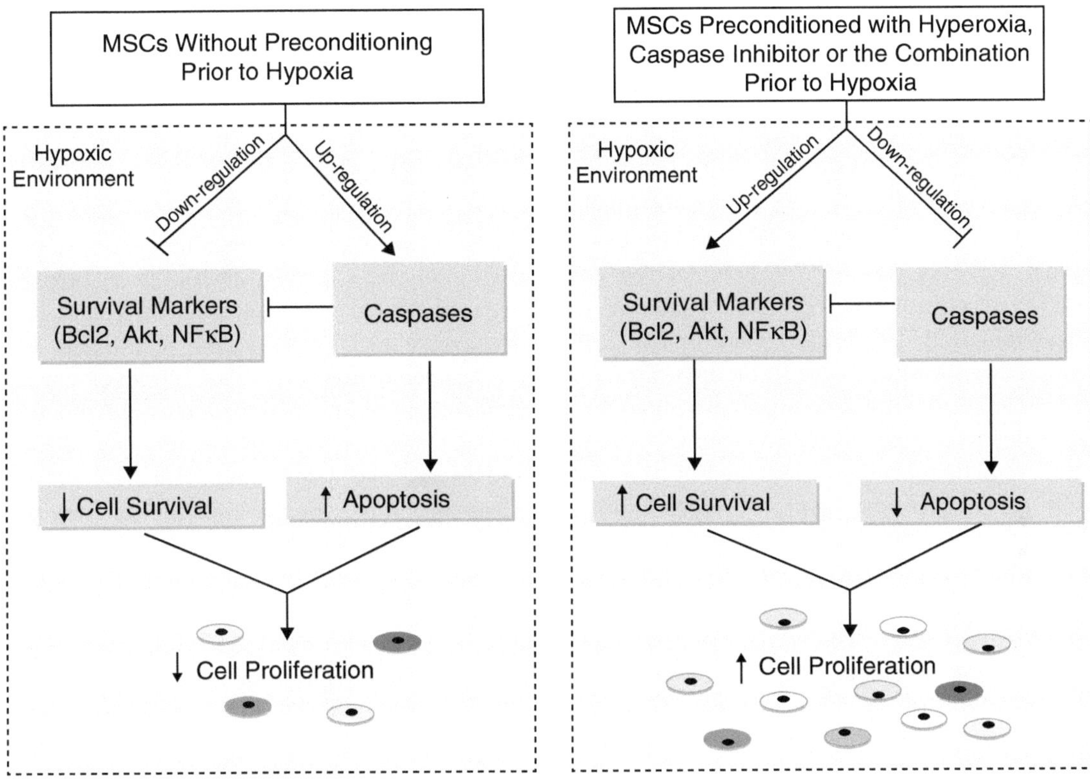

Fig. 1 Schematic diagram demonstrating the effect of preconditioning mesenchymal stem cells (MSC) with hyperoxia and/or pan-caspase inhibitor on apoptosis and cell survival prior to being placed into a hypoxic environment (From Ref. [57])

(hyperoxia) and/or pan-caspase inhibition significantly downregulated apoptosis and upregulated survival markers ultimately resulting in a significant increase in MSC proliferation in vitro when placed in a hypoxic environment [57] (Fig. 1). Studies have also demonstrated that treatment after stem cell transplantation can also have beneficial effects; post-myocardial infarction rats that were treated with MSC therapy followed by exposure to daily cycles of hyperoxygenation after transplantation resulted in enhanced stem cell engraftment, cardiac function, and increased endothelial nitric oxide synthase expression [58].

Angiogenesis

FGF-2 and vascular endothelial growth factor (VEGF) have been used to stimulate angiogenesis [59]. EPCs overexpressing VEGF were shown to possess enhanced proliferative and adhesive capabilities both in vitro and in vivo [46]. In addition, MSCs overexpressing VEGF were transplanted into the left ventricles of swine demonstrating improved cardiac function and neovascularization [60].

Scaffolds

In an attempt to improve stem cell adherence and engraftment during transplantation, scaffolds have been developed. Scaffolds are composed of a gelling substance that supports cell attachment, proliferation, and differentiation. The scaffolds can homogenously distribute cells and can be injected directly into areas of interest. Studies have shown that scaffolding used during stem cell transplantation in the infracted heart can increase cell survival and decrease infarct size, in turn improving left ventricular function [61, 62].

Barriers to Stem Cell Therapy

Several barriers exist for cell-based therapy limiting its potential effectiveness starting with stem cell delivery. Most of the stem cells delivered in the heart are lost immediately after transplantation [63–65]. In patients with acute myocardial infarction, only 1.2–3.6 % of bone marrow cells delivered intracoronary remained within the myocardium [65]. In addition, even though EPCs have been shown to home to the myocardial infarct border zone, the amount of cells found is relatively low, making up approximately 3 % of the infused cells [66]. Direct injection of stem cells into the myocardium during open-heart surgery also showed suboptimal results where approximately 32 % and 88 % of the cells were lost in non-beating and beating porcine hearts, respectively [67]. Even if the cells are retained in the myocardium after direct injection, cell survival is low. Myoblasts delivered into infracted hearts of mice resulted in approximately 7 % of grafted cells surviving after 72 h [68].

Certain types of stem cells like ESCs are subject to immune rejection. Parthenogenetically generated undifferentiated mouse ESCs displayed immune rejection of major histocompatibility complex (MHC) [69]. In addition, Swijnenburg et al. [70], demonstrated that robust cellular and humoral immune responses were triggered by transplanted human ESCs resulting in ESC rejection. Undifferentiated mouse iPSCs have also been shown to result in rejection upon transplantation [71].

Human ESCs and iPSCs are uniquely defined by their pluripotent differentiation potential and endless self-renewing ability, thus capable of being tumorigenic. Blum et al. (2009) [72], used global gene expression analysis to show that survivin, an antiapoptotic oncofetal gene, is highly expressed in human ESCs and continued expression of survivin upon differentiation in vivo may contribute to teratoma formation by these cells. In addition, there are concerns regarding the stability of ESCs and iPSCs during their prolonged culture and storage, and certain genomic or epigenomic abnormalities might affect the multiplication and differentiation potential of these cells [73]. Moreover, ethical issues regarding human ESCs have limited their use in the clinical practice.

Clinical Studies

Various stem cells and progenitor cells have been used in human clinical trials for the regeneration of cardiac tissue post-myocardial infarction. Randomized clinical studies using cell-based therapy in acute myocardial infarction or chronic ischemic cardiomyopathy patients are shown in Table 1. Human clinical studies have demonstrated variable results regarding the improvement of left ventricular function and regional wall motion abnormalities post-stem cell transplantation. While many randomized trials failed to demonstrate a significant improvement in cardiac function post-transplantation, several other randomized trials did show a significant improvement up to 12 % in left ventricular ejection fraction [74–91]; non-randomized trials ranged from no significant improvement in left ventricular ejection fraction to a significant improvement up to 14 % [92–99]. A meta-analysis involving 18 randomized and non-randomized trials with approximately 1,000 patients with acute myocardial infarction or chronic ischemic cardiomyopathy, transplantation of adult bone marrow-derived stem cells (mononuclear cells, MSCs, and circulating progenitor cells) when compared to control significantly improved left ventricular ejection fraction by 5.4 % ($P < 0.001$), decreased infarct scar size by 5.4 % ($P < 0.005$), and decreased left ventricular end-systolic volume by 4.8 mL ($P < 0.01$) [100]. The variability in clinical outcomes may be secondary not only to the type of stem cell injected or route of administration, but the quantity of cells injected; mononuclear bone marrow cell transplantation in post-myocardial infarction patients resulted in an improvement in left ventricular ejection fraction compared to control when a higher number of cells were injected, 10^8 compared to 10^7 [101]. An important challenge for many patients with chronic angina is to provide relief of their symptoms. Intramyocardial transplantation of CD34+ cells has shown to provide a decrease in the frequency

Table 1 Randomized clinical trials using cell based therapy in patients with acute myocardial infarction or chronic ischemic cardiomyopathy

Study/year	Clinical setting	N	Type of cells	Method of cell transplantation	Follow-up (months)	% Change in LVEF as compared to control
Makkar et al. (CADUCEUS)/2012	AMI	31	CDC	Intracoronary	6	Non-significant
Hirsch et al. (HEBE)/2011	AMI	200	MNC	Intracoronary	4	Non-significant
Traverse et al. (LateTIME)/2011	AMI	87	MNC	Intracoronary	6	Non-significant
Bolli et al. (SCIPIO)/2011	Ischemic CMP	23	CSC	Intramyocardial	12	+12.3 % ($p < 0.001$)
Menasché et al. (MAGIC)/2008	Ischemic CMP	97	SMB	Intramyocardial	6	Non-significant
Meluzin et al./2007	AMI	60	MNC	Intracoronary	12	+3.0 % ($p < 0.05$)
Meyer et al. (BOOST)/2006	AMI	60	MNC	Intracoronary	18	Non-significant
Assmus et al. (TOPCARE-CHD)/2006	Ischemic CMP	51	MNC	Intracoronary	3	+4.1 % ($p < 0.05$)
Schächinger et al. (REPAIR-AMI)/2006	AMI	204	MNC	Intracoronary	4	+2.5 % ($p < 0.05$)
Assmus et al. (TOPCARE-CHID)/2006	ICMP	47	CPC	Intracoronary	3	Non-significant
Ge et al. (TCT-STAMI)/2006	AMI	20	MNC	Intracoronary	6	Non-significant
Hendrikx et al./2006	Ischemic CMP	20	MNC	Intramyocardial	4	Non-significant
Janssens et al./2006	AMI	67	MNC	Intracoronary	4	Non-significant
Lunde et al. (ASTAMI)/2006	AMI	100	MNC	Intracoronary	6	−3.0 %; $p = 0.05$
Chen et al./2006	Ischemic CMP	45	MSC	Intracoronary	12	Non-significant
Kang et al./2006	AMI/Ischemic CMP	82	CPC	Intracoronary	6	Non-significant
Ruan et al./2005	AMI	20	BMC	Intracoronary	6	+9.2 %; $p < 0.05$
Erbs et al./2005	Ischemic CMP	26	CPC	Intracoronary	3	Non-significant
Chen et al./2004	AMI	69	MSC	Intracoronary	6	+12.0 % ($p < 0.05$)

AMI acute myocardial infarction, *BMC* bone marrow cells, *CDC* cardiosphere derived cells, *CMP* cardiomyopathy, *CPC* circulating progenitor cells, *CSC* cardiac stem cells, *LVEF* left ventricular ejection fraction, *MNC* bone marrow mononuclear cells, *MSC* mesenchymal stem cells, *NS* non-significant, *SMB* skeletal myoblasts

of angina, exercise times, and nitroglycerin use when compared to patients who did not receive cell therapy [102].

Summary

Myocardial infarction resulting from coronary artery occlusion may lead to ventricular remodeling, left ventricular dysfunction, and heart failure (4,5). To attenuate left ventricular remodeling and possibly restore cardiac function, stem cell therapy has been under extensive evaluation; however, only mild improvement at best in cardiac function has been seen to date. Several barriers exist in cell-based therapy limiting its ability to restore or rejuvenate myocardial tissue post-myocardial infarction. Meticulous and dedicated research will need to continue in order to overcome these obstacles and provide patients with a viable option for cardiac recovery post-myocardial infarction.

References

1. Boudoulas KD, Hatzopoulos AK (2009) Cardiac repair and regeneration: the Rubik's cube of cell therapy for heart disease. Dis Model Mech 2:344–358
2. Takemura G, Fujiwara H (2004) Role of apoptosis in remodeling after myocardial infarction. Pharmacol Ther 104:1–16
3. Gill C, Mestril R, Samali A (2002) Losing heart: the role of apoptosis in heart disease-a novel therapeutic target? FASEB J 16:135–146
4. Narula J, Pandey P, Arbustini E et al (1999) Apoptosis in heart failure: release of cytochrome c from mitochondria and activation of caspase-3 in human cardiomyopathy. Proc Natl Acad Sci 96:8144–8149
5. Sabbah HN, Sharov VG, Goldstein S (2000) Cell death, tissue hypoxia and the progression of heart failure. Heart Fail Rev 5:131–138
6. Pagidipati NJ, Gaziano TA (2013) Estimating deaths from cardiovascular disease: a review of global methodologies of mortality measurement. Circulation 127:749–756
7. Singla DK, Hacker TA, Ma L et al (2006) Transplantation of embryonic stem cells into the infarcted mouse heart: formation of multiple cell types. J Mol Cell Cardiol 40:195–200
8. Christoforou N, Oskouei BN, Esteso P et al (2010) Implantation of mouse embryonic stem cell-derived cardiac progenitor cells preserves function of infarcted murine hearts. PLoS One 5:e11536
9. Liu Y, Ye X, Mao L et al (2013) Transplantation of parthenogenetic embryonic stem cells ameliorates cardiac dysfunction and remodeling after myocardial infarction. Cardiovasc Res 97:208–218
10. Burt RK, Chen YH, Verda L et al (2012) Mitotically inactivated embryonic stem cells can be used as an in vivo feeder layer to nurse damaged myocardium after acute myocardial infarction: a preclinical study. Circ Res 111:1286–1296
11. Takahashi K, Yamanaka S (2006) Induction of pluripotent stem cells from mouse embryonic and adult fibroblast cultures by defined factors. Cell 126:663–676
12. Stadtfeld M, Maherali N, Breault DT et al (2008) Defining molecular cornerstones during fibroblast to iPS cell reprogramming in mouse. Cell Stem Cell 2:230–240
13. Xiong Q, Ye L, Zhang P et al (2013) Functional consequences of human induced pluripotent stem cells therapy: myocardial ATP turnover rate in the in vivo swine hearts with post-infarction remodeling. Circulation 127(9):997–1008
14. Gu M, Nguyen PK, Lee AS et al (2012) Microfluidic single-cell analysis shows that porcine induced pluripotent stem cell-derived endothelial cells improve myocardial function by paracrine activation. Circ Res 111:882–893
15. Deng J, Han Y, Yan C et al (2010) Overexpressing cellular repressor of E1A-stimulated genes protects mesenchymal stem cells against hypoxia- and serum deprivation-induced apoptosis by activation of PI3K/Akt. Apoptosis 15:463–473
16. Yamanaka S (2012) Induced pluripotent stem cells: past, present, and future. Cell Stem Cell 10:678–684
17. Gore A, Li Z, Fung HL, Young JE et al (2011) Somatic coding mutations in human induced pluripotent stem cells. Nature 471:63–67
18. Hussein SM, Batada NN, Vuoristo S et al (2011) Copy number variation and selection during reprogramming to pluripotency. Nature 471:58–62
19. Zhao T, Zhang Z-N, Rong Z et al (2011) Immunogenicity of induced pluripotent stem cells. Nature 474:212–215
20. Deten A, Volz HC, Clamors S et al (2004) Hematopoietic stem cells do not repair the infarcted mouse heart. Cardiovasc Res 65:52–63
21. Muller-Sieburg CE, Cho RH, Thoman M et al (2002) Deterministic regulation of hematopoietic stem cell self-renewal and differentiation. Blood 100:1302–1309
22. Jackson KA, Majka SM, Wang H et al (2001) Regeneration of ischemic cardiac muscle and vascular endothelium by adult stem cells. J Clin Invest 107:1395–1402
23. Segers VF, Lee RT (2008) Stem-cell therapy for cardiac disease. Nature 451:937–942
24. Young PP, Vaughan DE, Hatzopoulos AK (2007) Biologic properties of endothelial progenitor cells and their potential for cell therapy. Prog Cardiovasc Dis 49:421–429
25. Hristov M, Erl W, Weber PC (2003) Endothelial progenitor cells: mobilization, differentiation, and homing. Arterioscler Thromb Vasc Biol 23:1185–1189
26. Dzau VJ, Gnecchi M, Pachori AS (2005) Enhancing stem cell therapy through genetic modification. J Am Coll Cardiol 46:1351–1353
27. Shantsila E, Watson T, Lip GY (2007) Endothelial progenitor cells in cardiovascular disorders. J Am Coll Cardiol 49:741–752
28. Uccelli A, Moretta L, Pistoia V (2008) Mesenchymal stem cells in health and disease. Nat Rev Immunol 8:726–736
29. Dominici M, Le Blanc K, Mueller I et al (2006) Minimal criteria for defining multipotent mesenchymal stromal cells. The international society for cellular therapy position statement. Cytotherapy 8:315–317
30. Wang S, Qu X, Zhao RC (2012) Clinical applications of mesenchymal stem cells. J Hematol Oncol 5:19
31. Grauss RW, Winter EM, van Tuyn J et al (2007) Mesenchymal stem cells from ischemic heart disease patients improve left ventricular function after acute myocardial infarction. Am J Physiol Heart Circ Physiol 293:H2438–H2447

32. Otto Beitnes J, Oie E, Shahdadfar A et al (2012) Intramyocardial injections of human mesenchymal stem cells following acute myocardial infarction modulate scar formation and improve left ventricular function. Cell Transplant 21:1697–1709
33. Grinnemo KH, Månsson-Broberg A, Leblanc K et al (2006) Human mesenchymal stem cells do not differentiate into cardiomyocytes in a cardiac ischemic xenomodel. Ann Med 38:144–153
34. Hosoda T (2012) C-kit-positive cardiac stem cells and myocardial regeneration. Am J Cardiovasc Dis 2:58–67
35. Barile L, Messina E, Giacomello A et al (2007) Cardiac stem cells: isolation, expansion and experimental use for myocardial regeneration. Nat Clin Pract Cardiovasc Med 4:S9–S14
36. Boyle AJ, Schulman SP, Hare JM (2006) Is stem cell therapy ready for patients? Stem cell therapy for cardiac repair. Circulation 114:339–352
37. Beltrami AP, Barlucchi L, Torella D et al (2003) Adult cardiac stem cells are multipotent and support myocardial regeneration. Cell 114:763–776
38. Oh H, Bradfute SB, Gallardo TD et al (2003) Cardiac progenitor cells from adult myocardium: homing, differentiation, and fusion after infarction. Proc Natl Acad Sci 100:12313–12318
39. Hamdi H, Furuta A, Bellamy V et al (2009) Cell delivery: intramyocardial injections or epicardial deposition? A head-to-head comparison. Ann Thorac Surg 87(4):1196–1203
40. Asahara T, Murohara T, Sullivan A et al (1997) Isolation of putative progenitor endothelial cells for angiogenesis. Science 275:964–967
41. Kang SK, Shin IS, Ko MS et al (2012) Journey of mesenchymal stem cells for homing: strategies to enhance efficacy and safety of stem cell therapy. Stem Cells Int 2012:342968
42. Perin EC, López J (2006) Methods of stem cell delivery in cardiac diseases. Nat Rev Cardiol 3:S110–S113
43. Murry CE, Soonpaa MH, Reinecke H et al (2004) Haematopoietic stem cells do not transdifferentiate into cardiac myocytes in myocardial infarcts. Nature 428:664–668
44. Dimmeler S, Zeiher AM, Schneider MD (2005) Unchain my heart: the scientific foundations of cardiac repair. J Clin Invest 115:572–583
45. Penn MS, Mangi AA (2008) Genetic enhancement of stem cell engraftment, survival, and efficacy. Circ Res 102:1471–1482
46. Li W, Ma N, Ong LL et al (2007) Bcl-2 engineered MSCs inhibited apoptosis and improved heart function. Stem Cells 2:18–27
47. Mangi AA, Noiseux N, Kong D et al (2003) Mesenchymal stem cells modified with Akt prevent remodeling and restore performance of infarcted hearts. Nat Med 9:1195–1201
48. Hodgkinson CP, Gomez JA, Mirotsou M et al (2010) Genetic engineering of mesenchymal stem cells and its application in human disease therapy. Hum Gene Ther 21:1513–1526
49. Strauer BE, Steinhoff G (2011) 10 years of intracoronary and intramyocardial bone marrow stem cell therapy of the heart: from the methodological origin to clinical practice. J Am Coll Cardiol 58:1095–1104
50. Tsujimoto Y, Yunis J, Onorato-Showe L et al (1984) Molecular cloning of the chromosomal breakpoint of B-cell lymphomas and leukemias with the t(11;14) chromosome translocation. Science 224:1403–1406
51. Filip S, Mokry J, Horacek J et al (2008) Stem cells and the phenomena of plasticity and diversity: a limiting property of carcinogenesis. Stem Cells Dev 17:1031–1038
52. Smart N, Riley PR (2008) The stem cell movement. Circ Res 102:1155–1168
53. Chen J, Chemaly E, Liang L et al (2010) Effects of CXCR4 gene transfer on cardiac function after ischemia-reperfusion injury. Am J Pathol 176:1705–1715
54. Rombouts WJC, Ploemacher RE (2003) Primary murine MSC show highly efficient homing to the bone marrow but lose homing ability following culture. Leukemia 17:160–170
55. Tang YL, Tang Y, Zhang YC et al (2005) Improved graft mesenchymal stem cell survival in ischemic heart with a hypoxia-regulated heme oxygenase-1 vector. J Am Coll Cardiol 46:1339–1350
56. Pasha Z, Wang Y, Sheikh R et al (2008) Preconditioning enhances cell survival and differentiation of stem cells during transplantation in infarcted myocardium. Cardiovasc Res 77:134–142
57. Saini U, Gumina RJ, Wolfe B et al (2013) Preconditioning mesenchymal stem cells with caspase inhibition and hyperoxia prior to hypoxia exposure increases cell proliferation. J Cell Biochem 114:2612
58. Khan M, Meduru S, Gogna R et al (2012) Oxygen cycling in conjunction with stem cell transplantation induces NOS3 expression leading to attenuation of fibrosis and improved cardiac function. Cardiovasc Res 93:89–99
59. Ng Y-S, D'Amore PA (2001) Therapeutic angiogenesis for cardiovascular disease. Curr Control Trials Cardiovasc Med 2:278–285
60. Wang X, Hu Q, Mansoor A et al (2006) Bioenergetic and functional consequences of stem cell-based VEGF delivery in pressure-overloaded swine hearts. Am J Physiol Heart Circ Physiol 290: H1393–H1405
61. Christman KL, Vardanian AJ, Fang Q et al (2004) Injectable fibrin scaffold improves cell transplant survival, reduces infarct expansion, and induces neovasculature formation in ischemic myocardium. J Am Coll Cardiol 44:654–660
62. Lu WN, Lü SH, Wang HB et al (2009) Functional improvement of infarcted heart by co-injection of embryonic stem cells with temperature-responsive chitosan hydrogel. Tissue Eng Part A 15:1437–1447

63. Scudellari M (2009) The delivery dilemma. Nat Rep Stem Cells. doi:10.1038/stemcells.2009.104
64. Kang WJ, Kang HJ, Kim HS et al (2006) Tissue distribution of 18F-FDG-labeled peripheral hematopoietic stem cells after intracoronary administration in patients with myocardial infarction. J Nucl Med 47:1295–1301
65. Hofmann M, Wollert KC, Meyer GP et al (2005) Monitoring of bone marrow cell homing into the infarcted human myocardium. Circulation 111: 2198–2202
66. Aicher A, Zeiher AM, Dimmeler S (2005) Mobilizing endothelial progenitor cells. Hypertension 45:321–325
67. Teng CJ, Luo J, Chiu RC et al (2006) Massive mechanical loss of microspheres with direct intramyocardial injection in the beating heart: implications for cellular cardiomyoplasty. J Thorac Cardiovasc Surg 132:628–632
68. Suzuki K, Murtuza B, Beauchamp JR et al (2004) Dynamics and mediators of acute graft attrition after myoblast transplantation to the heart. FASEB J 10:1153–1155
69. Kim K, Lerou P, Yabuuchi A et al (2007) Histocompatible embryonic stem cells by arthenogenesis. Science 315:482–486
70. Swijnenburg RJ, Schrepfer S, Govaert JA et al (2008) Immunosuppressive therapy mitigates immunological rejection of human embryonic stem cell xenografts. Proc Natl Acad Sci 105: 12991–12996
71. Zhao Y, Li T, Wei X et al (2012) Mesenchymal stem cell transplantation improves regional cardiac remodeling following ovine infarction. Stem Cells Transl Med 1:685–695
72. Blum B, Bar-Nur O, Golan-Lev T et al (2009) The anti-apoptotic gene survivin contributes to teratoma formation by human embryonic stem cells. Nat Biotechnol 27:281–287
73. Martins-Taylor K, Xu R-H (2012) Concise review: genomic stability of human induced pluripotent stem cells. Stem Cells 30:22–27
74. Makkar RR, Smith RR, Cheng K et al (2012) Intracoronary cardiosphere-derived cells for heart regeneration after myocardial infarction (CADUCEUS): a prospective, randomized phase 1 trial. Lancet 379:895–904
75. Hirsch A, Nijveldt R, van der Vleuten PA et al (2011) Intracoronary infusion of mononuclear cells from bone marrow or peripheral blood compared with standard therapy in patients after acute myocardial infarction treated by primary percutaneous coronary intervention: results of the randomized controlled HEBE trial. Eur Heart J 32:1736–1747
76. Traverse JH, Henry TD, Ellis SG et al (2011) Effect of intracoronary delivery of autologous bone marrow mononuclear cells 2 to 3 weeks following acute myocardial infarction on left ventricular function the late TIME randomized trial. JAMA 306:2110–2119
77. Bolli R, Chueh AR, D'Amario D et al (2011) Cardiac stem cells in patients with ischaemic cardiomyopathy (SCIPIO): initial results of a randomized phase 1 trial. Lancet 378:1847–1857
78. Menasché P, Alfieri O, Janssens S et al (2008) The myoblast autologous grafting in ischemic cardiomyopathy (MAGIC) trial: first randomized placebo-controlled study of myoblast transplantation. Circulation 117:1189–1200
79. Meluzin J, Mayer J, Groch J et al (2006) Autologous transplantation of mononuclear bone marrow cells in patients with acute myocardial infarction: the effect of the dose of transplanted cells on myocardial function. Am Heart J 152:e9–e15
80. Meyer GP, Wollert KC, Lotz J et al (2006) Intracoronary bone marrow cell transfer after myocardial infarction: eighteen months' follow-up data from the randomized, controlled BOOST (Bone marrow transfer to enhance ST-elevation infarct regeneration) trial. Circulation 113: 1287–1294
81. Assmus B, Honold J, Schachinger V et al (2006) Transcoronary transplantation of progenitor cells after myocardial infarction. N Engl J Med 355: 1222–1232
82. Schächinger V, Erbs S, Elsasser A et al (2006) Intracoronary bone marrow–derived progenitor cells in acute myocardial in acute myocardial infarction. N Engl J Med 355:1210–1221
83. Ge J, Li Y, Qian J et al (2006) Efficacy of emergent transcatheter transplantation of stem cells for treatment of acute myocardial infarction (TCT-STAMI). Heart 92:1764–1767
84. Hendrikx M, Hensen K, Clijsters C et al (2006) Recovery of regional but not global contractile function by the direct intramyocardial autologous bone marrow transplantation: results from a randomized controlled clinical trial. Circulation 114(suppl): I101–I107
85. Janssens S, Dubois C, Bogaert J et al (2006) Autologous bone marrow–derived stem-cell transfer in patients with ST-segment elevation myocardial infarction: double-blind, randomized controlled trial. Lancet 367:113–121
86. Lunde K, Solheim S, Aakhus S et al (2006) Intracoronary injections of mononuclear bone marrow cells in acute myocardial infarction. N Engl J Med 355:1199–1209
87. Chen S, Liu S, Tian SN et al (2006) Intracoronary transplantation of autologous bone marrow mesenchymal stem cells for ischemic cardiomyopathy due to isolated chronic occluded left anterior descending artery. J Invasive Cardiol 18:552–556
88. Kang H, Lee H, Na S et al (2006) Differential effect of intracoronary infusion of mobilized peripheral blood stem cells by granulocyte colony stimulating factor on left ventricular function and remodeling in patients with acute myocardial infarction versus old myocardial infarction: the MAGIC

Cell-3-DES randomized, controlled trial. Circulation 114(suppl):I145–I151
89. Ruan W, Pan C, Huang G, Li Y et al (2005) Assessment of left ventricular segmental function after autologous bone marrow stem cells transplantation in patients with acute myocardial infarction by tissue tracking and strain imaging. Chin Med J (Engl) 118:1175–1181
90. Erbs S, Linke A, Adams V et al (2005) Transplantation of blood-derived progenitor cells after recanalization of chronic coronary artery occlusion: first randomized and placebo-controlled study. Circ Res 97:756–762
91. Chen SL, Fang WW, Ye F et al (2004) Effect on left ventricular function of intracoronary transplantation of autologous bone marrow mesenchymal stem cell in patients with acute myocardial infarction. Am J Cardiol 94:92–95
92. Choi JH, Choi J, Lee WS et al (2007) Lack of additional benefit of intracoronary transplantation of autologous peripheral blood stem cell in patients with acute myocardial infarction. Circ J 71: 486–494
93. Gavira JJ, Herreros J, Perez A et al (2006) Autologous skeletal myoblast transplantation in patients with nonacute myocardial infarction: 1-year follow-up. J Thorac Cardiovasc Surg 131:799–804
94. Ince H, Petzsch M, Rehders TC et al (2004) Transcatheter transplantation of autologous skeletal myoblasts in postinfarction patients with severe left ventricular dysfunction. J Endovasc Ther 11:695–704
95. Katritsis DG, Sotiropoulou PA, Karvouni E et al (2005) Transcoronary transplantation of autologous mesenchymal stem cells and endothelial progenitors into infarcted human myocardium. Catheter Cardiovasc Interv 65:321–329
96. Mocini D, Staibano M, Mele L et al (2006) Autologous bone marrow mononuclear cell transplantation in patients undergoing coronary artery bypass grafting. Am Heart J 151:192–197
97. Strauer BE, Brehm M, Zeus T, Köstering M et al (2002) Repair of infracted myocardium by autologous intracoronary mononuclear bone marrow cell transplantation in humans. Circulation 106: 1913–1918
98. Strauer BE, Brehm M, Zeus T et al (2005) Regeneration of human infarcted heart muscle by intracoronary autologous bone marrow cell transplantation in chronic coronary artery disease: the IACT Study. J Am Coll Cardiol 46:1651–1658
99. Tatsumi T, Ashihara E, Yasui T et al (2007) Intracoronary transplantation of non-expanded peripheral blood-derived mononuclear cells promotes improvement of cardiac function in patients with acute myocardial infarction. Circ J 71:1199–1207
100. Abdel-Latif A, Bolli R, Tleyjeh IM et al (2007) Adult bone marrow-derived cells for cardiac repair: a systematic review and meta-analysis. Arch Intern Med 167:989–997
101. Meluzín J, Janousek S, Mayer J et al (2008) Three-, 6-, and 12-month results of autologous transplantation of mononuclear bone marrow cells in patients with acute myocardial infarction. Int J Cardiol 128:185–192
102. Losordo DW, Henry TD, Davidson C et al (2011) Intramyocardial, autologous CD34+ cell therapy for refractory angina. Circ Res 109:428–436

Endothelial Progenitor Cells: Application in Vascular Medicine

Venkata Naga Srikanth Garikipati, Prasanna Krishnamurthy, Suresh Kumar Verma, and Raj Kishore

Introduction

For more than a decade now, biology and therapeutic efficacy of endothelial progenitor cells (EPCs) were largely driven by the first observations of Ashara et al. in identifying EPCs in adult peripheral blood (PB) [1] and were shown to derive from bone marrow (BM) further migrating and incorporating into foci of physiological or pathological neovascularization [2, 3]. As a matter of fact, postnatal neovascularization was believed to be established by the mechanism of "angiogenesis," by in situ proliferation and migration of preexisting endothelial cells (ECs) [4]. However, the finding that EPCs can home to sites of neovascularization and differentiate into ECs in situ is consistent with "vasculogenesis"; a critical paradigm has been demonstrated in embryonic neovascularization [5] and also shown recently for the adult organism in which a pool of progenitor cells contributes to postnatal neovascular formation [6]. The discovery of EPCs has therefore considerably changed our understanding of adult blood vessel formation. Furthermore, we and other groups envisage the potential of EPC to improve the clinical applicability in the fight against cardiovascular diseases.

This book chapter will focus on the potential value of EPCs for therapeutic vasculogenesis in ischemic diseases, particularly on "circulating EPCs" in terms of vascular medicine.

V.N.S. Garikipati
Center for Translational Medicine, Temple University School of Medicine, MERB-983 3500 N Broad St, Philadelphia, PA, 19104, USA
e-mail: tuf70539@temple.edu

P. Krishnamurthy
Department of Cardiovascular Sciences, Houston Methodist Research Institute, 6670 Bertner Ave, MS: F8-046, Houston, TX-77030, USA

S.K. Verma
Center for Translational Medicine, Temple University School of Medicine, MERB-954 3500 N Broad St, Philadelphia, PA, 19104, USA

R. Kishore, Ph.D., FAHA (✉)
Center for Translational Medicine, Temple University School of Medicine, MERB-953 3500 N Broad St, Philadelphia, PA, 19104, USA
e-mail: raj.kishore@temple.edu

Characterization of EPCs

EPCs in circulation can be broadly subdivided mainly into two categories: hematopoietic lineage EPCs and non-hematopoietic lineage EPCs. The hematopoietic EPCs originate from bone marrow and represent a pro-vasculogenic subpopulation of hematopoietic stem cells (HSCs) [7–9]. The non-hematopoietic EPCs are blood- or tissue-derived entities, exhibiting EC-like phenotype [10] or the ability to differentiate into EC-like cells [11]. EPCs and HSCs have been shown to express variety of cell surface markers, including membrane receptors like CD34, CD133, Flk- 1/KDR, CXCR4, and CD105 (Endoglin) for human samples and receptors like c-Kit, Sca-1, and CD34 in combination with Flk-1 (vascular

I. Somasundaram (ed.), *Stem Cell Therapy for Organ Failure*,
DOI 10.1007/978-81-322-2110-4_7, © Springer India 2014

endothelial growth factor receptor (VEGFR2) in case of mouse samples [12–14]. Nevertheless, identifying a particular combination of markers for isolation of these cells has been a big challenge for the researchers in this field.

The origin of non-hematopoietic EPCs remains to be clarified, but they are generally thought to be derived from non-hematopoietic tissue-resident lineage stem cells or organ blood vessels but not likely from HSC [15]. Non-hematopoietic EPCs have been shown to be derived from non-hematopoietic tissue stem cells. These newer EPCs are also known as endothelial outgrowth cell (EOC). These EOCs are derived from the endothelial colony formation assay system developed and reported by Yoder's group and others [16, 17]. Unlike EPC, EOC readily and easily forms tubelike structures in culture. However, the origin and role of EOC are still a controversial area of research, but it appears to be a potential cell type for neovascularization [18, 19]. Recently, Aicher et al. [20] have shown circulating cells from BM and non-BM aid in vasculature, demonstrated by the mobilization of c-kit+/CD45- progenitor by a parabiosis model from the liver and small intestine.

Wojakowski et al. have recently identified very small embryonic-like cells (7–8 nm) from BM expressing various ESC-specific markers mobilized into PB in patients following acute myocardial infarction (AMI). VSELs also represent a potential cell type to enhance neovascularization by functioning as EPC in ischemic diseases [21, 22].

What Is EPC Contributing to Neovascularization?

Direct method: In the context of EPC biology, vasculogenesis comprises the de novo formation of blood vessel via in situ migration, proliferation, differentiation, and/or incorporation of BM-derived EPCs into regenerating vasculature (Fig. 1). This phenomenon of BM-derived EPCs incorporation into foci of physiological and pathological neovascularization has been validated in various animal models [3]. Nevertheless, one well-established model, allowing the detection of BM-derived EPCs, uses transplantation of BM cells from transgenic mice in which LacZ is expressed under the regulation of an EC lineage-specific promoter, such as Flk-1 or Tie-2 (Flk-1/LacZ/BMT, Tie-2/LacZ/BMT), to wild-type control mice, followed by their use as a base for several different ischemic injury models. Using this model, it has been demonstrated that BM-derived Flk-1- and/or Tie-2-expressing endothelial lineage cells can localize to vascular structures during tumor growth [3, 23], wound healing [24], skeletal [3] and cardiac ischemia [12, 13], corneal neovascularization [25], and endometrial remodeling following hormone-induced ovulation [3, 23]. Regardless of the origin of EPCs, they undoubtedly play a significant role in contributing to neovascularization via vasculogenesis in ischemic tissues.

The tissue-resident EPCs secrete a variety of pro-angiogenic cytokines and growth factors in a paracrine fashion, thereby actively promoting proliferation and migration of preexisting ECs, thus triggering angiogenesis and contributing in an indirect mode to vascular regeneration to restore tissue homeostasis [4]. In fact, observations of the preclinical studies remind us the evidence of enhanced intrinsic recipient angiogenesis by extrinsic factors derived from transplanted EPCs in myocardial ischemia models [26, 27].

This paracrine aspect of EPC activity toward indirect contribution to neovascularization was established by several groups, identifying EPC secretome with the presence of various cytokines and other pro-angiogenic factors such as VEGF, hepatic growth factor (HGF), angiopoietin (Ang-1, stroma-derived factor (SDF)-1a, insulin-like growth factor (IGF)-1, and endothelial nitric oxide synthase (eNOS [28, 29]. Therefore, EPCs can also mediate tissue-protective effects and contribute to neovascularization in ischemic tissues via production of several important indirect support factors.

Since EPCs were first described more than a decade ago, we and other groups focused especially on the regenerative potential of these progenitor cells, trying to understand the biology and functions with the ultimate

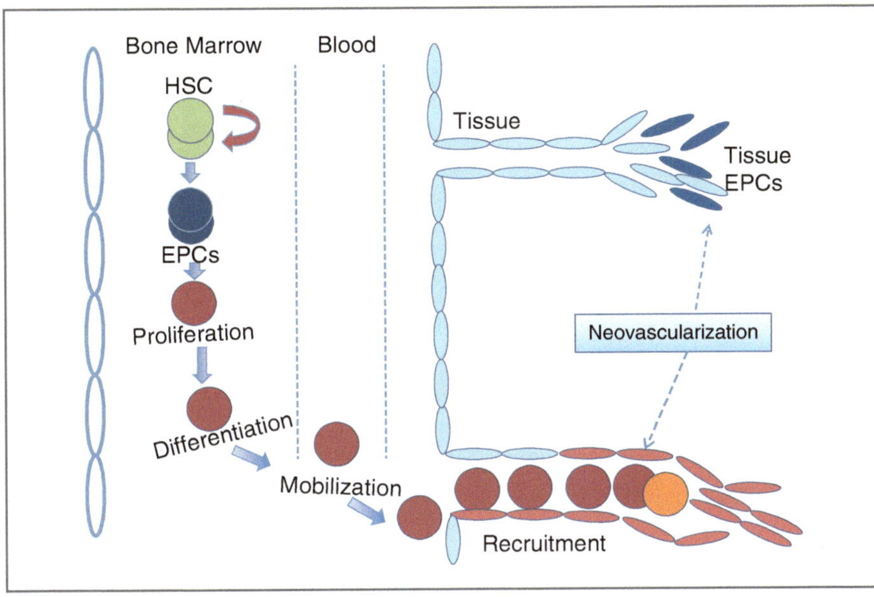

Fig. 1 Mechanism of EPC-mediated postnatal neovascularization. Circulating EPCs mobilized from BM are recruited into foci of neovascularization and contribute to new blood vessel formation

goal to translate to bedside, as this cell type has enormous potential to treat cardiovascular/ischemic diseases [5, 11, 30, 31]. The preclinical/clinical transplantation of blood-/BM-derived vasculogenic progenitor cells, of EPCs, opened a new door for the treatment of ischemic diseases.

EPC Transplantation in Animal Model

It was shown that transplantation of EPCs to be safe and enhanced neovascularization and regeneration of ischemic tissues. Therapeutic neovascularization was first reported during transplantation of human PB-derived cultured EPCs into immunodeficient mice with hind limb ischemia. Interestingly, culture-expanded EPCs restored neovascularization in murine hind limb ischemia model [32]. Murohara et al. reported similar benefits with the administration of EPCs isolated from cord blood in a nude rat model of hind limb ischemia [33]. In an other study, human EPCs transplantation resulted in their migration to sites of ischemic myocardium and differentiated into ECs in the site of neovascularization, in a nude rat myocardial ischemia model. These findings were in corroboration with the observed preservation of left ventricular (LV) function and a reduction in infarction size [27, 34]. Another study in which human cord blood-derived EPCs were transplanted in a nude rat hind limb ischemia model also demonstrated similar findings with enhanced neovascularization in ischemic tissues [33].

Experiments performed in our lab evaluated EPC therapy along with IL-10 enhanced EPC survival in ischemic microenvironment (Fig. 2). Intramyocardial injection of EPC along with IL-10 inhibited fibrosis and preserved function in mice MI and also enhanced neovascularization [13].

In the recent times, several groups have tested the therapeutic potential of CD34+ cells as a possible EPC-enriched fraction. As mentioned earlier, a clear demarcation between HSCs and EPCs with the existing methodology in the literature is difficult; however, CD34+ (hematopoietic marker) is widely used as a surface marker to isolate/enrich EPC. Schatteman et al.'s [35] transplantation of human CD34+ cells into diabetic nude mice with hind limb ischemia resulted in significant blood flow recovery in ischemic limbs.

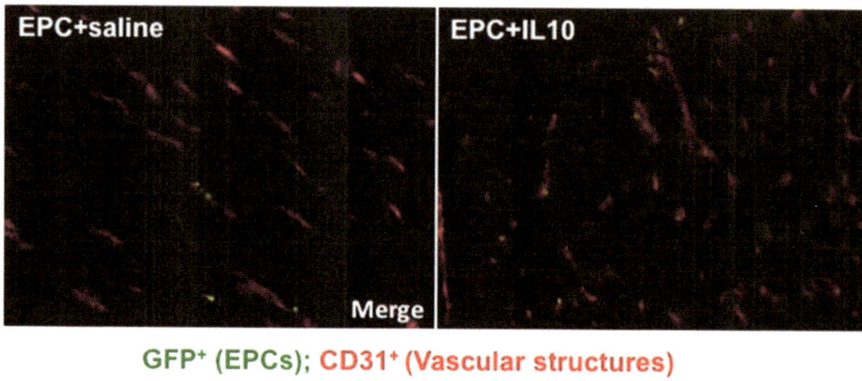

Fig. 2 EPC-mediated neovascularization in border zone of LV infarct at 28 days after MI. Engraftment of EPC (GFP+, green fluorescence) into vascular structures (CD31 staining for capillaries, red fluorescence) is seen as yellow structures

Kocher et al. [31] transplanted human CD34+ cells into a nude rat model of myocardial ischemia and observed preservation of LV function and inhibition of cardiac apoptosis.

We have shown that modifying human CD34+ EPCs with small molecules such as valproic acid and 5-azacytidine induces cardiomyogenic differentiation and histological fibrosis area, and echocardiographic functional parameters were better persevered in SCID mice model of MI [12].

EPC Transplantation in Clinical Trials

Numerous clinical trials have been initiated and are currently in progress to understand the safety issues and therapeutic efficacy of EPCs [40–48] observed in animal models on ischemic diseases [36] (Table 1). Our group performed a phase I/II, randomized, placebo-controlled, dose-ranging, clinical trial to evaluate the intramyocardial transplantation of G-CSF-mobilized CD34+ cells in 24 patients with intractable angina pectoris [37]. Favorable trends in angina frequency, exercise tolerance, and perfusion defect were observed in patients administered with CD34+ cells compared with patients who received placebo.

As for the evaluation of safety issues, neither death nor life-threatening adverse events were observed in therapy group; the results from phase II study support the safety and efficacy of intramyocardially injected autologous CD34+ cells for symptom reduction and improved exercise capacity in "no-option" patients with refractory angina. Larger-scale studies are warranted to verify these effects and to refine the methods for collecting and administering CD34+ cells to patients with disabling angina symptoms.

Lessons from Clinical EPC Transplantation

Our animal studies as well as other published reports suggest that EPC (heterologous) transplantation requires systemic injections of 0.5–2.0×10^4 EPCs/g body weight of the recipient animal to achieve a satisfactory improvement of hind limb and myocardial ischemia models [12, 13, 32, 34, 38, 39]. In general, cultured EPCs obtained from healthy human volunteers yield 5×10^6 cells/100 ml of PB on day 7. Based on these data in human, it was estimated that 12 l of blood will be required to obtain sufficient cell number for treatment of patients with CLI. On the top of it, several other factors such as aging [49], diabetes [50], hypercholesterolemia [51], hypertension [52], and smoking [53, 54] affect the number and biological activity of circulating and BM EPCs and thus limit the success of EPC transplantation. In fact, most patients undergoing EPC therapy for ischemic diseases have background diseases as mentioned above.

Table 1 Endothelial progenitor cells: its applications in vascular medicine

Trail name	Disease type	Patients (T/C)	EPC type	Outcome	References
Kuroda et al. (2014)	NUF	7	G-CSF-PB-CD34+	Safe and bone reunion ↑	[40]
Fujita et al. (2014)	CLI	11	G-CSF-PB-CD34+	Rest pain scale ↑ Physiological parameters ↑	[41]
Kuroda et al. (2010)	NUF	4/0	G-CSF-PB-CD34+	Bone reunion ↑	[42]
Burt et al. (2010)	CLI	9/0	G-CSF-PB-CD34+/CD133+	Physical component score ↑ Leg amputation ↓	[43]
TOPCARE-AMI. (2009)	AMI	30/29	PB/BM derived EPCs	LVEF ↑ Perfusion ↑ Infarct size ↓	[44]
Losordo et al. (2007)	AP	18/6	G-CSF-PB-CD34+	LVEF ↑ CCS class ↓	[37]
Li et al. (2007)	AMI	35/35	CD133+	LVEF ↑	[45]
Boyle et al. (2006)	OMI	5/0	G-CSF-PB-CD34+	Myocardial neovascularization ↑	[46]
Bartunek et al. (2005)	AMI	19/16	G-CSF-PB-CD34+	LVEF ↑ Perfusion ↑	[47]
Stamm et al. (2003)	RMI	46/9	G-CSF-PB-CD34+	LVEF ↑ Perfusion ↑	[48]

Abbreviations: *AMI* acute myocardial infarction, *AP* angina pectoris, *BM* bone marrow, *CLI* critical limb ischemia, *EPCs* endothelial progenitor cells, *G-CSF* granulocyte colony-stimulating factor, *NUF* nonunion fracture, *OMI* old myocardial infarction, *PB* peripheral blood, *RMI* recent myocardial infarction, *RT* randomized trial, *T/C* treatment/control

Therefore, considering autologous EPC therapy, certain technical improvements that may help to overcome the shortcomings of EPCs should include (1) local delivery of EPCs; (2) endogenous EPC mobilization, that is, cytokine/growth factor supplementation to promote BM-derived EPC mobilization; (3) enrichment procedures, that is, apheresis or BM aspiration; (4) enhancement of EPC functions by gene transduction; or (5) culture expansion of EPCs from self-renewable primitive stem/progenitor cells isolated from BM or other sources. It is of paramount importance to improve strategies that will salvage EPC dysfunction and improve the bioactivity of these cells for the successful treatment of ischemic diseases, especially in light of the current findings implicating that EPC function and mobilization may be impaired in certain diseases. Thus, additional investigations designed to optimize techniques for EPC isolation, expansion, mobilization, recruitment, and transplantation are needed to continue the advancement of this novel therapeutic modality.

References

1. Asahara T, Murohara T, Sullivan A, Silver M, van der Zee R, Li T et al (1997) Isolation of putative progenitor endothelial cells for angiogenesis. Science 275(5302):964–967. PubMed PMID: 9020076
2. Shi Q, Rafii S, Wu MH, Wijelath ES, Yu C, Ishida A et al (1998) Evidence for circulating bone marrow-derived endothelial cells. Blood 92(2):362–367. PubMed PMID: 9657732
3. Asahara T, Masuda H, Takahashi T, Kalka C, Pastore C, Silver M et al (1999) Bone marrow origin of endothelial progenitor cells responsible for postnatal vasculogenesis in physiological and pathological neovascularization. Circ Res 85(3):221–228. PubMed PMID: 10436164
4. Folkman J, Shing Y (1992) Angiogenesis. J Biol Chem 267(16):10931–10934. PubMed PMID: 1375931
5. Risau W, Sariola H, Zerwes HG, Sasse J, Ekblom P, Kemler R et al (1988) Vasculogenesis and angiogenesis in embryonic-stem-cell-derived embryoid bodies. Development 102(3):471–478. PubMed PMID: 2460305
6. Isner JM, Asahara T (1999) Angiogenesis and vasculogenesis as therapeutic strategies for postnatal neovascularization. J Clin Invest 103(9):1231–1236. PubMed PMID: 10225965. Pubmed Central PMCID: 408362

7. Bailey AS, Jiang S, Afentoulis M, Baumann CI, Schroeder DA, Olson SB et al (2004) Transplanted adult hematopoietic stems cells differentiate into functional endothelial cells. Blood 103(1):13–19. PubMed PMID: 12958072
8. Pelosi E, Valtieri M, Coppola S, Botta R, Gabbianelli M, Lulli V et al (2002) Identification of the hemangioblast in postnatal life. Blood 100(9):3203–3208. PubMed PMID: 12384418
9. Grant MB, May WS, Caballero S, Brown GA, Guthrie SM, Mames RN et al (2002) Adult hematopoietic stem cells provide functional hemangioblast activity during retinal neovascularization. Nat Med 8(6):607–612. PubMed PMID: 12042812
10. Ingram DA, Caplice NM, Yoder MC (2005) Unresolved questions, changing definitions, and novel paradigms for defining endothelial progenitor cells. Blood 106(5):1525–1531. PubMed PMID: 15905185
11. Aicher A, Brenner W, Zuhayra M, Badorff C, Massoudi S, Assmus B et al (2003) Assessment of the tissue distribution of transplanted human endothelial progenitor cells by radioactive labeling. Circulation 107(16):2134–2139. PubMed PMID: 12695305
12. Thal MA, Krishnamurthy P, Mackie AR, Hoxha E, Lambers E, Verma S et al (2012) Enhanced angiogenic and cardiomyocyte differentiation capacity of epigenetically reprogrammed mouse and human endothelial progenitor cells augments their efficacy for ischemic myocardial repair. Circ Res 111(2):180–190. PubMed PMID: 22589372. Pubmed Central PMCID: 3406600
13. Krishnamurthy P, Thal M, Verma S, Hoxha E, Lambers E, Ramirez V et al (2011) Interleukin-10 deficiency impairs bone marrow-derived endothelial progenitor cell survival and function in ischemic myocardium. Circ Res 109(11):1280–1289. PubMed PMID: 21959218. Pubmed Central PMCID: 3235675
14. Timmermans F, Plum J, Yoder MC, Ingram DA, Vandekerckhove B, Case J (2009) Endothelial progenitor cells: identity defined? J Cell Mol Med 13(1):87–102. PubMed PMID: 19067770
15. Timmermans F, Van Hauwermeiren F, De Smedt M, Raedt R, Plasschaert F, De Buyzere ML et al (2007) Endothelial outgrowth cells are not derived from CD133+ cells or CD45+ hematopoietic precursors. Arterioscler Thromb Vasc Biol 27(7):1572–1579. PubMed PMID: 17495235
16. Lin Y, Weisdorf DJ, Solovey A, Hebbel RP (2000) Origins of circulating endothelial cells and endothelial outgrowth from blood. J Clin Invest 105(1):71–77. PubMed PMID: 10619863. Pubmed Central PMCID: 382587
17. Ingram DA, Mead LE, Tanaka H, Meade V, Fenoglio A, Mortell K et al (2004) Identification of a novel hierarchy of endothelial progenitor cells using human peripheral and umbilical cord blood. Blood 104(9):2752–2760. PubMed PMID: 15226175
18. Hur J, Yoon CH, Kim HS, Choi JH, Kang HJ, Hwang KK et al (2004) Characterization of two types of endothelial progenitor cells and their different contributions to neovasculogenesis. Arterioscler Thromb Vasc Biol 24(2):288–293. PubMed PMID: 14699017
19. Sieveking DP, Buckle A, Celermajer DS, Ng MK (2008) Strikingly different angiogenic properties of endothelial progenitor cell subpopulations: insights from a novel human angiogenesis assay. J Am Coll Cardiol 51(6):660–668. PubMed PMID: 18261686
20. Aicher A, Rentsch M, Sasaki K, Ellwart JW, Fandrich F, Siebert R et al (2007) Nonbone marrow-derived circulating progenitor cells contribute to postnatal neovascularization following tissue ischemia. Circ Res 100(4):581–589. PubMed PMID: 17272807
21. Wojakowski W, Kucia M, Kazmierski M, Ratajczak MZ, Tendera M (2008) Circulating progenitor cells in stable coronary heart disease and acute coronary syndromes: relevant reparatory mechanism? Heart 94(1):27–33. PubMed PMID: 17395668
22. Losordo DW, Kishore R (2009) A big promise from the very small identification of circulating embryonic stem-like pluripotent cells in patients with acute myocardial infarction. J Am Coll Cardiol 53(1):10–12. PubMed PMID: 19118717. Pubmed Central PMCID: 2664705
23. Masuda H, Kalka C, Takahashi T, Yoshida M, Wada M, Kobori M et al (2007) Estrogen-mediated endothelial progenitor cell biology and kinetics for physiological postnatal vasculogenesis. Circ Res 101(6):598–606. PubMed PMID: 17656679
24. Bauer SM, Goldstein LJ, Bauer RJ, Chen H, Putt M, Velazquez OC (2006) The bone marrow-derived endothelial progenitor cell response is impaired in delayed wound healing from ischemia. J Vasc Surg 43(1):134–141. PubMed PMID: 16414400
25. Murayama T, Tepper OM, Silver M, Ma H, Losordo DW, Isner JM et al (2002) Determination of bone marrow-derived endothelial progenitor cell significance in angiogenic growth factor-induced neovascularization in vivo. Exp Hematol 30(8):967–972. PubMed PMID: 12160849
26. Urbich C, Aicher A, Heeschen C, Dernbach E, Hofmann WK, Zeiher AM et al (2005) Soluble factors released by endothelial progenitor cells promote migration of endothelial cells and cardiac resident progenitor cells. J Mol Cell Cardiol 39(5):733–742. PubMed PMID: 16199052
27. Kawamoto A, Gwon HC, Iwaguro H, Yamaguchi JI, Uchida S, Masuda H et al (2001) Therapeutic potential of ex vivo expanded endothelial progenitor cells for myocardial ischemia. Circulation 103(5):634–637. PubMed PMID: 11156872
28. Jujo K, Ii M, Losordo DW (2008) Endothelial progenitor cells in neovascularization of infarcted myocardium. J Mol Cell Cardiol 45(4):530–544. PubMed PMID: 18755197. Pubmed Central PMCID: 2628572
29. Miyamoto Y, Suyama T, Yashita T, Akimaru H, Kurata H (2007) Bone marrow subpopulations contain distinct types of endothelial progenitor cells and angiogenic cytokine-producing cells. J Mol Cell Cardiol 43(5):627–635. PubMed PMID: 17900610

30. Kawamoto A, Asahara T, Losordo DW (2002) Transplantation of endothelial progenitor cells for therapeutic neovascularization. Cardiovasc Rad Med 3(3–4):221–225. PubMed PMID: 12974378
31. Kocher AA, Schuster MD, Szabolcs MJ, Takuma S, Burkhoff D, Wang J et al (2001) Neovascularization of ischemic myocardium by human bone-marrow-derived angioblasts prevents cardiomyocyte apoptosis, reduces remodeling and improves cardiac function. Nat Med 7(4):430–436. PubMed PMID: 11283669
32. Kalka C, Masuda H, Takahashi T, Kalka-Moll WM, Silver M, Kearney M et al (2000) Transplantation of ex vivo expanded endothelial progenitor cells for therapeutic neovascularization. Proc Natl Acad Sci U S A 97(7):3422–3427. PubMed PMID: 10725398. Pubmed Central PMCID: 16255
33. Murohara T, Ikeda H, Duan J, Shintani S, Sasaki K, Eguchi H et al (2000) Transplanted cord blood-derived endothelial precursor cells augment postnatal neovascularization. J Clin Invest 105(11):1527–1536. PubMed PMID: 10841511. Pubmed Central PMCID: 300847
34. Kawamoto A, Iwasaki H, Kusano K, Murayama T, Oyamada A, Silver M et al (2006) CD34-positive cells exhibit increased potency and safety for therapeutic neovascularization after myocardial infarction compared with total mononuclear cells. Circulation 114(20):2163–2169. PubMed PMID: 17075009
35. Schatteman GC, Hanlon HD, Jiao C, Dodds SG, Christy BA (2000) Blood-derived angioblasts accelerate blood-flow restoration in diabetic mice. J Clin Invest 106(4):571–578. PubMed PMID: 10953032. Pubmed Central PMCID: 380249
36. Dai Y, Ashraf M, Zuo S, Uemura R, Dai YS, Wang Y et al (2008) Mobilized bone marrow progenitor cells serve as donors of cytoprotective genes for cardiac repair. J Mol Cell Cardiol 44(3):607–617. PubMed PMID: 18221754. Pubmed Central PMCID: 2685071
37. Losordo DW, Schatz RA, White CJ, Udelson JE, Veereshwarayya V, Durgin M et al (2007) Intramyocardial transplantation of autologous CD34+ stem cells for intractable angina: a phase I/IIa double-blind, randomized controlled trial. Circulation 115(25):3165–3172. PubMed PMID: 17562958
38. Hung HS, Shyu WC, Tsai CH, Hsu SH, Lin SZ (2009) Transplantation of endothelial progenitor cells as therapeutics for cardiovascular diseases. Cell Transplant 18(9):1003–1012. PubMed PMID: 19650968
39. Iwasaki H, Kawamoto A, Ishikawa M, Oyamada A, Nakamori S, Nishimura H et al (2006) Dose-dependent contribution of CD34-positive cell transplantation to concurrent vasculogenesis and cardiomyogenesis for functional regenerative recovery after myocardial infarction. Circulation 113(10):1311–1325. PubMed PMID: 16534028
40. Kuroda R, Matsumoto T, Niikura T, Kawakami Y, Fukui T, Lee SY et al (2014) Local transplantation of granulocyte colony stimulating factor-mobilized CD34+ cells for patients with femoral and tibial nonunion: pilot clinical trial. Stem Cells Transl Med 3(1):128–134. PubMed PMID: 24307697. Pubmed Central PMCID: 3902290
41. Fujita Y, Kinoshita M, Furukawa Y, Nagano T, Hashimoto H, Hirami Y et al (2014) Phase II clinical trial of CD34+ cell therapy to explore endpoint selection and timing in patients with critical limb ischemia. Circ J 78(2):490–501. PubMed PMID: 24257136
42. Kuroda R, Matsumoto T, Miwa M, Kawamoto A, Mifune Y, Fukui T et al (2011) Local transplantation of G-CSF-mobilized CD34(+) cells in a patient with tibial nonunion: a case report. Cell Transplant 20(9):1491–1496. PubMed PMID: 21176407
43. Burt RK, Testori A, Oyama Y, Rodriguez HE, Yaung K, Villa M et al (2010) Autologous peripheral blood CD133+ cell implantation for limb salvage in patients with critical limb ischemia. Bone Marrow Transplant 45(1):111–116. PubMed PMID: 19448678
44. Schachinger V, Assmus B, Britten MB, Honold J, Lehmann R, Teupe C et al (2004) Transplantation of progenitor cells and regeneration enhancement in acute myocardial infarction: final one-year results of the TOPCARE-AMI Trial. J Am Coll Cardiol 44(8):1690–1699. PubMed PMID: 15489105
45. Li ZQ, Zhang M, Jing YZ, Zhang WW, Liu Y, Cui LJ et al (2007) The clinical study of autologous peripheral blood stem cell transplantation by intracoronary infusion in patients with acute myocardial infarction (AMI). Int J Cardiol 115(1):52–56. PubMed PMID: 16822566
46. Boyle AJ, Whitbourn R, Schlicht S, Krum H, Kocher A, Nandurkar H et al (2006) Intra-coronary high-dose CD34+ stem cells in patients with chronic ischemic heart disease: a 12-month follow-up. Int J Cardiol 109(1):21–27. PubMed PMID: 15970342
47. Bartunek J, Vanderheyden M, Vandekerckhove B, Mansour S, De Bruyne B, De Bondt P et al (2005) Intracoronary injection of CD133-positive enriched bone marrow progenitor cells promotes cardiac recovery after recent myocardial infarction: feasibility and safety. Circulation 112(9 Suppl):I178–I183. PubMed PMID: 16159812
48. Stamm C, Westphal B, Kleine HD, Petzsch M, Kittner C, Klinge H et al (2003) Autologous bone-marrow stem-cell transplantation for myocardial regeneration. Lancet 361(9351):45–46. PubMed PMID: 12517467
49. Heiss C, Kleinbongard P, Dejam A, Perre S, Schroeter H, Sies H et al (2005) Acute consumption of flavanol-rich cocoa and the reversal of endothelial dysfunction in smokers. J Am Coll Cardiol 46(7):1276–1283. PubMed PMID: 16198843
50. Ii M, Takenaka H, Asai J, Ibusuki K, Mizukami Y, Maruyama K et al (2006) Endothelial progenitor thrombospondin-1 mediates diabetes-induced delay in reendothelialization following arterial injury. Circ Res 98(5):697–704. PubMed PMID: 16484619

51. Vasa M, Fichtlscherer S, Aicher A, Adler K, Urbich C, Martin H et al (2001) Number and migratory activity of circulating endothelial progenitor cells inversely correlate with risk factors for coronary artery disease. Circ Res 89(1):E1–E7. PubMed PMID: 11440984
52. Imanishi T, Moriwaki C, Hano T, Nishio I (2005) Endothelial progenitor cell senescence is accelerated in both experimental hypertensive rats and patients with essential hypertension. J Hypertens 23(10):1831–1837. PubMed PMID: 16148606
53. Kondo T, Hayashi M, Takeshita K, Numaguchi Y, Kobayashi K, Iino S et al (2004) Smoking cessation rapidly increases circulating progenitor cells in peripheral blood in chronic smokers. Arterioscler Thromb Vasc Biol 24(8):1442–1447. PubMed PMID: 15191940
54. Michaud SE, Dussault S, Haddad P, Groleau J, Rivard A (2006) Circulating endothelial progenitor cells from healthy smokers exhibit impaired functional activities. Atherosclerosis 187(2):423–432. PubMed PMID: 16288934

Part IV

Stem Cells and Organ Systems: Musculo Skeletal System

Bone Marrow Versus Dental Pulp Stem Cells in Osteogenesis

Mohammad Mahboob Kanafi, Sireesha Ganneru, Dhanasekaran Marappagounder, Padmanav Behera, and Ramesh R. Bhonde

Introduction

Although the consideration of stem cells is currently approaching its hundredth year as one of the organizing principles of developmental biology, it demonstrates no sign of losing its youthful luster. A range of sources of stem cells have been identified that has the potential to self-renewal and capacity to form multiple lineages. Regardless of the discovery of existence of stem cells in various tissues and body fluids, bone marrow has been potentially considered as a persuasive and primeval source of stem cells for treating a wide horizon of disease [1, 2]. Although bone marrow-derived MSCs could be differentiated into mesodermal and non-mesodermal lineages [3–5], osteoblasts, responsible for osteogenesis, and hematopoietic cells, for hematopoiesis are closely associated with the bone marrow, suggesting a reciprocal relationship between the two [6]. Much of the work in MSCs found within the bone marrow stroma on its in vitro and in vivo applications involved in osteogenesis, adipogenesis, cartilage, and muscle formation including osteoblast, osteocytes, adipocytes, chondrocytes, myoblast, and myocytes are gaining importance due to its inherent bone formation capacity [7]. Hence, bone marrow resident stem cells made them the most primitive and promising source from ancient days for treating bone-related diseases. Nevertheless, it is unfortunate that these sources could have not been effective in treatment of all possible diseases due to various disadvantages of BM-MSCs; one of the main drawbacks is that osteogenic potential of bone marrow cells decreases with age [8], and hence, the search for alternate sources of adult stem cells is also underway. It has been demonstrated that stromal adipocytes in bone marrow cavity increases as age increases. In other words, adipocyte accumulation in the human bone marrow stroma correlates with trabecular bone loss with aging [9–12]. Thus, adipose stromal cells both isolated either from bone marrow or from adipose tissue itself has evolved as a contemporary source for bone regeneration [13, 14]. However, we predict that identifying a source that will be similar to the characteristics of bone marrow, possessing

Mohammad Mahboob Kanafi and Sireesha Ganneru have contributed equally.

M.M. Kanafi • R.R. Bhonde (✉)
School of Regenerative Medicine, Manipal Institute of Regenerative Medicine, Manipal University, GKVK Post, Bellary Road, Allalasandra, Near Hotel Royal Orchid, Yelahanka, Bangalore 560065, India
e-mail: mahboob13591980@gmail.com; rr.bhonde@manipal.edu

S. Ganneru • P. Behera
Department of Stem Cell Research, National Institute of Nutrition (ICMR), Secunderabad, Hyderabad 500007, India
e-mail: sirisha_bc@yahoo.com; saipadmanav@gmail.com

D. Marappagounder
Stem Cell Banking & Research, Ree Laboratories Private Limited, Mumbai 400053, India
e-mail: dhanasekarbio@gmail.com

I. Somasundaram (ed.), *Stem Cell Therapy for Organ Failure*,
DOI 10.1007/978-81-322-2110-4_8, © Springer India 2014

inherent bone-forming capacity might be more valuable in bone tissue engineering, repair, and regeneration.

One such source is stem/progenitor cells isolated from dental pulp tissue. Dental pulp tissue has the ability to regenerate dentin in response to dental disorders such as caries. Dentin regeneration is occurred by committed precursor cells called the odontoblasts. It is thus very likely that pulp tissue contains a loaded source of stem cells or progenitor cells which have the capacity to form odontoblasts. These stem/progenitor cells can also be used to induce differentiation resulting in various lineages, apart from dentin, such as osteocytes, adipocytes, chondrocytes, and neurons.

The concept of dental pulp stem cell banking has gained concentration in terms that the cells isolated from the deciduous teeth can be stored and utilized for future treatments since it is well known that dental pulp has stem cells that are multi-lineage [15, 16]. At this point, it is essential to identify the characteristics of deciduous and adult dental pulp mesenchymal stem cells and their differential potential especially across this specific lineage. Recently, stem cells have been isolated and expanded from pulp tissue of permanent teeth, deciduous teeth, periodontal ligament, and apical papilla from an immature tooth, and it has been reported that they generated dentin-like tissue. Subsequent regenerative procedures include the development of guided tissue regeneration (GTR) procedures for osteogenesis, chondrogenesis, neurogenesis, and also dentinogenesis. Gronthos et al. in his study reported that the deciduous tooth contained multipotent stem cells which were extremely proliferative and clonogenic capable of differentiating, into variety of cell types including neural cells, adipocytes and odontoblasts. After in vivo transplantation, they were able to induce bone, generate dentin, and stay alive in mouse brain along with expression of neural markers. Although it could serve a better tool in treating multitude of diseases, its role in osteogenesis and its therapeutic efficacy in bone-related diseases are promising due to its inherent bone formation capacity similar to that of bone marrow stem cells.

Therefore, the present study reviews the need, significance, and advantages of bone marrow stem cells in bone repair and regeneration in the former part. However, during crisis with the negative attributes of bone marrow cells, as mentioned, we demonstrate the dental pulp stem cells might be an ideal alternative source for in vivo bone regeneration capacity in latter half, due to its similarity of inherent bone-forming capacity as that of bone marrow.

Bone Marrow Stem Cells

Bone marrow has traditionally been seen to be composed of two main distinct lineages, the hematopoietic compartment, and the nonhematopoietic cells including the reticular, fat, and endothelial cells, fibroblasts, osteoblasts, and mesenchymal progenitors. Among the cells that compose the stromal tissue, the mesenchymal stem/progenitor cells represents the key component, which are able to differentiate into its progeny, such as osteocytes, chondrocytes, and adipocytes. The mesenchymal stem/progenitor cells have received, in the past years, plenty of attention from the scientific community for their ability to differentiate into different lineages. The first detailed functional description of the tri-lineage differentiation potential of BM-MSCs was provided by Pittenger and colleagues in 1999 [17]. They isolated populations of human BM-MSCs from the bone marrow taken from the iliac crests with a frequency ranging from 1 out of 10.000–100.000, analyzing their immunophenotype and the in vitro differentiation potential. They showed that some of the clones, which they obtained, were able to differentiate in osteoblast, chondroblasts, and adipocytes, pointing out a multipotent ability for these clones. Not every clone retained these abilities. Out of six colonies analyzed, all of them were able to differentiate in osteoblasts, five were able to differentiate in chondrocytes, and only two were able to differentiate in adipocytes, confirming that these populations were very heterogeneous and made up of multipotent stem cells and progenitors already committed towards a specific cell lineage [17].

The most encouraging differentiation trait of BM-MSCs is osteoblastic differentiation in vitro and bone formation in vivo using beta glycerophosphate, dexamethasone, ascorbic acid, throughout the period of 2–3 weeks. Thus, development of osteoblast differentiation from human bone marrow in combination with biomimetic scaffolds provides the possibility of tissue engineering for bone and cartilage [18]. Osteoblasts are responsible not only for forming bone in the normal adult bone remodeling process but also for providing a specific niche microenvironment for hematopoietic stem cells (HSCs) governed by bone morphogenetic protein (BMP), parathyroid hormone (PTH), and Tie2/angiopoietin-1 signaling pathways [19]. These emerging evidences suggest a functional role of osteogenic cells for controlling HSC niches in vivo [2]. This demonstrates the efficacy of osteogenic potential of BM-MSCs.

Why BM-MSCs in Osteogenesis?

Bone tissue is capable of regeneration, yet the natural bone healing process is in some cases insufficient. This is because regeneration of damaged bone is related to four fundamental processes such as osteogenesis, osteoinduction, osteoconduction, and osteopromotion [20]. Excessive loss of bone due to trauma, tumor resection, nonhealing fractures, and so on are cases which lose natural regeneration process, thus requiring transplantation of large bone tissues or substitutes to restore the structural integrity [21]. The use of autologous and allogenic bone grafts was into clinical practice. Although autogenous, spongeous bone graft was considered the "golden standard" among tissue transplants supporting bone regeneration, but it does not suit ideal to clinical practice. Grafts were associated with donor site morbidity and the possible transmission of diseases [21]. Besides, lack of an adequate supply of autologous bone grafts and the unsuitability of allografts do exists. Attention of contemporary research is therefore directed to the finding of an optimum substitute for the standard bone grafts used. BMPs combined with the osteoconductive materials such as hydroxyapatite, tricalcium phosphate, and so on [22, 23] showed promising results. The application of BMPs has yielded positive results in supporting bone regeneration, yet their exclusively osteoinductive property presents a certain strategic limitation.

Thus, there has been some impetus to use MSCs to encourage repair and regenerate bones. Bone regeneration using transplantation of MSCs alone or combined with biomaterials has become the interesting areas of recent research due to its successful healing of the particular bone defect [24]. Despite the availability of stem cells from various adult tissues [25, 26], bone marrow is gaining consensus from ancient times, due to the fact that these stem cells are abundant in the bone marrow which is their suitable source [27]. When bone integrity is damaged (e.g. after fracture), under normal circumstances, MSCs from bone marrow play an important role in its healing. This is due to its inherent bone formation capacity. During fetal bone development, a part of MSC population in bone marrow remains unchanged and forms the source of undifferentiated stem cells [25]. Repair mechanism takes place by chemoattractive property of MSCs through release of cytokines from the damaged bone matrix. MSCs from periosteum and bone marrow are transferred to the damage site, where they continue to multiply and differentiate into its respective lineages to heal the repair. Some study explains a mechanism where bone regeneration occurs through the migration of distant MSCs from peripheral blood to the site of bone injury were they reinforce the healing potential of local MSCs [28]. Bone regeneration is analogous to embryonic development of the skeleton. It is provided by a sum of cellular, humoral, and mechanical factors involved in the formation of new bone in which MSCs play an important role. It is, thus, validated that MSCs from bone marrow serves an appropriate source of bone regeneration.

Applications of BM-MSCs in Bone Regeneration

Formation of new bones during repair is dependent on the quality of MSCs which is directly proportional to the source of osteogenic lines of cells capable of forming bone matter. In this sense, the strategy of using the MSCs that possess more osteogenic potential transplanted into the bone defect appears promising. Studies on murine model showed very promising results especially for bone repair and metabolic bone disorders [24]. Since their first use in 1951, MSCs have been successfully applied for bone regeneration. The subject of intensive research in the field of tissue engineering is the application of MSCs alone or in combination with suitable scaffolds in order to achieve bone tissue regeneration. For successful tissue engineering approaches, implantation of MSCs will require the use of growth and differentiation factors that will favor differentiation and maintenance of bone or chondrocyte phenotype together with an appropriate scaffold to provide a three-dimensional environment. Defining the optimal combination of stem cells, growth factors and scaffolds is thus essential to provide functional bone and cartilage [21]. This would be a contribution for clinical practice in patients with extensive bone defects (tumor resection, traumatic injuries with bone loss, complicated fractures) or in cases of decreased healing ability of bone tissue (older age, osteoporosis) or genetic diseases of the skeleton (osteogenesis imperfecta) [29].

A number of studies have been performed on the use of growth factors and biomaterials to improve tendon-to-bone healing. Besides, interest in using MSCs for tissue engineering has been validated in numerous preclinical models and is under evaluation in clinics. Several clinical trials are recruited for the therapeutic application of MSCs for cartilage defects, osteoporosis, bone fracture, or osteonecrosis. Methods have also been developed for the expansion of bone marrow osteoprogenitors, which indicates the possibility of using autologous human stromal progenitors in the regeneration of large bone defects [30]. Several researchers have described the purification and expansion of bone marrow cells from mice, rats, rabbits, dogs, and humans, and their repair and functional recovery of diaphyseal defects/segmental bone defects have also been reported with the use of osteoprogenitor cells grown on scaffolds of macroporous of hydroxyapatites or other carriers. Other applications of using MSCs as a vehicle for gene delivery approaches have also been demonstrated [31–34]. After successful BM-MSCs, transplantation donor cells actively form bone on the surface of the carrier vehicle, and the recipient cells are induced to form hematopoietic marrow elements, leading to bone/marrow organ structure (craniofacial). Thus, the use of this cell-based tissue-engineering approach to treat patients with large bone defects is also underway [22], thereby leading to substantial improvement in our ability to repair large defects in long bones.

Apart from tissue-engineering-based approach, several clinical investigators from various parts of the world have reported on the safety and therapeutic effect of direct BM-MSCs administration in patients with osteoarthritis and other bone diseases [22, 35]. As an example, Nejadnik and colleagues [36, 37] compared the efficacy of first-generation autologous chondrocyte implantation with that of autologous BM-MSCs, and identified BM-MSCs for cartilage repair showed a better outcome. Besides, a number of studies on the direct use of MSCs to improve the repair of tendon defects have been carried [38, 39]. As an example, Lim et al. studied the role of MSCs at the tendon-bone junction during reconstruction of the ACL in the rabbit [40].

However, as compared to uncommitted BM-SCs, freshly isolated heterogenous bone marrow cell transplantation has not proven successful. This is because uncommitted BM-MSCs were identified to express many osteogenic markers such as CBFA 1/Runx2, osterix, osteopontin, parathyroid hormone receptor, and osteocalcin which are not expressed by freshly isolated BM-MSCs [41]. Interestingly, study reported that a subset of high-proliferating single colony-derived BM-MSCs clones (approximately 60 %) was capable of forming ectopic bone upon in vivo transplantation into immunocompromised mice [42]. Ex vivo expanded BM-MSCs successful repair of bone defects has been achieved in both calvaria and long bone in various animal models [43–47].

Dental Stem Cells

The quest for MSC-like cells in different tissues has led to identification of a variety of stem cells in all organs and tissues in the body in the past decades. Dental-tissue-derived MSC-like populations are among many other stem cells residing in specialized tissues that have been isolated and characterized [48]. The first kind of stem cells was isolated from the human pulp tissue and termed "postnatal dental pulp stem cells" (DPSC) [49]. Later, four additional types of dental-MSC-like populations were recognized: stem cells from exfoliated deciduous teeth (SHED) [50], periodontal ligament stem cells (PDLSCs) [51], stem cells from apical papilla (SCAP) [52], and dental follicle precursor cells (DFPCs) [53].

Postnatal Human Dental Pulp Stem Cells (DPSCs)

The postnatal human dental pulp stem cells (DPSCs) were first isolated by Gronthos and colleagues from pulp of permanent teeth and identified as clonogenic and rapidly proliferative stem cells [49]. Studies have demonstrated that multiple-colony-derived DPSCs can have a population doubling of more than 120, single-colony-derived strains of DPSCs proliferate 10–20 population doublings, and approximately two-thirds of the single-colony derived hDPSCs are able to form the same amount of dentin as multi-colony hDPSCs [54]. On comparison of these cells with BM-MSCs, the DPSCs were found to have an identical expression outline for a range of markers related to endothelium, smooth muscle, bone, and fibroblasts, as that for BM-MSCs [49]. The similarity between DPSCs and BM-MSCs was also confirmed by cDNA microarray profiling when DPSCs and BM-MSCs showed similar level of gene expression for more than 4000 known human genes. DPSCs expressed a high level of collagen type XVIII α-1, insulin-like growth factor-2 (IGF-2), discordin domain tyrosine kinase-2, NAD(P)H menadione oxidoreductase, homolog-2 of Drosophila large disk, and cyclin-dependent kinase-6, whereas the insulin-like growth factor binding protein-7 (IGFBP-7), and collagen type I α-2 genes are expressed in high levels in BM-MSCs. However, the functional roles of many of these genes in the development of dentin and bone can be an interesting concept for further study and research in future [55]. Since characterization studies revealed the mesenchymal stem-cell-like qualities of DPSCs such as self-renewal and multi-lineage differentiation potential, recently tri-lineage differentiation of DPSCs adipo-, osteo-, and chondro-differentiation became a common approach for identification of mesenchymal property of these cells in majority of published works [56]. Spontaneous differentiation of STRO-1+ DPSCs into odontoblasts, osteoblasts, and chondrocytes has been also observed in vitro [57].

DPSCs in Osteogenesis

In vitro expanded DPSCs are capable of differentiating into dentin/pulp-like tissue in vivo. In a study, in vitro expanded DPSCs were transplanted into immunocompromised mice with hydroxyapatite/tricalcium phosphate powder (HA/TCP). Six weeks after transplantation, dentin-like structures were observed lining the surface of the hydroxyapatite/tricalcium phosphate particles. Dentin matrix protein markers like bone sialoprotein, osteocalcin, and DSPP were found to be expressed in the DPSC transplants, and generated dentin was found to thicken over time [49]. DPSCs also demonstrated their capability in differentiation into dentin-like structure by seeding onto human dentin surfaces and implanting into immunocompromised mice [58]. It has been shown that DPSCs from inflamed pulps (DPSCs-IPs) has a decreased osteo-/dentinogenic potential when compared with that of DPSCs normal pulps (DPSCs-NPs) [59]. Nevertheless, on transplantation of DPSCs-IPs into immunocompromised mice, pulp/dentin surfaces similar to that of DPSCs-NPs transplantation is formed [59]. Osteogenic differentiation potential of the human dental pulp cells was discovered when a subpopulation of these cells developed into bone-like tissue in vivo. The cells were termed as osteoblasts derived from human pulpar stem cells (ODHPSCs) [60].

In the field of orofacial and maxillofacial surgery, addition of mass to existing tissue is often essential to reconstruct largely damaged tissue. However, limited availability of autografts, and the inability of allografts to integrate with the surrounding tissue, confines their application [61, 62]. Under such scenarios, stem cell therapy and tissue-engineering technology, or its combination, have been found to be useful. Several studies have documented the in vitro and in vivo osteogenic potential of DPSCs [60–62]. Laino et al. demonstrated the formation of functional lamellar bone constructs in vivo from CD 44+/RunX 2+ (osteoblast precursor marker) differentiated DPSCs [63]. The potential of DPSCs to catalyze responses required to restore tooth structure and function following clinical procedures has been successfully utilized in regenerative endodontics [64]. Recent findings revealing the critical ability of DPSCs to vascularize engineered constructs has expanded its potential in hard tissue engineering [65–67].

Previous studies have also investigated the in vitro and in vivo behavior of DPSCs on 2-D and 3-D collagen, ceramic, and titanium scaffolds directed towards applications in orofacial tissue engineering [68–70]. Reports involving *3-D* porous HA/TCP showed bone-like hard tissue formation by STRO-1 selected DPSCs with distinct lamellae structure and bone marrow-like tissue [71]. In a clinical study, DPSCs were used in conjunction with a collagen sponge scaffold to repair alveolar bone defects caused due to wisdom tooth extraction [72]. Results from these investigations conclude that DPSCs in combination with a suitable scaffold system provide immense potential for the repair and regeneration of periodontal and maxillofacial tissues. However, the above-mentioned studies make use of porous scaffolds into which DPSCs are seeded for the purpose of reconstructing dental or bone tissue. Immobilization ensures sustainability and functionality of cells while avoiding physical stress and inflammatory responses caused at transplantation or delivery sites. Several types of biomaterials, synthetic polymers like polydimethylsiloxane (PDMS), polyethylene glycol (PEG), and natural materials like silk, collagen, and alginate have been adopted over the past decade to immobilize cells [73–75]. The material characteristics of alginate allow its mechanical strength, permeability, and degradability to be tailored to application requirements. The diffusibility of alginate under physiological conditions allows timely release of cells as well as replacement of the biomaterial with regenerated tissue. In addition, the hydrophilic nature of cross-linked alginate provides a framework similar to the extracellular matrix in which cells proliferate, differentiate, and form a functional tissue [76]. Recent studies have investigated the potential of periodontal ligament stem cells (PDLSCs) and gingival mesenchymal stem cells (GMSCs) encapsulated in oxidized alginate micro beads for applications in bone tissue engineering [77].

Results obtained from these studies investigating the morphology, growth, proliferation, immunophenotype, and genotype expression is crucial in determining the potential of utilizing immobilized DPSCs to deliver stem cells and to engineer functional native tissue constructs for oral and maxillofacial bone regeneration applications. Results from a study clearly emphasized the significance of immobilization of DPSCs in 3-D calcium alginate microspheres, leading to consistent cell survivability and functionality. In the same study, the improved osteogenic differentiation of immobilized DPSCs was evidenced by enhanced mineralization, protein secretion, and an upregulated osteo-related gene profile, and interestingly, it was also shown that immobilization triggered osteogenic differentiation of DPSCs without any use of conventional induction factors.

Stem Cells from Human Exfoliated Deciduous Teeth (SHED)

Primary teeth also contain stem cells that referred as SHED [50]. Gronthos et al. in his study reported that the deciduous tooth contained multipotent stem cells which were extremely proliferative and clonogenic capable of differentiating into variety of cell types including neural cells, adipocytes, and odontoblasts. After in vivo transplantation, they were able to induce bone, generate dentin, and stay alive in mouse brain along with expression of neural markers. Although it could serve a better tool in treating multitude of diseases, its role in osteogenesis and its therapeutic efficacy in bone-related diseases are promising due to its similar characteristics inherent bone formation capacity similar to that of bone marrow stem cells.

Isolation of high-quality human postnatal stem cells from accessible resources is usually a priority in the field of stem cell research. As every child loses milk teeth, the obtaining of SHED from them becomes a simple and convenient when compared with other sources of stem cells like as BM-MSCs; hence, this property has given a considerable advantage to SHED among other type of stem cells. In comparison to BM-MSCs, SHED are found to have a higher cell proliferation rate and show a higher number of single colony clusters (CFU-F) [78]. They are also able to proliferate more than 140 population doublings, which is significantly higher than BM-MSCs and DPSCs [50]. SHED are CD34-, CD45-, STRO-1+, SSEA4+, CD73+, CD105+, CD146+, and CD166+. These cells show significant higher levels of STRO-1 and CD146 and lower levels of CD105 [78]. Immature DPSCs (IDPSCs), stem cells isolated from deciduous teeth, have embryonic stem cell markers like Oct4, Nanog, stage-specific embryonic antigens (SSEA-3, SSEA-4), and tumor recognition antigens (TRA-1-60 and TRA-1-81) [79]. Heterogeneous population of SHED has molecular similarity with neural crest cells and stem cells in vitro.

SHED in Osteogenesis

SHED, on osteogenic induction medium, form alizarin red positive nodules, and various bone markers like CBFA1, ALP, MEPE, and bone sialoprotein get upregulated, indicating calcium accumulation and the ability of SHED to differentiate into odontoblastic lineage in vitro [50]. One-month-old culture of SHED-derived osteoblasts secreted extracellular mineralized matrix which went on to develop into 3D woven bone samples in vitro. These cells were positive for alkaline phosphates (ALP), alizarin red, and calcium and to specific antibodies [80]. Myogenic and chondrogenic potentials of SHED have also been demonstrated [79].

In vivo odontoblastic differentiation potential of SHED was demonstrated by transplanting the ex vivo expanded SHED into immunocompromised mice, where these cells developed into human-specific Alu-positive odontoblasts directly associated with a dentin-like structure, while the regenerated dentin being immune reactive to dentin sialophosphoprotein (DSPP), a dentin-specific antibody [70]. However, in vivo complete dentin pulp-like complex regeneration of SHED is not possible [50]. Although SHED are not able to differentiate directly into osteoblasts, but they found to be capable of inducing recipient murine cells to osteocytes, when transplanted into immunocompromised mice [50]. One-fourth of the single-colony-derived SHED clones exhibited the ability to generate ectopic dentin-like tissue equivalent to that generated by multi-colony-derived SHED, while all the single-colony-derived SHED clones were capable of inducing bone formation in immunocompromised mice. Therapeutic potential of SHED was discovered when SHED were found to be able to repair bone defects. In vivo transplantation of SHED-derived bone samples into immune-suppressed rats gave rise to lamellar bone containing entrapped osteocytes [80]. Another study revealed that in the process of SHED-mediated osteogenesis, the hematopoietic marrow elements often found in bone marrow mesenchymal stem cell-generated bone were absent, while mesenchymal stem cell markers like CC9/MUC18/CD146, with an array of growth factor receptors such as transforming growth factor receptors I and II, fibroblast growth factor receptors I and III, and vascular endothelial growth factor receptor I were co-expressed, implying their comprehensive differentiation potential [81]. In vivo transplantation of SHED into immunocompromised mice demonstrated dense engraftment of these cells in various tissues and organs like the liver, spleen, and kidney; hence, the relative ease of recovery and the expression profiles of various markers justify further investigation of SHED for treatment of diseases [79].

Bony defects in the craniomaxillofacial skeleton remain a major and challenging health concern. Maxillofacial surgeons have been trying for centuries to restore functionality and aesthetic appearance applying different strategy including cell-based and protein-based therapies as new strategies without entirely satisfactory results. Nowadays SHED has been proved as a potential source of stem cells to be used in plastic surgery, particularly among craniofacial anomalies. The results of an investigation in the field of stem cell therapy which has been conducted to assess potential of SHED in reconstruction of large-sized cranial bone defects in non-immune-suppressed rats were shown that these stem cells with collagen membrane are able to induce new bone formation at the site of defects without stimulation of the allogenic graft rejection by recipient organism [82].

The curative efficacy of SHED in orofacial bone defects has also been proved when isolated stem cells from miniature pig deciduous teeth, engrafted into pre-generated critical-size bone defects in swine mandible models. Results of this study indicated that stem cells from miniature pig deciduous teeth are able to engraft and regenerate bone to repair critical-size mandibular defects [83]. Recently scientists have suggested that the tissue-engineered bone complex with nano-hydroxyapatite/collagen/poly (L-lactide) nHAC/PLA, recombinant human bone morphogenetic protein 2 (rhBMP-2), and autologous DPSCs might be a better alternative to autologous bone for the clinical reconstruction of periodontal bone defects. In this connection, the capacity of a tissue-engineered bone complex of rhBMP-2 mediated DPSCs and nHAC/PLA to reconstruct critical-size alveolar bone defects in rabbit was evaluated. Findings of this study indicated that nHAC/PLA is an acceptable scaffold for autologous DPSC seeding, proliferation, and differentiation and rhBMP-2 promotes osteogenic capability of DPSCs as a potential cell source for periodontal bone regeneration [84].

DPSCs and Osteogenesis: Our Short Experience

The abovementioned studies make use of porous scaffolds into which DPSCs are seeded for the purpose of reconstructing dental or bone tissue. As such, approaches involving DPSC immobilization aimed at stem cell delivery and hard tissue engineering need to be investigated. With regard to these discussed literatures, we investigated and examined the impact of immobilization on viability and osteogenic differentiation of DPSCs. Morphological analyses correlate with current literature that DPSCs assume spherical shape when immobilized due to matrix tension [77, 85–87]. It can also be inferred from our results that parameters governing immobilization like the concentration of alginate used, the choice of the cross-linker, its molar concentration, and the cell density (2×10^6 cells/mL of alginate) did not hinder the viability of cells (Fig. 1). Initial decrease in the viability of immobilized DPSCs could be attributed to cell shock observed due to the change in the microenvironment (*2-D to 3-D*). Immobilization causes mechanical stress on the cytoskeleton thereby influencing cell behavior. When exposed to sufficiently high stress, it is possible that cells undergo programmed death [88, 89].

The viability and proliferation of immobilized DPSCs were also compared to DPSCs grown in *2-D*. Results show that cells proliferated at a higher rate in *2-D* and reached saturation on day

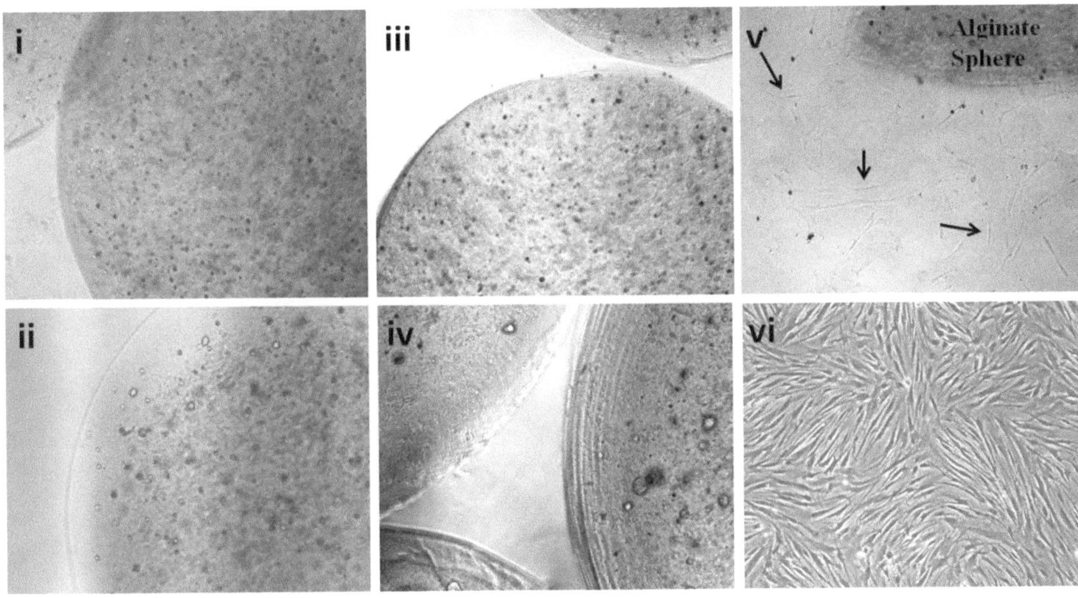

Fig. 1 Micrographs of DPSCs immobilized in alginate microspheres and supplied with conventional MSC medium (control spheres) on days 1 (i) and 10 (ii) supplied with osteogenic induction medium – induced spheres – on days 1 (iii) and 10 (iv) (*arrows* point towards cell aggregates). Immobilized DPSCs adhering to the culture dish upon release (v) (*arrows* point towards DPSC attaining fibroblast-like morphology). DPSCs in passage 3 in 2-D culture on reaching confluency (vi)

4 upon occupation of available surface area in the culture dish. Observably, the percentage viability of these cells reduced after day 4. Immobilized cells interact with the matrix, respond to mechanical cues, and undergo controlled proliferation. It can be concluded that while an immobilization matrix provides a more natural environment for the cells to grow in, it does not provide a platform for the expansion of cells. Immunophenotype analysis of the immobilized cells in *control spheres* on day 10 of culture revealed the expression of cell surface marker CD 73 and CD 90. This confirms that the stem cell characteristics of DPSC had not changed due to immobilization. Similarly, DPSCs in *induced spheres* marked for osteocalcin, a late marker of osteoblastic differentiation, showed maximum expression in aggregations of cells. Minimal osteocalcin expression was also observed in control spheres although the cells did not form aggregates. Calcium quantification analysis showed high calcium content in *induced spheres* as compared to DPSCs differentiated in *2-D*. Osteocalcin expression and the presence of calcium indicate that the alginate matrix provides optimal support for DPSCs to form aggregates, secrete bone-related proteins, and calcify the matrix when differentiation is induced. Alizarin red staining showed significant mineralization in *induced spheres* on days 14 and 21. Data concerning Alizarin red staining of matrix mineralization in immobilization systems are rare. However, reports of the use of alizarin red staining to support osteo-differentiation data have emerged lately [90]. As alginate hydrogels are capable of supplying nutrients to cells, the stain can also be diffused to interact with any mineralization. However, several wash steps need to be incorporated to remove unspecific binding of the stain. In this study, we found that results from Alizarin red staining corroborates with immunocytochemical and calcium quantification data indicating Alizarin red staining as yet another technique for determining mineralization.

It is likely that a *3-D* immobilization matrix, as against 2D, caused similar responses in DPSCs thereby elevating the production of the matrix proteins. Surprisingly, the expression of osteo-specific genes were also elevated in DPSC *control*

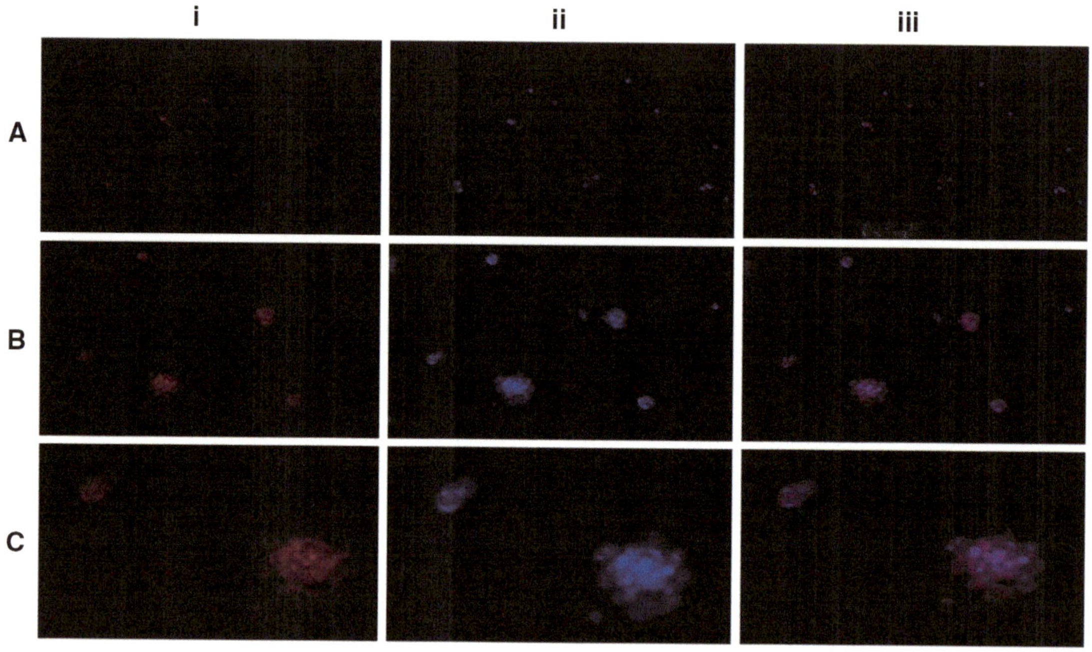

Fig. 2 Immunocytochemical evaluation of osteogenesis of immobilized DPSCs using osteoblastic specific marker, osteocalcin on day 21. (A in 10x) Individual DPSCs in control spheres (without any supplementation of induction medium) expressing osteocalcin. (B in 10x) Osteocalcin expression in cell aggregates in differentiation-induced spheres. (C in 20x) Osteocalcin expression can be observed clearly in cell aggregates in an induced sphere. PE-conjugated osteocalcin is expressed in (i), DAPI used to counterstain the nuclei of the cells is expressed in (ii), and the composite of the images is presented as (iii)

spheres supplemented with conventional MSC media suggesting correlation of the ectodermal origin of DPSCs with osteogenesis (Fig. 2). Stem cells from the human bone marrow (BM-MSCs), without supplementation of induction factors, have been shown to express osteogenic associated markers like OCN, osteopontin (BSP1), and ALP when grown on the surface of unmodified alginate [91]. DPSCs, like BM-MSCs, are equally capable of differentiating into osteoblasts by responding to specific environmental signals. As such, the presence of the markers could be due to (a) the innate quality of DPSCs being naturally prone to differentiate along the osteo-lineage owing to the source from which they are obtained and (b) the 3-D environment created by the alginate scaffold that allows cell-cell interaction imitating the physiological environment.

Results from this study clearly exhibited the significance of immobilization of DPSCs in 3-D calcium alginate microspheres leading to consistent cell survivability and functionality. The improved osteogenic differentiation of immobilized DPSCs was evidenced by enhanced mineralization, protein secretion, and an upregulated osteo-related gene profile. Interestingly, it was also shown that immobilization triggered osteogenic differentiation of DPSCs without any use of conventional induction factors. Collectively, our results demonstrate the potential of immobilized DPSCs to be utilized in stem cell delivery and hard tissue regeneration.

Why DPSCs Are Better than BM-MSCs at Instances

Irrespective of its prehistoric source, bone marrow-derived stem cells were not promising in attempting curative therapies for all diseases.

It became acknowledged from the advancement occurred in bone marrow stem cells by understanding the basic biology and molecular pathways. The first and foremost of the disadvantages put forward is the frequency of lesser number of nucleated cells obtained from large quantity of sample [92]. The second important disadvantage of BM-MSCs is that the proliferation and differentiation capacity of MSC decline with age, reducing their therapeutic potential [92, 93]. Additionally, low frequency of mesenchymal stem cell and the heterogeneity of mononuclear cells with granulocyte interface might create a threat for cell migration and engraftment [94, 95].

In concert with the decreased osteogenic potential of bone marrow cells with age, adipocytes accumulate in the bone marrow stroma. In neonates, adipocytes are barely present in the bone marrow stroma, but the number and size of stromal adipocytes increase with aging, and more than 90 % of the bone marrow cavity is occupied by adipocytes in the aged bone [96]. Interestingly, adipocyte accumulation in the human bone marrow stroma correlates with trabecular bone loss with aging [96–98]. Mice with premature aging (SAMP 6 strain) also show decreased bone formation and increased number of adipocytes in the bone marrow stroma. In hypokinetic rats, bone loss resulting from a decreased osteoblast number [99] is associated with increased adipocyte number and size in the bone marrow cavity. This inverse correlation between the two differentiation processes suggests the disadvantage of bone marrow.

Conclusion

This volume, thus, demands alternative valuable source of stem cell similar to that of bone marrow without compromising its quality. This opens the interesting possibility of promoting dental pulp stem cells. Discovery and advances in dental pulp stem cell biology and behavior have blazed new hopes and promises in the field of regenerative medicine. Although dental pulp stem cells are easily accessible and very good resource of MSCs, [100] conflicting results, possibly due to donor-associated variability, reduce its potential applicability [101], thereby, demanding further tremendous amount of work in order to complement the recent advances in bone tissue engineering. Some important hurdles need to be addressed include multi-differentiation potential, bioscaffolds, and inductive factors that implants and integrates into the surrounding environment for the reconstruction of functional complex organ systems.

References

1. Kumar AA, Kumar SR, Narayanan R, Arul K, Baskaran M (2009) Autologous bone marrow derived mononuclear cell therapy for spinal cord injury: a phase I/II clinical safety and primary efficacy data. Exp Clin Transplant 7:241–248
2. Miura Y, Gao Z, Miura M, Seo B-M, Sonoyama W, Chen W, Gronthos S, Zhang L, Shi S (2006) Mesenchymal stem cell-organized bone marrow elements: an alternative hematopoietic progenitor resource. Stem Cells 24:2428–2436
3. Hattan N, Kawaguchi H, Ando K, Kuwabara E, Fujita J, Murata M, Suematsu M, Mori H, Fukuda K (2005) Purified cardiomyocytes from bone marrow mesenchymal stem cells produce stable intracardiac grafts in mice. Cardiovasc Res 65:334–344
4. Mimeault M, Batra SK (2006) Concise review: recent advances on the significance of stem cells in tissue regeneration and cancer therapies. Stem Cells 24:2319–2345
5. Oh S-H, Muzzonigro TM, Bae S-H, LaPlante JM, Hatch HM, Petersen BE (2004) Adult bone marrow-derived cells trans-differentiating into insulin-producing cells for the treatment of type I diabetes. Lab Invest 84:607–617
6. Wilson A, Trumpp A (2006) Bone-marrow haematopoietic-stem-cell niches. Nat Rev Immunol 6:93–106
7. Kopp H-G, Avecilla ST, Hooper AT, Rafii S (2005) The bone marrow vascular niche: home of HSC differentiation and mobilization. Physiology 20:349–356
8. Rebelatto CK, Aguiar AM, Moreto MP, Senegaglia AC, Hansen P, Barchiki F, Oliveira J, Martins J, Kuligovski C, Mansur F, Christofis A, Amaral VF, Brofman PS, Goldenberg S, Nakao LS, Correa A (2008) Dissimilar differentiation of mesenchymal stem cells from bone marrow, umbilical cord blood, and adipose tissue. Exp Biol Med 233: 901–913
9. Meunier P, Aaron J, Edouard C, Vignon G (1971) Osteoporosis and the replacement of cell

populations of the marrow by adipose tissue. A quantitative study of 84 iliac bone biopsies. Clin Orthop 80:147–154
10. Verma S, Rajaratnam JH, Denton J, Hoyland JA, Byers RJ (2002) Adipocytic proportion of bone marrow is inversely related to bone formation in osteoporosis. J Clin Pathol 55:693–698
11. Burkhardt R, Kettner G, Bohm W, Schmidmeier M, Schlag R, Frisch B, Mallmann B, Eisenmenger W, Gilg T (1987) Changes in trabecular bone, hematopoiesis and bone marrow vessels in aplastic anemia, primary osteoporosis, and old age: a comparative histomorphometric study. Bone 8:157–164
12. Jilka RL, Weinstein RS, Takahashi K, Parfitt AM, Manolagas SC (1996) Linkage of decreased bone mass with impaired osteoblastogenesis in a murine model of accelerated senescence. J Clin Invest 97:1732–1740
13. Rodriguez AM, Elabd C, Amri E-Z, Ailhaud G, Dani C (2005) The human adipose tissue is a source of multipotent stem cells. Biochimie 87:125–128
14. Zuk PA, Zhu M, Ashjian P, De Ugarte DA, Huang JI, Mizuno H, Alfonso ZC, Fraser JK, Benhaim P, Hedrick MH (2002) Human adipose tissue is a source of multipotent stem cells. Mol Biol Cell 13:4279–4295
15. Kanafi MM, Pal R, Gupta PK (2013) Phenotypic and functional comparison of optimum culture conditions for upscaling of dental pulp stem cells. Cell Biol Int 37(2):126–136
16. Kanafi M, Ramesh A, Gupta P, Bhonde R (2014) Dental pulp stem cells immobilized in alginate microspheres for applications in bone tissue engineering. Int Endod J 47(7):687–697
17. Pittenger MF, Mackay AM, Beck SC, Jaiswal RK, Douglas R, Mosca JD, Moorman MA, Simonetti DW, Craig S, Marshak DR (1999) Multilineage potential of adult human mesenchymal stem cells. Science 284:143–147
18. Kamran Kaveh, Rashid Ibrahim, Md. Zuki Abu Bakar, Tengku Azmi Ibrahim (2011) Mesenchymal stem cells, osteogenic lineage and bone tissue engineering: a review. J Anim Vet Adv 10: 2317–2330
19. Arai F, Ohneda O, Miyamoto T, Zhang XQ, Suda T (2002) Mesenchymal stem cells in perichondrium express activated leukocyte cell adhesion molecule and participate in bone marrow formation. J Exp Med 195:1549–1563
20. Albrektsson T, Johansson C (2001) Osteoinduction, osteoconduction and osteointegration. Eur Spine J 10:S96–S101
21. Patel DM, Shah J, Srivastava AS (2013). Therapeutic potential of mesenchymal stem cells in regenerative medicine. Stem Cells Int:496218
22. Quarto R, Mastrogiacomo M, Cancedda R, Kutepov SM, Mukhachev V, Lavroukov A, Kon E, Marcacci M (2001) Repair of large bone defects with the use of autologous bone marrow stromal cells. N Engl J Med 344(5):385–386
23. Marcacci M, Kon E, Moukhachev V, Lavroukov A, Kutepov S, Quarto R, Mastrogiacomo M, Cancedda R (2003) Stem cells associated with macroporous bioceramics for long bone repair: 6- to 7-year outcome of a pilot clinical study. Tissue Eng 13(5): 947–955
24. Griffin M, Iqbal SA, Bayat A (2011) Exploring the application of mesenchymal stem cells in bone repair and regeneration. J Bone Joint Surg Br 93(4):427–434
25. Stocum DL (2000) Stem cells in regenerative biology and medicine. Wound Repair Regen 9(6): 429–442
26. Fuchs E, Segre JA (2000) Stem cells: a new lease on life. Cell 100:143–155
27. Li M, Ikehara S (2013) Bone-marrow-derived mesenchymal stem cells for organ repair. Stem Cells Int:132642
28. Devine MJ, Mierisch CM, Jang E, Anderson PC, Balian G (2002) Transplanted bone marrow cells localize to fracture callus in a mouse model. J Orthop Res 20:1232–1239
29. Horwitz EM, Prockop DJ, Fitzpatrick LA, Koo WWK, Gordon PL, Neel M, Sussman M, Orchard P, Marx JC, Pyeritz RE, Brenner MK (1999) Transplantability and therapeutic effects of bone marrow-derived mesenchymal cells in children with osteogenesis imperfecta. Nat Med 5:309–313
30. Ahdjoudj S, Lasmoles F, Holy X, Zerath E, Marie PJ (2002) Transforming growth factor beta2 inhibits adipocyte differentiation induced by skeletal unloading in rat bone marrow stroma. J Bone Miner Res 17:668–677
31. Goshima J, Goldberg VM, Caplan AI (1991) Osteogenic potential of culture-expanded rat marrow cells as assayed in vivo with porous calcium phosphate ceramic. Biomaterials 12:253–258
32. Gundle R, Joyner CJ, Triffitt JT (1995) Human bone tissue formation in diffusion chamber culture in vivo by bone-derived cells and marrow stromal fibroblastic cells. Bone 16:597–601
33. Bruder SP, Kurth AA, Shea M et al (1998) Bone regeneration by implantation of purified culture-expanded human mesenchymal stem cells. J Orthop Res 16:155–162
34. Jaiswal N, Haynesworth SE, Caplan AI, Bruder SP (1997) Osteogenic differentiation of purified, culture-expanded human mesenchymal stem cells in vitro. J Cell Biochem 64:295–312
35. Warnke PH, Springer IN, Wiltfang J et al (2004) Growth and transplantation of a custom vascularised bone graft in a man. Lancet 364:766–770
36. Nejadnik H, Hui JH, Feng Choong EP, Tai B-C, Lee EH (2010) Autologous bone marrow-derived mesenchymal stem cells versus autologous chondrocyte implantation: an observational cohort study. Am J Sports Med 38:1110–1116
37. Gupta PK, Das AK, Chullikana A, Majumdar AS (2012) Mesenchymal stem cells for cartilage repair in osteoarthritis. Stem Cell Res Ther 3(4):25

38. Young RG, Butler DL, Weber W et al (1998) Use of mesenchymal stem cells in a collagen matrix for Achilles tendon repair. J Orthop Res 16:406–413
39. Awad HA, Boivin GP, Dressler MR et al (2003) Repair of patellar tendon injuries using a cell-collagen composite. J Orthop Res 21:420–431
40. Lim JK, Hui J, Li L et al (2004) Enhancement of tendon graft osteointegration using mesenchymal stem cells in a rabbit model of anterior cruciate ligament reconstruction. Arthroscopy 20:899–910
41. Gronthos S, Zannettino ACW, Hay SJ, Shi S, Graves SE, Kortesidis A, Simmons PJ (2003) Molecular and cellular characterisation of highly purified stromal stem cells derived from human bone marrow. J Cell Sci 116:1827–1835
42. Kuznetsov SA, Friedenstein AJ, Robey PG (1997) Factors required for bone marrow stromal fibroblast colony formation in vitro. Br J Haematol 97:561–570
43. Bruder SP, Horowitz MC, Mosca JD, Haynesworth SE (1997) Monoclonal antibodies reactive with human osteogenic cell surface antigens. Bone 21:225–235
44. Krebsbach PH, Mankani MH, Satomura K, Kuznetsov SA, Robey PG (1998) Repair of craniotomy defects using bone marrow stromal cells. Transplantation 66:1272–1278
45. Kon E, Muraglia A, Corsi A et al (2000) Autologous bone marrow stromal cells loaded onto porous hydroxyapatite ceramic accelerate bone repair in critical-size defects of sheep long bones. J Biomed Mater Res 49:328–337
46. Petite H, Viateau V, Bensaid W et al (2000) Tissue-engineered bone regeneration. Nat Biotechnol 18:959–963
47. Mankani MH, Krebsbach PH, Satomura K, Kuznetsov SA, Hoyt R, Robey PG (2001) Pedicled bone flap formation using transplanted bone marrow stromal cells. Arch Surg 136:263–270
48. Huang GT-J, Gronthos S, Shi S (2009) Mesenchymal stem cells derived from dental tissues vs. those from other sources: their biology and role in regenerative medicine. J Dent Res 88(9):792–806
49. Gronthos S, Mankani M, Brahim J, Robey PG, Shi S (2000) Postnatal human dental pulp stem cells (DPSCs) in vitro and in vivo. Proc Natl Acad Sci U S A 97(25):13625–13630
50. Miura M, Gronthos S, Zhao M, Lu B, Fisher LW, Robey PG et al (2003) SHED: stem cells from human exfoliated deciduous teeth. Proc Natl Acad Sci U S A 100(10):5807–5812
51. Seo B-M, Miura M, Gronthos S, Bartold PM, Batouli S, Brahim J et al (2004) Investigation of multipotent postnatal stem cells from human periodontal ligament. Lancet 364(9429):149–155
52. Sonoyama W, Liu Y, Yamaza T, Tuan RS, Wang S, Shi S et al (2008) Characterization of the apical papilla and its residing stem cells from human immature permanent teeth: a pilot study. J Endod 34(2):166–171
53. Morsczeck C, Götz W, Schierholz J, Zeilhofer F, Kühn U, Möhl C et al (2005) Isolation of precursor cells (PCs) from human dental follicle of wisdom teeth. Matrix Biol 24(2):155–165
54. Suchanek J, Soukup T, Visek B, Ivancakova R, Kucerova L, Mokry J (2009) Dental pulp stem cells and their characterization. Biomed Pap Med Fac Univ Palacký Olomouc Czech 153(1):31–35
55. Yamada Y, Fujimoto A, Ito A, Yoshimi R, Ueda M (2006) Cluster analysis and gene expression profiles: a cDNA microarray system-based comparison between human dental pulp stem cells (hDPSCs) and human mesenchymal stem cells (hMSCs) for tissue engineering cell therapy. Biomaterials 27(20):3766–3781
56. Koyama N, Okubo Y, Nakao K, Bessho K (2009) Evaluation of pluripotency in human dental pulp cells. J Oral Maxillofac Surg 67(3):501–506
57. Yu J, He H, Tang C, Zhang G, Li Y, Wang R et al (2010) Differentiation potential of STRO-1+ dental pulp stem cells changes during cell passaging. BMC Cell Biol 11:32
58. Batouli S, Miura M, Brahim J, Tsutsui TW, Fisher LW, Gronthos S et al (2003) Comparison of stem-cell-mediated osteogenesis and dentinogenesis. J Dent Res 82(12):976–981
59. Alongi DJ, Yamaza T, Song Y, Fouad AF, Romberg EE, Shi S et al (2010) Stem/progenitor cells from inflamed human dental pulp retain tissue regeneration potential. Regen Med 5(4):617–631
60. Carinci F, Papaccio G, Laino G, Palmieri A, Brunelli G, D'Aquino R et al (2008) Comparison between genetic portraits of osteoblasts derived from primary cultures and osteoblasts obtained from human pulpar stem cells. J Craniofac Surg 19(3):616–625, discussion 626–627
61. Liu H-C EL-L, Wang D-S, Su F, Wu X, Shi Z-P et al (2011) Reconstruction of alveolar bone defects using bone morphogenetic protein 2 mediated rabbit dental pulp stem cells seeded on nano-hydroxyapatite/collagen/poly(L-lactide). Tissue Eng Part A 17(19-20):2417–2433
62. Mori G, Centonze M, Brunetti G, Ballini A, Oranger A, Mori C et al (2010) Osteogenic properties of human dental pulp stem cells. J Biol Regul Homeost Agents 24(2):167–175
63. Laino G, d'Aquino R, Graziano A, Lanza V, Carinci F, Naro F et al (2005) A new population of human adult dental pulp stem cells: a useful source of living autologous fibrous bone tissue (LAB). J Bone Miner Res 20(8):1394–1402
64. Huang GT-J (2011) Dental pulp and dentin tissue engineering and regeneration: advancement and challenge. Front Biosci (Elite Ed) 3:788–800
65. Marchionni C, Bonsi L, Alviano F, Lanzoni G, Di Tullio A, Costa R et al (2009) Angiogenic potential of human dental pulp stromal (stem) cells. Int J Immunopathol Pharmacol 22(3):699–706
66. Hazaimeh-NAL, Beattie J, Duggal M, Yang XB (2012). Angiogenic potential of human dental pulp

stem cells for skeletal tissue engineering. J Bone Joint Surg:94-B (SUPP XXXVI 43)

67. Gandia C, Armiñan A, García-Verdugo JM, Lledó E, Ruiz A, Miñana MD et al (2008) Human dental pulp stem cells improve left ventricular function, induce angiogenesis, and reduce infarct size in rats with acute myocardial infarction. Stem Cells Dayt Ohio 26(3):638–645
68. Zhang W, Walboomers XF, van Kuppevelt TH, Daamen WF, Bian Z, Jansen JA (2006) The performance of human dental pulp stem cells on different three-dimensional scaffold materials. Biomaterials 27(33):5658–5668
69. Prescott RS, Alsanea R, Fayad MI, Johnson BR, Wenckus CS, Hao J et al (2008) In vivo generation of dental pulp-like tissue by using dental pulp stem cells, a collagen scaffold, and dentin matrix protein 1 after subcutaneous transplantation in mice. J Endod 34(4):421–426
70. Cordeiro MM, Dong Z, Kaneko T, Zhang Z, Miyazawa M, Shi S et al (2008) Dental pulp tissue engineering with stem cells from exfoliated deciduous teeth. J Endod 34(8):962–969
71. Yang X, Walboomers XF, van den Beucken JJJP, Bian Z, Fan M, Jansen JA (2009) Hard tissue formation of STRO-1-selected rat dental pulp stem cells in vivo. Tissue Eng Part A 15(2):367–375
72. D' Aquino R, De Rosa A, Lanza V, Tirino V, Laino L, Graziano A et al (2009) Human mandible bone defect repair by the grafting of dental pulp stem/progenitor cells and collagen sponge biocomplexes. Eur Cell Mater 18:75–83
73. Bünger CM, Tiefenbach B, Jahnke A, Gerlach C, Freier T, Schmitz KP et al (2005) Deletion of the tissue response against alginate-pll capsules by temporary release of co-encapsulated steroids. Biomaterials 26(15):2353–2360
74. Kasoju N, Bora U (2012) Silk fibroin based biomimetic artificial extracellular matrix for hepatic tissue engineering applications. Biomed Mater Bristol Engl 7(4):045004
75. Munoz-Pinto DJ, Jimenez-Vergara AC, Hou Y, Hayenga HN, Rivas A, Grunlan M et al (2012) Osteogenic potential of poly (ethylene glycol)-poly(dimethylsiloxane) hybrid hydrogels. Tissue Eng Part A 18(15–16):1710–1719
76. Drury JL, Mooney DJ (2003) Hydrogels for tissue engineering: scaffold design variables and applications. Biomaterials 24(24):4337–4351
77. Moshaverinia A, Chen C, Akiyama K, Ansari S, Xu X, Chee WW et al (2012) Alginate hydrogel as a promising scaffold for dental-derived stem cells: an in vitro study. J Mater Sci Mater Med 23(12): 3041–3051
78. Yamaza T, Kentaro A, Chen C, Liu Y, Shi Y, Gronthos S et al (2010) Immunomodulatory properties of stem cells from human exfoliated deciduous teeth. Stem Cell Res Ther 1(1):5
79. Kerkis I, Kerkis A, Dozortsev D, Stukart-Parsons GC, Gomes Massironi SM, Pereira LV et al (2006) Isolation and characterization of a population of immature dental pulp stem cells expressing OCT-4 and other embryonic stem cell markers. Cells Tissues Organs 184(3–4):105–116
80. Laino G, Graziano A, d' Aquino R, Pirozzi G, Lanza V, Valiante S et al (2006) An approachable human adult stem cell source for hard-tissue engineering. J Cell Physiol 206(3):693–701
81. Seo BM, Sonoyama W, Yamaza T, Coppe C, Kikuiri T, Akiyama K et al (2008) SHED repair critical-size calvarial defects in mice. Oral Dis 14(5):428–434
82. De Mendonça CA, Bueno DF, Martins MT, Kerkis I, Kerkis A, Fanganiello RD et al (2008) Reconstruction of large cranial defects in nonimmunosuppressed experimental design with human dental pulp stem cells. J Craniofac Surg 19(1):204–210
83. Zheng Y, Liu Y, Zhang CM, Zhang HY, Li WH, Shi S et al (2009) Stem cells from deciduous tooth repair mandibular defect in swine. J Dent Res 88(3): 249–254
84. Liu H-C, L-L E, Wang D-S, Su F, Wu X, Shi Z-P et al (2011) Reconstruction of alveolar bone defects using bone morphogenetic protein 2 mediated rabbit dental pulp stem cells seeded on nano-hydroxyapatite/collagen/poly (L-lactide). Tissue Eng Part A 17(19–20):2417–2433
85. Aquino D', Riccardo ADR, Lanza V, Tirino V, Laino L, Graziano A, Desiderio V, Laino G, Papaccio G (2009) Human mandible bone defect repair by the grafting of dental pulp stem/progenitor cells and collagen sponge biocomplexes. Eur Cell Mater 18:75–83
86. Markusen JF, Mason C, Hull DA, Town MA, Tabor AB, Clements M et al (2006) Behavior of adult human mesenchymal stem cells entrapped in alginate-GRGDY beads. Tissue Eng 12(4):821–830
87. Nuttelman CR, Tripodi MC, Anseth KS (2004) In vitro osteogenic differentiation of human mesenchymal stem cells photoencapsulated in PEG hydrogels. J Biomed Mater Res A 68(4):773–782
88. Fletcher DA, Mullins RD (2010) Cell mechanics and the cytoskeleton. Nature 463(7280):485–492
89. Discher DE, Mooney DJ, Zandstra PW (2009) Growth factors, matrices, and forces combine and control stem cells. Science (New York, NY) 324(5935):1673–1677
90. Penolazzi L, Tavanti E, Vecchiatini R, Lambertini E, Vesce F, Gambari R et al (2010) Encapsulation of mesenchymal stem cells from Wharton's jelly in alginate microbeads. Tissue Eng Part C Methods 16(1):141–155
91. Wang L, Shelton RM, Cooper PR, Lawson M, Triffitt JT, Barralet JE (2003) Evaluation of sodium alginate for bone marrow cell tissue engineering. Biomaterials 24(20):3475–3481
92. Rebelatto CK, Aguiar AM, Moreto MP, Senegaglia AC, Hansen P, Barchiki F, Oliveira J, Martins J, Kuligovski C, Mansur F, Christofis A, Amaral VF, Brofman PS, Goldenberg S, Nakao LS, Correa A (2008) Dissimilar differentiation of mesenchymal

stem cells from bone marrow, umbilical cord blood, and adipose tissue. Exp Biol Med 233:901–913

93. Stolzing A, Jones E, McGonagle D, Scutt A (2008) Age-related changes in human bone marrow-derived mesenchymal stem cells: consequences for cell therapies. Mech Ageing Dev 129:163–173
94. Kuethe F, Richartz BM, Kasper C, Sayer HG, Hoeffken K, Werner GS, Figulla HR (2005) Autologous intracoronary mononuclear bone marrow cell transplantation in chronic ischemic cardiomyopathy in humans. Int J Cardiol 100: 485–491
95. Wagner W, Wein F, Seckinger A, Frankhauser M, Wirkner U, Krause U, Blake J, Schwager C, Eckstein V, Ansorge W, Ho AD (2005) Comparative characteristics of mesenchymal stem cells from human bone marrow, adipose tissue, and umbilical cord blood. Exp Hematol 33: 1402–1416
96. Rozman C, Feliu E, Berga L, Reverter JC, Climent C, Ferran MJ (1989) Age-related variations of fat tissue fraction in normal human bone marrow depend both on size and number of adipocytes: a stereological study. Exp Hematol 17:34–37
97. Charbord P, Tavian M, Humeau L, Peault B (1996) Early ontogeny of the human marrow from long bones: an immunohistochemical study of hematopoiesis and its microenvironment. Blood 87:4109–4119
98. Kirkland JL, Tchkonia T, Pirtskhalava T, Han J, Karagiannides I (2002) Adipogenesis and aging: does aging make fat go MAD? Exp Gerontol 37:757–767
99. Machwate M, Zerath E, Holy X, Hott M, Modrowski D, Malouvier A, Marie PJ (1993) Skeletal unloading in rat decreases proliferation of rat bone and marrow-derived osteoblastic cells. Am J Physiol 26:E790–E799
100. Alge DL, Zhou D, Adams LL, Wyss BK, Shadday MD, Woods EJ, Gabriel Chu TM, Scott Goebel W (2010) Donor-matched comparison of dental pulp stem cells and bone marrow-derived mesenchymal stem cells in a rat model. J Tissue Eng Regen Med 4(1):73–81
101. Shi S, Bartold PM, Miura M, Seo BM, Robey PG, Gronthos S (2005) The efficacy of mesenchymal stem cells to regenerate and repair dental structures. Orthod Craniofac Res 8:191–199

Adipose-Derived Stem Cells: In Musculoskeletal Disorders

Dhanasekaran Marappagounder, Sandeep Kumar Kotturu, Padmanav Behera, Rajanna Ajumeera, and Ravikumar Rajappa

Introduction

Conceptually and from a practical standpoint, bone marrow has been the most influential source of stem cells that offers a possibility of being used in a wide range of therapeutics. Clinical situations frequently demand stem cells with dependable quality and quantity to treat disorders of cellular degeneration. Challenges to bring advances to the clinical mount have expanded rapidly, engendering new perspectives concerning the identity, origin, and full therapeutic potential of various tissue-specific stem cells. Recent progress in stem cell biology has allowed researchers to investigate distinct stem cell populations in such divergent mammalian tissues and organs. Taking stem cells adaptable for regenerative medicine applications in adequate quantities at the right time is a challenge. In this respect, an emerging body of literature suggests that redundant adipose tissue serves as an abundant, accessible, and reliable source of stem cells that can be readily harvested with minimal risk to the patients. Rapidly accumulating evidence suggests that adipose tissue-derived stem cells (ADSC), especially from white adipose tissue, possess a far wider property of self-renewal and multilineage differentiation capacity, thereby highlighting their importance and effectiveness in regenerative medicine [1–5]. Despite literature supporting the capacity and plasticity of ADSC for regenerative medicine, there are functional and heterogeneous discrepancies associated with it, thus presenting ADSC research a difficult and challenging task. Promising strides are continuously being made to unravel these challenges and realize the potential of ADSC. While much progress on adipose-derived stem cells has been made in the last few years, there remain a lot to be explored.

This chapter focuses on the overview of adipose-derived stem cells. Further insight into the current knowledge on the advances in the applications of this adipose tissue-derived stem cell in musculoskeletal disorders has also been explored.

Dhanasekaran Marappagounder and Sandeep Kumar Kotturu have contributed equally.

D. Marappagounder (✉)
Stem Cell Banking & Research, Ree Laboratories Private Limited, Mumbai 400053, India
e-mail: dhanasekarbio@gmail.com

S.K. Kotturu
Molecular Biology Unit, National Institute of Nutrition (ICMR), Secunderabad, Hyderabad 500007, India
e-mail: sandeep.kotturu@gmail.com

P. Behera • R. Ajumeera
Department of Stem Cell Research, National Institute of Nutrition (ICMR), Secunderabad, Hyderabad 500007, India
e-mail: saipadmanav@gmail.com; arajanna@gmail.com

R. Rajappa
Harvard Medical School, Dr. Spector's Lab-VA Medical Center, Jamaica Plain, MA 02130, USA
e-mail: cra.dr.ravikumar@gmail.com

I. Somasundaram (ed.), *Stem Cell Therapy for Organ Failure*,
DOI 10.1007/978-81-322-2110-4_9, © Springer India 2014

Adipose-Derived Stem Cells

Attention in considering adipose tissue as a reservoir of stem cells was really undertaken only after the findings of Zuk and his coworkers, in the year 2001 [5, 6]. The translation of his findings associated with the easy sampling of adipose tissue with its low risk and morbidity attracted many new investigators. Subsequently, increasing evidences are accumulating on the pivotal role of subcutaneous fat-derived stem cells owing to their proliferative capacity and multilineage differentiation ability [3, 4, 7–11]. To investigate multipotency, several researchers had demonstrated the multilineage differentiation ability of subcutaneous adipose-derived stem cells [2, 3, 6, 12, 13]. For instance, Rodriguez and his coworkers [14] created a single clone from fast-adherent ASCs and proved that two out of 12 clones were able to undergo multilineage differentiation [15]. The remaining ten clones had bipotent capacity. These findings indicate that a high percentage of ASCs have multipotential as well as pluripotential capacity in vitro to differentiate into the major mesodermal and ectodermal lineages.

Adipose tissue-derived mesenchymal stem cells naturally differentiate into mature adipocytes [2, 3, 6, 12, 13]. In adipogenic induction medium such as 3-isobutyl-1-methylxanthine (IBMX), dexamethasone (DEX), indomethacin, insulin, and so on, ASCs were found to develop intracellular lipid vacuoles which coalesce and give rise to a single, cytoplasm-filling vacuole. Besides this definite marker of adipogenesis [14], comprising glycerol-3-phosphate dehydrogenase (GPDH), lipoprotein lipase, peroxisome proliferator-activated receptor γ (PPARγ), leptin, adipocyte fatty-acid-binding protein (aP2)11, CCAAT/enhancer binding protein (C/EBP), and glucose transporter 4 (Glut4) are found to be expressed. Similarly, ASCs differentiation towards the osteogenic cell lineage is well established for in vitro as well as for in vivo animal tissue engineering models [2, 3, 12, 13, 16, 41]. A clinical observation is in part responsible for the discovery of the osteogenic differentiation capacity of ASCs. A rare disorder named "progressive osseous heteroplasia" together with the capacity of MSCs to convert into the osteogenic lineage led to the assumption that ASCs are likewise able to differentiate into osteocytes [17]. Osteogenic induction of ASCs can be achieved by similar culture conditions as used in MSCs, including supplementation with ascorbic acid together with 1-alpha,25- dihydroxyvitamin D3 (1,25(OH)2D3), the hormonal metabolite of vitamin D, or dexamethasone [18]. Under osteogenic differentiation medium, ASCs are capable of expressing diverse genes and proteins found in the osteoblast's phenotype: type I collagen, alkaline phosphatase, osteocalcin, osteonectin, osteopontin, parathyroid hormone (PTH) receptor, bone morphogenetic protein-2 (BMP-2), BMP-4, BMP receptors I and II, bone sialoprotein, and RunX-1 [5, 18, 19].

Besides, the efficacy of retention capacity of its characteristics at prolonged culture condition of both rat and human adipose tissue-derived mesenchymal stem cells had also been demonstrated. Both murine and human stem cells were found to retain their properties evidence for the possibility of their characteristics until prolonged culturing. They were able to preserve their long-term stem cell characteristics and differentiation potential even at longer passages [10, 20, 21]. The human subcutaneous adipose tissue showed high telomerase activity that could be maintained for more than 100 population doublings. Thus, evidence is available proving that subcutaneous adipose tissues possess properties of true stem cells, which were retained even after extended in vitro culturing, thereby rewarding a prerequisite for possible successful cell-based therapies [20]. Besides, secretion of various growth factors that control and manage damaged neighboring cells has been an essential function of ADSC [22–24]. It is deduced that ADSCs may exert their beneficial effects via complex paracrine mechanisms in addition to a building-block function. This paracrine effect has been well demonstrated, thereby making ADSCs an attractive therapeutic tool.

Applications of ADSC in Musculoskeletal Disorders

The basic, experimental, and clinical research on SVF/ADSC has expanded exponentially over the past decade. Cell-based therapy using ADSCs presents a unique opportunity for their use in tissue repair and regeneration. The important experimental findings using SVF/ASC in recent years in treating wide range of diseases are increasing, thereby laying a blueprint for ADSC in cellular replacement and regenerative medicine. Adipose-derived stem cells are an abundant, easily accessible, and reproducible cell source for musculoskeletal regenerative medicine applications; however, in vivo repair processes continue to present major challenges. Musculoskeletal defects due to acute trauma, congenital malformations, degenerative diseases, and neoplasia are potential targets for cell-based regenerative therapies. But, the use of collagenase in isolation of ADSCs has made FDA to declare its use as a "Drug" and demands a laborious time-consuming process to acquire permission. This might warrant further investigations into other methods of separation.

Soft Tissue Defects

One of the most intuitive uses of ADSCs is for the replacement of adipose tissue itself. Large soft tissue defects are a common problem following trauma, burns, and oncological resections, such as mastectomy. The regeneration and augmentation of soft tissues requires long-term maintenance of aesthetic results. In order to develop more physiological alternatives for soft tissue reconstruction, several laboratories have investigated the possibility of creating tissue-engineered cell-seeded scaffolds for the generation of de novo adipose tissue. Current therapies are limited, and biomaterials, which include collagen, hyaluronic acid, silicon, and other filler materials. However, the influence of variables such as porosity, biomaterial composition, and seeding density has been under continuous investigations for the optimization of the constructs to improve adipogenesis. Besides, they have several disadvantages such as high cost, immunogenicity and allergenicity, and the risk of transmitting infectious diseases. Although, autologous fat grafts including ADSC [25] are in current clinical practice, their long-term graft retention [26] might be still controversial.

Yoshimura and coworkers used adipose-derived SVF cells for soft tissue augmentation by a novel strategy called cell-assisted lipotransfer (CAL) for treatment of facial lipoatrophy and breast augmentation. It was identified that ASC supplementation has improved its efficacy and improved facial contour, with no adverse effects, although there are no statistically significant difference identified. Furthermore, breast tissue augmentation and reconstruction trial had been reported successful [27–30]. Alternatively, Kim and his coworkers were successful in transplanting the differentiating adipocytes from ADSC for treating depressed scar with up to 75 % success rate [31]. Developing strategies to reconstruct larger tissue defects, however, remains a formidable challenge. Furthermore, although the preliminary studies proved efficient in retention and volume-restoring capabilities of transplanted fat, further attention in these techniques is of utmost importance to draw any conclusion.

Muscular Dystrophies

Muscular dystrophies are a clinically and genetically heterogeneous group of disorders characterized by progressive degeneration and loss of skeletal muscles. The continuous and gradual muscle degeneration in progressive muscular dystrophies leads to depletion of satellite cells, and consequently, the capacity to restore the skeletal muscle is lost. Knowledge of the genetic and molecular mechanisms underlying muscular dystrophies (MDs) has advanced in recent times [32]. However, congenital muscular dystrophies (CMD) are disabling and often lethal disorders. The CMDs share the same muscle pathology status similar to other traditional muscular dystrophies, of which Duchenne and

Becker muscular dystrophies are the major forms. However, the mechanisms leading to the muscle pathologies (sarcolemma instability, degeneration and regeneration of muscle cells, apoptosis and fibrosis) differ between the common CMD types and other muscular dystrophies. Stem cell-based therapy holds promise for treating genetic diseases and has been utilized in animal models and human clinical trials for different types of muscular dystrophies, in particular Duchenne muscular dystrophy [33]. Although initial efforts on myoblast transfer have proven successful, poor cell survival, immune rejection, and poor migration of transplanted cells limited its applications. Cell-based therapies for muscular disease evolved out of interest to restore dystrophin levels in patients with Duchenne muscular dystrophy (DMD). The discovery of muscle-derived stem cells has led to new investigations not only in skeletal muscle disease but also in other applications [19, 34, 35]. However, muscle-derived stem cells are limited and difficult to obtain; hence, alternative way of using adult stem cells from non-muscle tissues to replace damaged muscle fibers for treatments is underway. However, effective cellular therapy for ECM-related CMDs rests on the ability of the therapeutic cells to secrete normal ECM proteins that can prevent muscle cell degeneration rather than on the potential of these cells to differentiate into muscle fibers.

The regenerative potential of BMSCs has been promising. Cells injected into cardiotoxin-damaged muscle were shown to engraft and incorporate into regenerating myofibers. Further trials in a transgenic murine model were also effective [36]. This demonstrates the effect of BMSC to repair damage following acute injury as well as in a degenerative model. Emerging results with ASCs point to the possibility of a similar therapeutic potential. Although bone marrow is the main source for MSC isolation, subcutaneous fat represents an alternative repository for stem cells and is currently a subject of intensive investigations [37]. Adipose cell lineage plays a positive role and is required for efficient muscle regeneration after acute injury. It would be interesting to examine whether an appropriate proportion of adipogenic cells is necessary for the normal growth and maintenance of the skeletal muscle. Another important question to address is at which level adipogenic cells regulate muscle regeneration. Conceivably, preadipocytes may interact directly with myogenic progenitors and regulate their differentiation.

ADSCs were demonstrated to possess myogenic differentiation capacity in vitro as well as in vivo [3, 9, 22, 38]. The in vitro myogenic differentiation potential was evident from the expression of characteristic markers and the formation of multinucleated myotubules. Degenerative diseases, on the other hand, are characterized by slow but progressive accumulation of damage. Cell-based therapies that may replenish the exhausted supply of satellite cells [36] may be particularly suited to prevent this decline. Vitali Alexeev et al. [39] and coworkers demonstrated that ADSCs cultured in vitro secrete a variety of ECM proteins, including collagen VI and, therefore, can provide therapeutic ECM proteins without cell differentiation in the muscle environment for the treatment of congenital muscular dystrophy. Significant work is needed to establish the methods necessary to treat progressive diseases; however, the demonstration that ASCs have myogenic potential both in vitro and in vivo is encouraging. Although these results were statistically significant, it remains to be seen whether they will result in clinically noticeable improvements.

Bone and Cartilage Defects

Both BMSCs and ADSCs have proven to be favorable candidates based on their osteogenic capacity in in vitro and in vivo studies [40–42]. Osteoblast differentiation represents a crucial event during skeletal tissue formation, bone repair, and bone remodeling. Craniofacial defects, in particular calvarial defects, have been of special interest in treatment with ADSC. Due to the restricted amount of BM available in patients with calvarial defects, ADSCs combined with milled autologous cancellous bone and fibrin glue were sued to repair a large calvarial defect [43]. In efforts to utilize this potential for tissue-engineered bone repairs, many laboratories have begun seeding osteogenically differentiated ASCs onto various scaffolds and biomaterials. The biomaterials that rendered the most significant result are PGA [12], atelocollagen [44], and hydroxyapatite/tricalcium phosphate (HA-TCP) [45]. Although evidence to date suggests that ASCs may one day be useful in the treatment of difficult osseous repairs, further investigations are needed to determine their ultimate safety and efficacy in the clinic.

Cartilage, particularly articular, primarily serves a structural and mechanical function in the body. Clinical cartilage repair has remained an elusive goal for some time. Autologous [46] and allogeneic [47] chondrocyte transplants have been used successfully, but are limited by donor site morbidity and the slow repairs seen, respectively, with these approaches. Recognition of the chondrogenic differentiation potential seen in many stem cells has led to the exploration of an alternative source of cells. Chondrogenic potential, described in vitro in ASCs, includes evidence of cell condensation into nodules and the production of an extracellular matrix rich in proteoglycans and collagen type II [48–51]. It has been proved successful for a minimum period of 12 weeks when implanted with alginate constructs subcutaneously in nude mice [52]. A direct comparison of the in vitro chondrogenic potential of ASCs and BMSCs examined similarities in histological staining and gene expression. In vivo experiments using ASC spheroids had been identified and were successful at generating cartilage-like tissue [53]. Induced spheroids were implanted between two muscle bellies in immunodeficient mice. At 6 weeks, the implants were harvested and found to have produced a cartilage-like tissue consisting of cells within lacunae surrounded by a gel-like extracellular matrix in the absence of any fibrous network. Although this study demonstrates an in vivo potential for differentiation intramuscularly, no physiological models of cartilage repair have yet been tested. To this end, one laboratory is investigating what effects scaffold material [52, 53], oxygen tension [53], and media composition have on the biomechanical properties of ASC-seeded constructs [54]. Clear differences are seen depending on the combination of factors used; however, nothing yet approaches the mechanical properties of mature cartilage. Thus far, results suggest that future in vivo models may demonstrate a potential for ASCs to enhance the healing of debilitating osteochondral diseases. Repairs with the resilience necessary for weight-bearing joints, however, will probably be more difficult to develop. However, ASC seem to have a greater chondropotential effect, and further enhanced work on the same might improve engineering of cartilage in vitro as well in vivo.

Conclusion

Adipose tissue is an abundant, easily accessible, and reproducible cell source for musculoskeletal regenerative medicine applications, especially in conditions where BMSCs could not be obtained, as discussed in the previous chapter. The advances that have been observed with ASC have provided evidence of their great potential and applicability in cell therapy, as well as in the enhancement of healing process. Although its utility is gaining importance in preclinical studies and clinical trials, the factors that drive the repair processes and the details of underlying science continue to present major challenges.

References

1. Aust L, Devlin B, Foster SJ et al (2004) Yield of human adipose-derived adult stem cells from liposuction aspirates. Cytotherapy 6:7–14
2. Gimble JM (2003) Adipose tissue-derived therapeutics. Expert Opin Biol Ther 3:705–713
3. Gimble JM, Guilak F (2003) Adipose-derived adult stem cells: isolation, characterization, and differentiation potential. Cytotherapy 5:362–369
4. Gimble JM, Katz AJ, Bunnell BA (2007) Adipose-derived stem cells for regenerative medicine. Circ Res 100:1249–1260
5. Zuk PA, Zhu M, Mizuno H et al (2001) Multilineage cells from human adipose tissue: implications for cell-based therapies. Tissue Eng 7:211–228

6. Zuk PA, Zhu M, Ashjian P et al (2002) Human adipose tissue is a source of multipotent stem cells. Mol Biol Cell 13:4279–4295
7. Jurgens W, Oedayrajsingh-Varma M, Helder M et al (2008) Effect of tissue-harvesting site on yield of stem cells derived from adipose tissue: implications for cell-based therapies. Cell Tissue Res 332:415–426
8. Katz AJ, Tholpady A, Tholpady SS et al (2005) Cell surface and transcriptional characterization of human adipose-derived adherent stromal (hADAS) cells. Stem Cells 23:412–423
9. Mitchell JB, McIntosh K, Zvonic S et al (2006) Immunophenotype of human adipose-derived cells: temporal changes in stromal-associated and stem cell–associated markers. Stem Cells 24:376–385
10. Zhu Y, Song K, Fan X et al (2008) Adipose-derived stem cell: a better stem cell than BMSC. Cell Biochem Funct 26:664–675
11. De Ugarte DA, Morizono K, Elbarbary A et al (2003) Comparison of multi-lineage cells from human adipose tissue and bone marrow. Cells Tissues Organs 174:101–109
12. Lee JA, Parrett BM, Conejero JA et al (2003) Biological alchemy: engineering bone and fat from fat-derived stem cells. Ann Plast Surg 50(6):610–617
13. Wosnitza M, Hemmrich K, Groger A et al (2007) Plasticity of human adipose stem cells to perform adipogenic and endothelial differentiation. Differentiation 75:12–23
14. Rodriguez A-M, Elabd C, Delteil FDR et al (2004) Adipocyte differentiation of multipotent cells established from human adipose tissue. Biochem Biophys Res Commun 315:255–263
15. Rodriguez AM, Elabd C, Amri E-Z et al (2005) The human adipose tissue is a source of multipotent stem cells. Biochimie 87:125–128
16. Elabd C, Chiellini C, Massoudi A et al (2007) Human adipose tissue-derived multipotent stem cells differentiate in vitro and in vivo into osteocyte-like cells. Biochem Biophys Res Commun 361:342–348
17. Bruder SP, Jaiswal N, Haynesworth SE (1997) Growth kinetics, self-renewal, and the osteogenic potential of purified human mesenchymal stem cells during extensive subcultivation and following cryopreservation. J Cell Biochem 64:278–294
18. Ogawa R, Mizuno H, Hyakusoku H et al (2004) Chondrogenic and osteogenic differentiation of adipose-derived stem cells isolated from GFP transgenic mice. J Nippon Med Sch 71(4):240–241
19. Halvorsen YD, Franklin D, Bond AL et al (2001) Extracellular matrix mineralization and osteoblast gene expression by human adipose tissue-derived stromal cells. Tissue Eng 7(6):729–741
20. Dhanasekaran M, Indumathi S, Rashmi M, Rajkumar JS, Sudarsanam D (2012) Unravelling the retention of proliferation ad differentiation potency in extensive culture of human subcutaneous fat-derived mesenchymal stem cells in different media. Cell Prolif 45(6):516–526
21. Maddox JR, Liao X, Li F, Niyibizi C (2009) Effects of culturing on the stability of the putative murine adipose derived stem cells markers. Open Stem Cell J 1:54–61
22. Kim WS, Park BS, Kim HK et al (2008) Evidence supporting antioxidant action of adipose-derived stem cells: protection of human dermal fibroblasts from oxidative stress. J Dermatol Sci 49(2):133–142
23. Kim WS, Park BS, Park SH et al (2009) Antiwrinkle effect of adipose-derived stem cell: activation of dermal fibroblast by secretory factors. J Dermatol Sci 53(2):96–102
24. Kim WS, Park SH, Ahn SJ et al (2008) Whitening effect of adipose-derived stem cells: a critical role of TGF-β1. Biol Pharm Bull 31(4):606–610
25. Gir P, Brown SA, Oni G et al (2012) Fat grafting: evidence-based review on autologous fat harvesting, processing, reinjection, and storage. Plast Reconstr Surg 130:249–258
26. Minn KW, Min KH, Chang H et al (2010) Effects of fat preparation methods on the viabilities of autologous fat grafts. Aesthetic Plast Surg 34(5):626–631
27. Yoshimura K, Sato K, Aoi N et al (2008) Cell-assisted lipotransfer for cosmetic breast augmentation: supportive use of adipose-derived stem/stromal cells. Aesthetic Plast Surg 32:48–55
28. Kamakura T, Ito K (2011) Autologous cell-enriched fat grafting for breast augmentation. Aesthetic Plast Surg 35:1022–1030
29. Wang L, Lu Y, Luo X et al (2012) Cell-assisted lipotransfer for breast augmentation: a report of 18 patients. Zhonghua Zheng Xing Wai Ke Za Zhi 28:1–6
30. Yoshimura K, Asano Y, Aoi N et al (2010) Progenitor-enriched adipose tissue transplantation as rescue for breast implant complications. Breast J 16:169–175
31. Kim M, Kim I, Lee SK et al (2011) Clinical trial of autologous differentiated adipocytes from stem cells derived from human adipose tissue. Dermatol Surg 37:750–759
32. Bonnemann CG (2011) The collagen VI-related myopathies: muscle meets its matrix. Nat Rev Neurol 7:379–390
33. Meregalli M, Farini A, Parolini D, Maciotta S, Torrente Y (2010) Stem cell therapies to treat muscular dystrophy: progress to date. BioDrugs 24:237–247
34. Hattori H, Sato M, Masuoka K et al (2004) Osteogenic potential of human adipose tissue-derived stromal cells as an alternative stem cell source. Cells Tissues Organs 178(1):2–12
35. Ogawa R, Mizuno H, Hyakusoku H et al (2004) Chondrogenic and osteogenic differentiation of adipose-derived stem cells isolated from GFP transgenic mice. J Nippon Med Sch 71(4):240–241
36. Dezawa M, Ishikawa H, Itokazu Y et al (2005) Bone marrow stromal cells generate muscle cells and repair muscle degeneration. Science 309(5732): 314–317

37. Fraser JK, Wulur I, Alsonso Z, Hedrick M (2006) Fat tissue: an underappreciated source of stem cells for biotechnology. Trends Biotechnol 24:150–154
38. Bacou F, El Andalousi RB, Daussin PA et al (2004) Transplantation of adipose tissue-derived stromal cells increases mass and functional capacity of damaged skeletal muscle. Cell Transplant 13(2): 103–111
39. Alexeev V, Arita M, Donahue A, Bonaldo P, Chu M-L, Igoucheva O (2014) Human adipose-derived stem cell transplantation as a potential therapy for collagen VI-related congenital muscular dystrophy. Stem Cell Res Ther 5:21
40. Wadagaki R, Mizuno D, Yamawaki-Ogata A et al (2011) Osteogenic induction of bone marrow-derived stromal cells on simvastatin-releasing, biodegradable, nano- to microscale fiber scaffolds. Ann Biomed Eng 39:1872–1881
41. Cowan CM, Shi YY, Aalami OO et al (2004) Adipose-derived adult stromal cells heal critical-size mouse calvarial defects. Nat Biotechnol 22:560–567
42. Haynesworth SE, Goshima J, Goldberg VM et al (1992) Characterization of cells with osteogenic potential from human marrow. Bone 13:81–88
43. Lendeckel S, Jodicke A, Christophis P et al (2004) Autologous stem cells (adipose) and fibrin glue used to treat widespread traumatic calvarial defects: case report. J Craniomaxillofac Surg 32:370–373
44. Hattori H, Sato M, Masuoka K et al (2004) Osteogenic potential of human adipose tissue-derived stromal cells as an alternative stem cell source. Cells Tissues Organs 178(1):2–12
45. Hicok KC, Du Laney TV, Zhou YS et al (2004) Human adipose-derived adult stem cells produce osteoid in vivo. Tissue Eng 10(3–4):371–380
46. Tissue Genesis Cell Isolation System. Tissue Genesis Incorporated. February 10th, 2009. Available at: http://www.tissuegenesis.com/
47. Cytori's Celution® 700 System to be Regulated as a Medical Device by U.S. FDA. 20 Jul 2009. Available at: http://www.medicalnewstoday.com/articles/158091.php
48. Jaiswal RK, Jaiswal N, Bruder SP et al (2000) Adult human mesenchymal stem cell differentiation to the osteogenic or adipogenic lineage is regulated by mitogen-activated protein kinase. J Biol Chem 275:9645–9652
49. Gonzalez-Rey E, Gonzalez MA, Varela N et al (2010) Human adipose-derived mesenchymal stem cells reduce inflammatory and T cell responses and induce regulatory T cells in vitro in rheumatoid arthritis. Ann Rheum Dis 69:241–248
50. Fraser JK, Wulur I, Alfonso Z et al (2007) Differences in stem and progenitor cell yield in different subcutaneous adipose tissue depots. Cytotherapy 9:459–467
51. Dragoo JL, Samimi B, Zhu M et al (2003) Tissue-engineered cartilage and bone using stem cells from human infrapatellar fat pads. J Bone Joint Surg Br 85(5):740–747
52. Guilak F, Awad HA, Fermor B et al (2004) Adipose-derived adult stem cells for cartilage tissue engineering. Biorheology 41(3-4):389–399
53. Awad HA, Wickham MQ, Leddy HA et al (2004) Chondrogenic differentiation of adipose-derived adult stem cells in agarose, alginate, and gelatin scaffolds. Biomaterials 25(16):3211–3222
54. Fang B, Song Y, Liao L et al (2007) Favorable response to human adipose tissue-derived mesenchymal stem cells in steroid-refractory acute graft-versus-host disease. Transplant Proc 39:3358–3362

Part V

Stem Cells and Organ Systems: Hematopoietic System

Fetal Blood Research: A Brief Perspective on the Present, Past, and Future of This Exciting Field

Niranjan Bhattacharya and Sanjukta Bhattacharya

Cord Blood Transfusion and Its Potential

All over the world, millions of people are saved every year as a result of blood transfusions. At the same time, many, particularly in developing countries, still die because of an inadequate supply of safe blood and blood products. A reliable supply of safe blood is essential to improve health standards at several levels, especially among women and children, and particularly in the poorer sections of society anywhere in the world. Half a million women still die of complications related to pregnancy and childbirth, and 99 % of these are in developing countries. Hemorrhage accounts for 25 % of the complications and is the most common cause of maternal death. Malnutrition, thalassemia, and severe anemia are prevalent diseases in children which require blood transfusion, apart from other complicated diseases.

Globally, over 80 million units of blood are collected every year, but the tragedy is that only 39% of this is collected in the developing world which contains 82 % of the global population.

There are about 100 million births in the world per year in a conservative estimate. In India alone, there are more than 20 million births or production of placenta. One of the products of the placenta is cord blood with its immense potentials. The placenta is a complex organ that regulates feto-maternal interactions. Many cytokines that influence the lymphohematopoietic environment are produced in the placenta in abundance. Therefore, placental umbilical cord blood contains beneficial substances other than the contents of normal adult blood. Further, because the fetus grows into a neonate within the safe confines of the womb where the placenta plays an important role in ensuring security, placental cord blood may be assumed to be safe due to the molecular screening intrinsic to the functional barrier of a healthy placenta.

We can estimate a production of 8,785,000 l of cord blood in the world per year on the basis of average 84–90 ml/placenta collection of cord whole blood. Transfusion of ABO screened and matched (HLA) randomized fetal blood has been found to be safe in anemia due to malaria, diabetes, thalassemia, leprosy, rheumatoid arthritis, tuberculosis, malignancy, and AIDS, only to name a few diseases that can cause anemia. Our group of medical scientists and clinicians began cord blood transfusion with assistance from the Department of Science and Technology Government of West Bengal, India [1–12].

N. Bhattacharya (✉)
Department of Regenerative Medicine and Translational Science, Calcutta School of Tropical Medicine and STM Cord Blood Bank, Kolkata, India
e-mail: sanjuktaniranjan@gmail.com

S. Bhattacharya
Department of International Relations, Jadavpur University, Kolkata, India

I. Somasundaram (ed.), *Stem Cell Therapy for Organ Failure*,
DOI 10.1007/978-81-322-2110-4_10, © Springer India 2014

Cord Blood and Stem Cell

Today, a microscopic section of the blood's mononuclear cells (.01 % nucleated cells) is used for transplantation purposes; the rest, i.e., 99.99 % goes to the trash/incinerator. The blood volume of a fetus at term is 80–85 ml/kg. The placental vessel at term contains approximately 150 ml of cord blood. Cord blood contains three types of hemoglobin, HbF, HbA, and HbA2, of which HbF constitutes the major fraction. HbA accounts for 15–40 % and HbA2 is present only in trace amounts at birth. HbF, which is the major component, has a greater oxygen-binding affinity than HbA.

Hematopoietic stem cells from cord blood are now harvested in many laboratories all over the world and stored in cord blood banks. Cord blood is globally known for its role as a source of hematopoietic cells, but apart from that, cord blood also contains potent angiogenesis-stimulating cells. $CD34^+$, $CD11b^+$ fraction is approximately less than half of the $CD34C^+$ fraction of cord blood. This was demonstrated to possess the ability to differentiate into functional endothelial cells in vitro and in vivo [13]. In addition, there are some mesenchymal stem cells in the cord blood which are classically defined as adherent to plastic and expressing a non-hematopoietic cell surface phenotype, consisting of $CD34^-$, $CD45^-$, and $HLA\text{-}DR^-$, while possessing markers such as STRO-1, VCAM, CD13, CD29, CD44, CD90, CD105, SH-3, and STRO-1 [14]. In addition cord blood cells with markers and activities resembling embryonic stem cells have been found [15]. Investigators have identified a population of $CD34^-$ cells expressing OCT-4, Nanog, SSEA-3, and SSEA-4, which could differentiate into cells of the mesoderm, ectoderm, and endoderm lineage [15].

The first widespread utilization of cord blood as a stem cell source was in the treatment of pediatric hematological malignancies after myeloablative conditioning. Outside the area of oncology, the clinical use of cord blood has expanded into various areas that range from reconstituting a defective immune system to correcting congenital hematological abnormalities, to inducing angiogenesis. In addition to current clinical use, cord blood is currently under intense experimental investigation in preclinical models of pathophysiologies that range from myocardial ischemia to stroke, to muscle regeneration [16–18] apart from indications of transfusion to combat anemia.

Differences in Adult RBC and Cord Blood RBC

The red cell collected from the newborn's cord blood differs from the adult RBC in many ways, viz., there is an increase of the immunoreactive myosin in red cell membrane [19] and the total value of lipid, phospholipid, and cholesterol are more in cord blood red cell than in adults [20]. Even the antigen expression of cord blood RBC differs from the adult RBC. A, B, S, and Lutheran antigens are expressed in lesser amount in cord blood than the adult blood; in addition there is a complete absence of Lewis antigen in the cord blood [21]. There is also fundamental metabolic difference between the cord blood and the adult blood, for example, the activities of phosphoglycerate kinase, enolase, glyceraldehyde-3-phosphate dehydrogenase, glucose phosphate isomerase, etc., of the Embden-Meyerhof pathway are definitely increased in cord blood [22] and even the non-glycolytic enzymes like carbonic anhydrase and acetylcholine esterase are distinctly different from the adult blood [23].

Future Research in Cord Blood

Cellular and acellular constituents of cord blood have potential for clinical applications. Recent advances in biology and medicine have introduced new technologies to study the mechanisms of genetic switching of the hemoglobin chain from alpha to beta during human fetal development, the site of hematopoiesis during fetal development, and its change from yolk sac eventually to bone marrow. This raises a number of pertinent questions: do the developing hematopoietic stem cells come from the same origin or from different sites during the shift of the place off its synthesis and turnover? What is the exact role of the stroma as a hematopoietic organ, and what is its interaction with hematopoietic progenitor cells still, remains an unsolved mystery.

Why do hematopoietic stem cells home to particular site in cases of amphibians, birds, and mammals? Knowledge and understanding of these issues may lead to the development of animal models, successful therapies, and novel tools to characterize clinical states and provide better care to patients.

We invite investigators to contribute to the understanding of the molecular mechanisms underlying the immunomodulation capability of cord blood, as well as the development of strategies to use this immunomodulation in clinical practice, and the evaluation of outcomes of the new modalities for the characterization of the components of fetal blood and the placenta and their use in therapy and in means for measuring outcomes from treatment trials. Pregnancy cytokines provide new insights into understanding the expression of different antigens, their presence or absence. Regulations using animal models, like human hematopoiesis in animal hosts following xenograft in SCID mouse system and fetal sheep systems, also appear to be very exciting. Similarly, the therapeutic potential of nucleated RBC, CFU, and Gower 1 and 2 hemoglobin collected from the developing human fetus, if applied in human and animal system to combat refractory anemia, may have fruitful clinical implication for futuristic medicine. These are areas that may be explored in the coming years.

Future research in this field may focus on the following aspects in particular:

1. Recent developments in autologous and allogenic cord blood transfusion from the pediatric to the geriatric group
2. Advances in serum constituent characterization therapy using cord blood serum, viz., cord blood biomarkers, IL-1 beta, IL-6, and IL-8 are selectively associated with fetal infection. These markers may be clinically useful indicators of extensive intrauterine infection associated with poor neonatal outcome.
3. Role of fetal blood in suppression of inflammation and role of cytokines like IL-3, G-CSF, M-CSF4, and GM-CSF in immunomodulation.
4. The mechanism of inflammation of the growing fetal blood using animal model or some other model.
5. Role of modifiers of inflammation contained (regulatory t cells)) in fetal blood
6. Emergency use of fetal blood in nuclear radiation disaster like the scenario in Japan.
7. Use of fetal blood in different indications as blood substitute when there is more demand than supply of blood for transfusion, as in the war against terror and even in peacetime emergencies (earthquake, Tsunami, others).
8. Use of serum from cord blood for any purpose, viz., cord blood serum to treat corneal xerosis and ulceration.
9. Antigen expression and metabolic differences of cord blood and adult blood with special reference to clinical implications.
10. Comparison of cord blood preservation and its functional variation and comparison with adult blood.
11. Adverse outcomes after cord blood transfusion or its constituent therapy, if it happens, how to prevent and combat the situation.

References

1. Bhattacharya N (2006) Placental umbilical cord blood transfusion: a new method of treatment of patients with diabetes and microalbuminuria in the background of anemia. Clin Exp Obstet Gynecol 33(3):164–168
2. Bhattacharya N (2006) Placental umbilical cord blood transfusion: a novel method of treatment of patients with malaria in the background of anemia. Clin Exp Obstet Gynecol 33(1):39–43
3. Bhattacharya N (2006) A preliminary study of placental umbilical cord whole blood transfusion in under resourced patients with malaria in the background of anaemia. Malar J 5:20
4. Bhttacharya N (2006) Placental umbilical cord whole blood transfusion to combat anemia in the

background of advanced rheumatoid arthritis and emaciation and its potential role as immunoadjuvant therapy. Clin Exp Obstet Gynecol 33(1):28–33
5. Bhattacharya N (2006) Transient spontaneous engraftment of CD34 hematopoietic cord blood stem cells as seen in peripheral blood: treatment of leprosy patients with anemia by placental umbilical cord whole blood transfusion. Clin Exp Obstet Gynecol 33(3):159–163
6. Bhattacharya N (2006) Placental umbilical cord whole blood transfusion to combat anemia in the background of tuberculosis and emaciation and its potential role as an immuno-adjuvant therapy for the under-resourced people of the world. Clin Exp Obstet Gynecol 33(2):99–104
7. Bhattacharya N (2006) A preliminary report of 123 units of placental umbilical cord whole blood transfusion in HIV-positive patients with anemia and emaciation. Clin Exp Obstet Gynecol 33(2):117–121
8. Bhattacharya N (2005) Placental umbilical cord blood transfusion in transfusion-dependent beta thalassemic patients: a preliminary communication. Clin Exp Obstet Gynecol 32(2):102–106
9. Bhattacharya N (2006) A study of placental umbilical cord whole blood transfusion in 72 patients with anemia and emaciation in the background of cancer. Eur J Gynaecol Oncol 27(2):155–161
10. Bhattacharya N (2005) Placental umbilical cord whole blood transfusion: a safe and genuine blood substitute for patients of the under-resourced world at emergency. J Am Coll Surg 200(4): 557–563
11. Bhattacharya N (2006) Spontaneous transient rise of CD34 cells in peripheral blood after 72 hours in patients suffering from advanced malignancy with anemia: effect and prognostic implications of treatment with placental umbilical cord whole blood transfusion. Eur J Gynaecol Oncol 27(3):286–290
12. Bhattacharya N et al (2001) A study report of 174 units of placental umbilical cord whole blood transfusion in 62 patients as a rich source of fetal hemoglobin supply in different indications of blood transfusion. Clin Exp Obstet Gynecol 28(1):47–52
13. Hildbrand P, Cirulli V, Prinsen RC, Smith KA, Torbett BE, Salomon DR, Crisa L (2004) The role of angiopoietins in the development of endothelial cells from cord blood CD34+ progenitors. Blood 104:2010–2019. doi:10.1182/blood-2003-12-4219
14. De Ugarte DA, Alfonso Z, Zuk PA, Elbarbary A, Zhu M, Ashjian P, Benhaim P, Hedrick MH, Fraser JK (2003) Differential expression of stem cell mobilization-associated molecules on multi-lineage cells from adipose tissue and bone marrow. Immunol Lett 89:267–270. doi:10.1016/S0165-2478(03)00108-1
15. Zhao Y, Wang H, Mazzone T (2006) Identification of stem cells from human umbilical cord blood with embryonic and hematopoietic characteristics. Exp Cell Res 312:2454–2464
16. Brzoska E, Grabowska I, Hoser G, Streminska W, Wasilewska D, Machaj EK, Pojda Z, Moraczewski J, Kawiak J (2006) Participation of stem cells from human cord blood in skeletal muscle regeneration of SCID mice. Exp Hematol 34:1262–1270. doi:10.1016/j.exphem.2006.05.009
17. Hu CH, Wu GF, Wang XQ, Yang YH, Du ZM, He XH, Xiang P (2006) Transplanted human umbilical cord blood mononuclear cells improve left ventricular function through angiogenesis in myocardial infarction. Chin Med J (Engl) 119:1499–1506
18. Leor J, Guetta E, Feinberg MS, Galski H, Bar I, Holbova R, Miller L, Zarin P, Castel D, Barbash IM, Nagler A (2006) Human umbilical cord blood-derived CD133+ cells enhance function and repair of the infarcted myocardium. Stem Cells 24:772–780. doi:10.1634/stemcells.2005-0212
19. Matovcik LM, Groschel-Stewart U, Schrier SL (1986) Myosin in adult and human erythrocyte membrane. Blood 67:1668
20. Tuan D, Feingold E, Newman M, Weissman SM, Forget BG (1983) Different 3′ end points of deletions causing delta beta-thalassemia and hereditary persistence of fetal hemoglobin: implications for the control of gamma-globin gene expression in man. Proc Natl Acad Sci U S A 80(22):6937–6941
21. Marsh WL (1987) Erythrocytes' blood groups in human. In: Nathan DG, Oski FA (eds) Hematology of infancy and childhood, 3rd edn. W. B. Saunders, Philadelphia
22. Travis SF, Kumar SP, Paez PC, Delivoria-Papadopoulos M (1980) Red cell metabolic alterations in postnatal life in term infants: glycolytic enzymes and glucose-6-phosphate dehydrogenase. Pediatr Res 14(12):1349–1352
23. Stevenson SS (1943) Carbonic anhydrase in new born infants. J Clin Invest 22:403

An Outline of Haematopoietic Stem Cell Transplantation

Revathy Raj

Introduction

The world's first successful human bone marrow transplantation was performed by Dr. E. Donnall Thomas of New York between two HLA identical twins in 1956. In 1968, Dr. Robert A. Good performed the first successful sibling transplant at the University of Minnesota. The recipient was a 4-month-old boy who had inherited severe combined immunodeficiency syndrome ("bubble baby syndrome"). He was transfused stem cells from his donor, his 8-year-old HLA-matched sister. The disease had previously killed 11 male children in the boy's family.

Haematopoietic stem cell transplantation (HSCT) is currently the treatment of choice for haematological disorders such as high risk leukaemias that cannot be treated with chemotherapy alone, primary immunodeficiency and genetic diseases such as thalassaemia major, sickle cell anaemia and Hurler syndrome. A basic outline of haematopoietic stem cell transplantation has been provided in this chapter.

R. Raj (✉)
Hematopoietic Stem Cell Transplantation Unit, Apollo Speciality Hospitals, 320 Padma Complex, Anna Salai, Chennai - 600035, India
e-mail: revaraj@yahoo.com

What Are Haematopoietic Stem Cells?

Haematopoietic stem cells make all cells of the blood, namely, red, white and platelet series every day of our life. Umbilical cord blood is a rich source of haematopoietic stem cells from a new born baby used for transplantation.

What Are the Types of Transplantation?

Transplants can be classified into several types based on the source of stem cells used:

1. In Autologous stem cell transplantation the patient acts as his own stem cell donor.
2. In Syngeneic transplantation the donor is an identical twin.
3. In Allogeneic transplantation the donor is a histocompatible sibling or family member (related) or from a donor registry (unrelated).

What Are the Common Indications for Allogeneic HSCT?

1. *Haematological malignancies*
   - Acute myeloid leukaemia [1–3] – presence of >5 % blasts after 1st course of chemotherapy, absence of favourable genetic abnormalities, presence of adverse genetic abnormalities, e.g. −5, −7, del (5q)

I. Somasundaram (ed.), *Stem Cell Therapy for Organ Failure*, DOI 10.1007/978-81-322-2110-4_11, © Springer India 2014

- Acute lymphoblastic leukaemia [4, 5] – high risk cases in first remission, relapsed leukaemia in second remission
- Juvenile chronic myeloid leukaemia (JCML) and juvenile myelomonocytic leukaemia [6] (JMML)

2. *Bone marrow failure – acquired and congenital*
   - Severe aplastic anaemia
   - Diamond-Blackfan anaemia
   - Fanconi's anaemia
   - Myelodysplastic syndromes
   - Shwachman-Diamond syndrome
   - Kostmann syndrome
3. *Genetic Disorders*
   - Immunodeficiency disorders like severe combined immunodeficiency, Wiskott-Aldrich syndrome and chronic granulomatous disease
   - Platelet disorders like Glanzmann's thrombasthenia
   - Red cell disorders like thalassaemia major and sickle cell anaemia
   - Lysosomal storage disorders such as Hurler syndrome and adrenoleukodystrophy
   - Osteoclast defect such as osteopetrosis
   - Macrophage defect such as haemophagocytic lymphohistiocytosis

How Is a Donor Chosen for Allogeneic Stem Cell Transplantation?

Choice and Assessment of the Donor

The first step is to perform HLA typing on the patient, sibling and parents as HLA matching is the key to successful transplantation. Transplantation is performed only if there is a full six antigen HLA A, B and DR matched related donor available. If there are no matched siblings who can serve as a donor, then an extended family search can be done in which parents or cousins are tested. The probability of finding a donor in a family is about 35 % and it may be higher if there is consanguinity. HSCT can be done even if the donor and recipient are not ABO blood group matched unlike solid organ transplants. We are now looking at ten antigens – HLA A, B, C, DRB1 and DQB1 before proceeding with transplantation.

HLA Typing

Encoded by a cluster of genes on the short arm of chromosome six, the Human Leukocyte Antigens play a central role in processes of recognition and self versus non-self discrimination. There are three classes of HLA genes:

- HLA class I genes which code for HLA A, B and C antigens, which are found on virtually all nucleated cells in the body
- HLA class IT genes which code for the DR, DQ and DP antigens, expressed primarily on cells like macrophages, dendritic cells and B lymphocytes
- HLA Class III genes which encode complement components and other molecules like tumour necrosis factor, not directly involved in immune recognition

Each biologic parent contributes a haplotype and two haplotypes constitute a genotype. Donors can be chosen only if they are a full match. HLA typing plays a pivotal role in graft rejection and graft versus host disease (Fig. 1).

- Unrelated donor banks are searched if there are no family donors. These stem cell donors are registered in various countries like the USA (NMDP), Germany (DKMS), and India (DATRI). In addition, public cord blood banks can also be used as a ready source of stem cells for an urgent transplantation.

What Is the Process of Transplantation for a Patient?

Admission to BMT Unit

Once HLA matching is done and patient and donor are fit to proceed, the patient is admitted to the transplant ward. A dual lumen Hickman

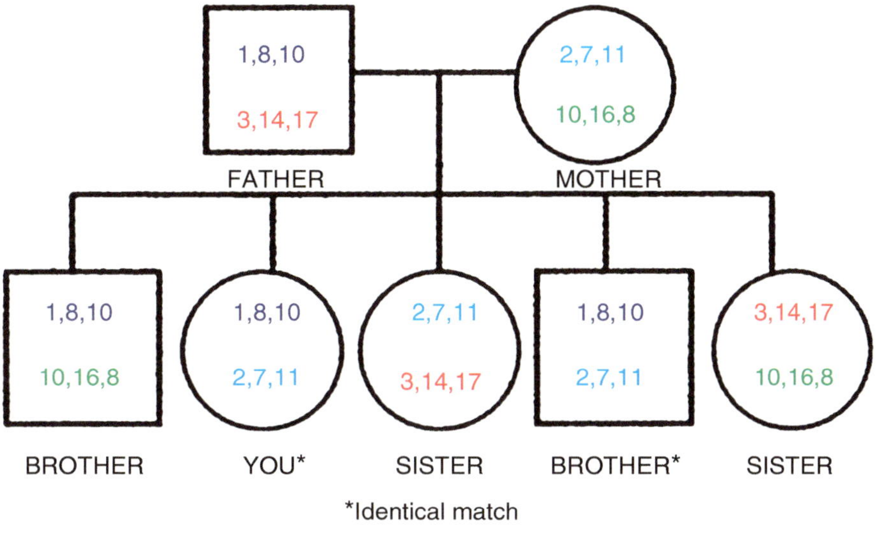

Fig. 1 HLA typing of a family

catheter is inserted under anaesthesia to help with blood tests, transfusions and administration of medication and fluids. The bone marrow transplant unit is a positive pressure HEPA (high efficiency particulate air)-filtered unit. HEPA filters are high efficiency filters with a size of 0.3 μm with a trapping efficiency of 99.97 %. The cooling unit is placed outside the sterile room, and the cooled air is allowed to enter into the room only after passing through these filters so that the air in the rooms is sterile. Additionally, air changes occur at the rate of 15–20 per hour so that the air blown in remains in the room for only 3–4 min. Since the rooms are under positive pressure, the air is pushed out into an adjoining room/sent out through vents, thus not allowed to recirculate in the room. The air quality is monitored using air quality monitors and air pressures are assessed using a pressure-testing device. Settle plate cultures are also done every month to look for microbiological contamination. The patient also needs one to one nursing to prevent infections during the period of prolonged neutropenia.

Conditioning

The patient needs to be prepared to receive new stem cells and the treatment given the week before stem cell infusion is called conditioning. Conditioning permits engraftment of donor stem cells and involves the creation of space by destroying the patient's own marrow (cytoreduction) and profound immunosuppression to prevent rejection. The choice of conditioning regimen for a given child is dictated by the underlying disease and donor characteristics. Chemotherapeutic agents [7] like busulfan orally at a dose of 16 mg/kg over 4 days followed by cyclophosphamide at 50 mg/kg for 2–4 days have been used as a standard conditioning regimen for benign disorders. Total body irradiation 12 Gy over 4 days with cyclophosphamide is used as conditioning in patients with leukaemia. If there is a high risk of rejection such as multiply transfused children with thalassaemia or aplastic anaemia, antilymphocyte globulin may be added to the conditioning regimen. In some diseases such as Fanconi's anaemia and myelodysplastic syndromes, reduced intensity conditioning transplants are being increasingly used. Drugs such as fludarabine are used to suppress the patient's immune system enough to allow engraftment of donor cells without use of megadose chemotherapy. Reduced intensity conditioning reduces immediate mortality and late effects and seems the way forward for transplanting children. However, graft rejection and relapse rates may increase with reducing conditioning drugs.

Collection of Stem Cells from the Donor

The donor is admitted a day before collection of stem cells. Under general anaesthesia, bone marrow is aspirated from the iliac bone using a bone marrow harvest needle in small aliquots of 5–10 ml. The volume of marrow harvested is dependent on the weight of the recipient. The total nucleated cell dose required is 300 million marrow cells/kg (3×10^8/kg) of the recipient. On average, 10 ml/kg of the recipient body weight is harvested to obtain adequate stem cells. When the donor is much smaller than the recipient, he/she may need a top up transfusion to be able to donate this volume of bone marrow. The donor is discharged the day after the harvest.

Stem cells can also be collected from the donor using an apheresis machine after administration of G-CSF for 4–5 days, and this is called a peripheral blood stem cell transplant [PBSCT] [8, 9].

Once the stem cells are harvested, they are directly infused into the child like a blood transfusion. If the blood groups are different, the cells may need to be red cell depleted and/or plasma depleted before infusion.

Stem Cell Infusion and Supportive Care

The harvested marrow looks just like blood and is infused into the child like a blood transfusion. Following infusion of donor marrow, the stem cells home in to the child's marrow and a rising white cell count after 15 days is evidence of engraftment. Granulocyte colony-stimulating factors (G-CSF, GM-CSF) are used to hasten engraftment in sick children. Supportive care required during these 2–4 weeks is of paramount importance and optimal care a difference and reduces transplant related mortality.Aggressive blood component support is crucial during the period of aplasia, whilst awaiting marrow recovery. Red cell transfusions are given to keep the haemoglobin levels above 9 g% and platelets transfusions given to maintain a platelet count above 10×109/L. All blood products are irradiated to prevent transfusion-associated GVHD [10]. Nutritional support to children who cannot eat due to nausea and mucositis from conditioning chemotherapy and antibiotic use as per hospital infection control policy is also required.

Engraftment

The patient is considered to be a full chimera if all the cells in the blood are donor in origin and a mixed chimera if residual host cells are present. Full donor chimerism is achieved 30–45 days following BMT if a myeloablative conditioning protocol is used. The child is transferred out of the BMT unit when his neutrophil count is over 500/cumm and discharged when there are no intravenous medications need to be administered.

What Are the Potential Complications Following Haematopoietic Stem Cell Transplantation?

Acute Complications

Acute graft versus host disease
Infections
Graft rejection
Sinusoidal obstruction syndrome
Haemorrhagic cystitis
Interstitial pneumonia

Chronic Complications

Chronic graft versus host disease
Relapse
Sterility
Cataract
Second malignancy

Graft Versus Host Disease (GVHD)

This is one of the most devastating complications of BMT and is termed acute if it occurs in the first 100 days following transplantation and chronic if it continues or develops after this period. Despite a six-antigen HLA match, the donor T lymphocytes on entry into the patient may recognise minor differences in the recipient's cells, recognising them as foreign, and attack the recipient. Any organ in the body may be affected, but the skin, intestine and liver usually bear the brunt of the attack by the donor lymphocytes in acute GVHD. Generalised erythema, maculopapular rash involving palms and soles and in severe cases toxic epidermal necrolysis (TEN) are the main cutaneous manifestations. Intestinal involvement usually manifests as diarrhoea, whilst upper gastrointestinal involvement may manifest with vomiting and loss of appetite. Progressive jaundice with minimal elevation of enzymes is seen in hepatic GVHD. Severe GVHD is associated with increased propensity to infections particularly with cytomegalovirus (CMV). Acute GVHD develops in about 30 % of patients transplanted with HLA identical marrow and may be an indirect cause of death in 20–30 % of affected individuals. Acute GVHD is graded 1–4 based on the degree of target organ involvement. Increased host age and transplants from female donors to male recipients particularly if the female is multiparous or has been transfused greatly increases the risk of GVHD.

Whilst acute GVHD presents as an inflammatory cytokine storm, chronic GVHD produces a picture similar to scleroderma. In the skin, manifestations range from dry patches or areas of variegated pigmentation to extensive dermal scarring that produces thickened atrophic skin. Gastrointestinal tract involvement may result in lichenoid lesions in the oral mucosa, xerostomia, dysphagia, diarrhoea or malabsorption. Chronic GVHD of the liver usually presents as a cholestatic process, which can progress to a syndrome similar to primary biliary cirrhosis. Other manifestations include sicca syndrome, pulmonary dysfunction and development of autoantibodies. Chronic GVHD is graded as either limited or extensive.

Prevention of GVHD is included in standard conditioning protocols with a combination of 2 drugs, methotrexate and cyclosporine. Methotrexate is given as an intravenous push in 3–4 doses over the first 10 days following infusion of stem cells. Cyclosporine is continued for a period of 6–12 months depending upon the disease and the presence of GVHD. This is unlike sold organ transplantation where the drug has to be continued lifelong. Once GVHD develops, the main stay of treatment is with corticosteroids. In refractory cases, other immunosuppressive agents like antithymocyte globulin (ATG), OKT3, mycophenolate, sirolimus and IL2 antibody, anti-CD20 antibody or TNF alpha blockade agents may be used. Chronic GVHD is treated with prednisolone along with cyclosporine.

Infections

In the initial period following transplantation, profound neutropenia, disruption of anatomic barriers secondary to mucositis and presence of central venous catheters are important risk factors resulting in bacterial and disseminated fungal infections with aspergillus and candida. At the earliest sign of infection or fever, the patient is started on intravenous antibiotics, usually a third- or fourth-generation cephalosporin to cover gram-negative organisms and a glycopeptide such as vancomycin or teicoplanin to cover gram-positive organisms. If there is persistent fever, carbapenems such as imipenem or meropenem are added. Antifungal agents like amphotericin and caspofungin are routinely administered even if fungal infection is not documented if there is persistent fever beyond 72 h of antibiotic coverage. Viral infections with herpes, CMV and adenoviruses contribute to the morbidity and mortality of transplantation after engraftment. The presence of GVHD, high intensity of immunosuppression and the use of drugs like fludarabine and antithymocyte globulin as part of conditioning are the risk factors for cytomegalovirus (CMV) and adenovirus infections. CMV infection commonly targets lungs, liver and intestine. Rapid early detection tests using PCR and prophylactic administration of ganciclovir to high-risk patients reduce the incidence of CMV infection. The most common manifestation of adenovirus infections is haemorrhagic cystitis, gastroenteritis, pneumonia and liver failure, and treatment consists of use of cidofovir. Patients with chronic GVHD are immunosuppressed and remain at risk for infections with encapsulated organisms such as pneumococcus and meningococcus. Prophylactic penicillin or trimethoprim sulpha for pneumocystis jirovecii prevention in the first-year posttransplantation is mandatory to prevent sepsis.

Interstitial Pneumonia

This is a complication observed in 30 % of patients where radiation is used for the conditioning and can be crippling. Busulfan used as a conditioning agent can also cause lung damage. GVHD, CMV and pneumocystis jirovecii are other agents that cause an interstitial pneumonia.

Regimen-Related Toxicity (RRT)

Complications seen primarily as a result of conditioning agents are termed as regimen-related toxicity.

Sinusoidal obstruction syndrome of the liver (SOS) was previously called venoocclusive disease (VOD) [11, 12]. This is a complication seen with conditioning regimens containing busulfan and cyclophosphamide and in patients heavily pretreated with chemotherapy or have iron overload in the liver. It is characterised by weight gain, ascites and tender hepatomegaly which appears in the first 1–3 weeks after transplant. If severe, it can progress to hepatic failure and death. Treatment is supportive with careful fluid management. Defibrotide is the drug of choice in established SOS. Prophylactic use of ursodeoxycholic acid has shown benefit.

Haemorrhagic cystitis is a well-known complication of cyclophosphamide but can also arise from infection with adenovirus or BK virus [13]. This is managed with supportive care with hydration and pain relief. Severe cases may require continuous bladder irrigation, cystoscopy and clot evacuation.

Graft Failure

Graft failure is defined as failure to achieve neutrophil recovery (ANC > 500) by day +30 after transplantation. It results from eradication of the incoming donor cells by recipient's immune system. Risk factors include inadequate conditioning, inadequate stem cell dose, T cell depletion and HLA disparity between donor and recipient.

What Is Autologous Stem Cell Transplantation?

Dose intensification of chemotherapeutic agents increases the response rate of chemosensitive tumours. However, haematopoietic toxicity is a limiting factor, and harvesting haematopoietic stem cells and then cryopreserving and reinfusing them after the dose of chemotherapy that would otherwise be lethal or require a prolonged period of recovery can help overcome this problem. This is the principle of autologous transplantation in solid tumours. The outcome of autograft is closely correlated with tumour burden at the time of transplantation. The current indications for autograft in paediatrics are as follows:

Neuroblastoma Stage 4 disease in children over 1 year of age and at relapse

Soft tissue sarcoma Primary refractory disease, stage 4 disease at diagnosis

Ewing's sarcoma Viable tumour after induction therapy, relapse

Wilm's tumour Primary refractory disease, relapse in case of unfavourable histology 2 or more relapses
Brain tumour Medulloblastoma in CR2
PNET In CR2
Lymphoma Relapsed/refractory lymphomas

What Does Alternate Donor Transplantation Mean?

Allogeneic transplantation requires the presence of a 6 antigen-matched HLA-identical donor. The chance that a sibling will be 6 antigens identical will be around 30 %, and this improves to about 40 % using an extended family search. This means that a large number of patients who require a transplant will not find a HLA identical donor within the family. In this situation, the options of transplant and donor source include:

(a) Matched unrelated donor (MUD)
(b) Cord blood transplants (CBT)
(c) Haploidentical transplants

These transplants may be associated with slower engraftment, higher rejection rates and higher incidence of GVHD, and hence, necessary modifications in conditioning protocols and manipulation of stem cells may be required to make the transplant a successful one.

Umbilical Cord Blood Transplant

Though there are currently more than eight million donors registered in marrow donor registries around the world, a substantial proportion of children who lack a sibling donor will never undergo transplantation from an HLA-matched unrelated donor either because such a donor cannot be found or because the time to identify a donor is too long [14, 15]. The use of haploidentical or half-matched family donors provides a potential source of haematopoietic stem cells for children who lack both a sibling and an unrelated donor. T-cell depletion of the graft can in part overcome the risk of severe GVHD, but it substantially increases the risk of severe and prolonged posttransplantation immunodeficiency [16, 17]. Haematopoietic stem cells from an unrelated cord blood (UCB) transplant can restore haematopoiesis and immune function after a myeloablative conditioning regimen. Umbilical cord is a rich source of stem cells which are pluripotential but are immunologically naive, and hence, these transplants are associated with a decreased risk of GVHD [18, 19]. This has led to the generation of both public and private cord blood banks (UCB) where cord blood units are stored for use in related and unrelated patients. UCB offers the advantage of significantly faster availability of banked cryopreserved UCB units compared with the availability of unrelated bone marrow grafts.

The first cord blood transplant (CBT) took place in 1988 for a child with Fanconi's anaemia. Since that time, CBT has been used for a multitude of illnesses, including haematologic and oncologic disorders, congenital immune deficiencies and inherited metabolic disorders. With the establishment of cord blood banks, more than 100,000 cord blood units have been made available for transplantation and facilitated more than 10,000 unrelated umbilical cord blood transplants (UCBT). Cord blood has potential advantages compared with bone marrow haematopoietic stern cells, namely, the rapid availability of cells and less stringent requirements for HLA matching between donor and recipient because of the lower risk of acute and chronic graft-versus-host disease (GVHD).

What Should a Patient Do for Posttransplant Care?

Following transplantation immunosuppression is given at full doses for 6 months after which it is tapered and stopped 1-year posttransplant. Penicillin and pneumocystis prophylaxis as well as supplemental folic acid is given for 1 year. The patient can return to normal activities within 6 months after a transplant and do not require any special treatment or care after 1-year post-transplant. This is unlike solid organ transplants, where they are on lifelong immunosuppression with its associated complications and high cost. Patients with GVHD need more careful follow-up and prolonged immunosuppression.

Immunisation

Immunisation programme should commence 12 months after a HLA identical sibling donor transplant and 18 months after any other allogeneic HSCT. The child should be off all immunosuppressive treatment including steroids and cyclosporine and should not have evidence of active chronic GVHD.

1 year

- Tetanus toxoid
- Inactivated polio vaccine
- Haemophilus influenza B
- Pneumococcal vaccine one dose (23 valent)

2 years – No graft versus host disease and not on immunosuppression

- Measles/mumps/rubella
- Hepatitis A and B
- Typhoid

What Is the Status of HSCT in India

The first allogeneic bone marrow transplant was performed in India in Tata Memorial Hospital Mumbai in 1985. Since then, transplant activity in the country has been gradually progressing. There are more than 30 transplant centres in India performing allogeneic and autologous transplants with a cumulative number of over 200 transplants. Approximately 30 % of these transplants will be for children. One of the reasons for the limited number of transplants in our country is the lack of resources or a suitable donor. The current cost of haematopoietic stem cell transplantation in the United States is about $150,000 (60 lakh rupees). In India, the average cost of transplantation is about Rs. 10 lakhs. The costs involved in alternate donor transplantation will be much higher and can range from Rs. 25–50 lakhs depending on the type of transplant. These costs will still be much lower than if the child were to travel abroad for a transplant [20]. Organising a national bone marrow and cord blood registry should be the answer to this problem, and several such registries have now begun in India heralding the dawn of a new era. Several voluntary organisations now assist families in providing financial aid for transplantation. There are 10,000 new babies with thalassaemia major being born in our country each year. Early transplantation will help reduce the burden of looking after these children long term with improved outcomes. Centres that cater only to children are also being developed in our country. Transplantation services help save valuable foreign exchange, upgrade the quality of medical care in tertiary referral institutions and provide a life-saving treatment for a child in his own country. Transplant-related mortality in our country is about 10–15 %.

Summary

Haematopoietic stem cell transplantation offers curative therapy for a number of malignancies and haematological diseases in children. Improvement in supportive care and better understanding of transplant immunology have led to dramatic improvement in transplant-related morbidity and mortality and has helped save lives. Novel experimental transplants are being introduced in order to make this procedure more accessible to all patients and improve safety of the procedure.

References

1. Meisel R, Enczmann J, Balzer S et al (2005) Similar survival following HLA-identical sibling transplantation for standard indication in children with haematologic malignancies: a single centre comparison of mobilized peripheral blood stem cell with bone marrow transplantation. Klin Padiatr 217(3):135–141
2. Pan L, Teshima T, Hill GR et al (1999) Granulocyte colony-stimulating factor-mobilized allogeneic stem cell transplantation maintains graft -versus-leukemia effects through a perforin-dependent

pathway while preventing graft-versus-host disease. Blood 93(12):4071–4078

3. Richardson P, Guinan E (1999) The pathology, diagnosis and treatment of hepatic veno occlusive disease; current status and novel approaches. Br J Haematol 107:485–493
4. Bearnian SI, Lee JL, Baron AB (1997) Treatment of hepatic verioocclusive disease with recombinant human tissue plasminogen activator, and heparin in 42 marrow transplant patients. Blood 89: 1501–1506
5. Hale GA, Rochester RI, Heslop HE et al (2003) Haemorrhagic cystitis after allogeneic bone marrow transplantation in children: clinical characteristics and outcome. Biol Blood Marrow Transplant 9:698–705
6. Ravindranath Y, Chang M, Steuber CP et al, Pediatric Oncology Group (2005) Pediatric Oncology Group (POG) studies of acute myeloid leukemia (AML): a review of four consecutive childhood AML trials conducted between 1981 and 2000. Leukemia 19(12):2101–2116
7. Alonzo TA, Wells RJ, Woods WG et al (2005) Postremission therapy for children with acute myeloid leukemia: the children's cancer group experience in the transplant era. Leukemia 19(6):965–970
8. Nemecek ER, Gooley TA, Woolfrey AB et al (2004) Outcome of allogeneic bone marrow transplantation for children with advanced acute myeloid leukemia. Bone Marrow Transplant 34(9):799–806
9. Badell I, Munoz A, Ortega JJ et al, Spanish Working Party for BMT in Children (GETMON) (2005) Long-terrn outcome of allogeneic or autologous haemopoietic cell transplantation for acute lymphoblastic leukaemia in second remission in children. GETMON experience 1983–1998. Bone Marrow Transplant 35(9):895–901
10. Shah AJ, Lenarsky C, Kapoor N et al (2004) Busulfan and cyclophosphamide as a conditioning regimen for pediatric acute lymphoblastic leukemia patients undergoing bone marrow transplantation. J Pediatr Hematol Oncol 26(2):91–97
11. Sharathkumar A, Saunders EF, Dror Y et al (2004) Allogeneic bone marrow transplantationvs chemotherapy for children with Philadelphia chromosome-positive acute lymphoblastic leukemia. Bone Marrow Transplant 33(1):39–45
12. Yusuf U, Frangoul HA, Gooley TA et al (2004) Allogeneic bone marrow transplantation in children with myelodysplastic syndrome or juvenile myelomonocytic leukemia: the Seattle experience. Bone Marrow Transplant 33(8):805–814
13. Dini G, Cancedda R, Locatelli F, Bacigalupo A et al (2001) Unrelated donor marrow transplantation: an update of the experience of the Italian Bone Marrow Group (GITMO). Haematologica 86(5):451–456
14. Yumura-Vagi K, Inoue M, Sakata N et al (2005) Unrelated donor bone marrow transplantation for 100 pediatric patients: a single institute's experience. Bone Marrow Transplant 36(4):307–313
15. Sakata N, Kawa K, Kato K et al (2004) Unrelated donor marrow transplantation for congenital immunodeficiency and metabolic disease: an update of the experience of the Japan Marrow Donor Program. Int J Hematol 80(2):174–182
16. Wall DA, Carter SL, Kernan NA et al, COBLT Steering Committee (2005) Busulfan! melphalan! antithymocyte globulin followed by unrelated donor cord blood transplantation for treatment of infant leukemia and leukemia in young children: the Cord Blood Transplantation study (COBLT) experience. Biol Blood Marrow Transplant 11(8): 637–646
17. Gluckman E, Rocha V (2004) Cord blood transplantation for children with acute leukaemia: a Eurocord registry analysis. Blood Cells Moi Dis 33(3):271–273
18. Styczynski J, Cheung YK, Garvin J et al (2004) Outcomes of unrelated cord blood transplantation in pediatric recipients. Bone Marrow Transplant 34(2):129–136
19. Lang P, Greil J, Bader P, Handgretinger R et al (2004) Long-term outcome after haploidentical stem cell transplantation in children. Blood Cells Mol Dis 33(3):281–287
20. Veys P, Amrolia P, Rao K (2003) The role of haploidentic.al stem cell transplantation in the management of children with haematological disorders. Br J Haematol 123(2):193–206

Placental Umbilical Cord Whole Blood and Nuclear Disaster Management: A Possibility

Niranjan Bhattacharya and Sanjukta Bhattacharya

In 1958, six physicists were exposed to large doses of neutron irradiation following an accident at Veneza, Yugoslavia. They were sent to France and treated with multiple transfusions of homologous bone marrow. Four of the victims had successful temporary bone marrow grafts as part of the treatment [1]. In serious radiation injury, graft-versus-host disease (GVHD) may occur, particularly in patients who have their marrow completely ablated, but this would be difficult to detect in the first 72 h. GVHD, however, can usually be adequately treated with cord blood because of its very minor nature.

In this connection, it may be mentioned that scientists, in an experiment, kept SJL/J mice alive for several years with the use of human umbilical cord blood after they received lethal levels of irradiation. Further, it is a known fact that under certain conditions, human cord blood does not have to be HLA matched to facilitate rescue from irradiation [2]. In addition, there are reports of unmatched HLA cord blood being used successfully for marrow transplantation. Clinical observations from China suggest that HLA-mismatched umbilical cord blood can be engrafted successfully in children with malignant disease [3]. Another investigator has reported that if female A/J mice (27–30 weeks old) are exposed to an absorbed dose of 9–10 Gy of (137)Cs γ-rays delivered acutely to the whole body, and subsequently treated either with 1×10^8 or 2×10^8 human umbilical cord blood (HUCB) mononucleated cells 24–52 h after the irradiation, the survival is more on the higher dosage and this is significant (($P < 0.01$) [4]. The point that is being made here is that human umbilical cord blood has the potential to treat patients in case of accidental radiation injury because of some of its properties.

Brief Pathophysiology of Radiation Injury

Victims of acute radiation-induced illness require prompt diagnosis and specific treatment. Radiation dosage may be estimated by automated biodosimetry, apart from clinical correlations by time, to start emesis, lymphocytic depletion kinetics, and assessing the resultant clinical condition by multiple organ-specific biochemical tests [5]. As a matter of fact, all cases of suspected radiation-induced injury should be considered as having multiple organ dysfunctions which may eventually lead to multiple organ failure. Acute radiation syndrome (ARS) occurs when there is high level external radiation involving the whole body or at least 60 % of the body surface area. High dosage exposure means a dosage greater than 1 Gy delivered at a relatively high dosage rate. The clinical manifestations of ARS occur

N. Bhattacharya (✉)
Department of Regenerative Medicine and Translational Science, Calcutta School of Tropical Medicine and STM Cord Blood Bank, Kolkata, India
e-mail: sanjuktaniranjan@gmail.com

S. Bhattacharya
Department of International Relations, Jadavpur University, Kolkata, India

I. Somasundaram (ed.), *Stem Cell Therapy for Organ Failure*,
DOI 10.1007/978-81-322-2110-4_12, © Springer India 2014

within an hour of exposure and may last for several weeks. Subclinical ARS may result from less than 1 Gy exposure. In case of whole body irradiation, there may have a prominent hematopoietic impact within the radiation exposure range from 1 to 8 Gy: for gastrointestinal impact, the dosage should be 6–20+ Gy, and for neurovascular impact, the dosage should be 20–50+ Gy. The average time from exposure to definite diagnosis takes about 3 weeks or so when the victim presents with features of neutropenia and pancytopenia [6].

Blast cells or premature cells replicate rapidly and have a long mitotic feature. These cells are found to be generally radiosensitive. There are other cells like spermatogonia, lymphocytes, small intestine, colon, stomach epithelium, and hematopoietic cells which are also generally radiosensitive in nature. All radiation injury processes must have a prodromal phase, which starts within a few hours of the radiation exposure. If the exposure is more than 2–3 Gy, it may lead to neutropenia and pancytopenia. A dosage within a range of 3–8 Gy will cause significant damage to the bone marrow, and an eventual 50 % death rate will ensue due to bone marrow structural or functional failure. 3–4 Gy exposure will result in exponential biological death involving bone marrow stem and progenitor cells. As mentioned, a little early prodromal symptoms with high dosage radiation will begin with nausea and vomiting within a few to 72 h [7]. Early onset of nausea means high exposure to radiation. On an average, 50 % of the patients who are exposed to 1.6–2.4 Gy of radiation will experience nausea or vomiting. The actual exposure of the victim depends on several factors like critical radiation exposure, distance from the focal point of radiation emission, shielding, reservoir of stem cells in the body, genetic background, sensitivity of the organ/tissue thus exposed, interface modification, preexisting diseases like diabetes, and bone marrow refraction due to any cause just to name a few variables. After the exposure, gastrointestinal and hematopoietic syndromes appear simultaneously [8]. In case of neurovascular syndrome, there is a burning sensation which appears within minutes followed by nausea and vomiting within hours and eventually loss of balance (ataxia), confusion, prostration, and severe watery diarrhea which begins within hours leading to electrolyte imbalance, which in turn triggers a multiorgan failure syndrome. Sometimes there is a combined injury involving radiation, trauma, and burn [9].

Treatment for Radiation Injury

Treatment for radiation injury may follow the suggested guideline given below:

(a) Grading and staging of the injury with meticulous history taking to assess the preexisting load of the disease after calculating the age of the victim and individualized organ-specific impact taking into consideration 1 % structural or functional cellular loss/year, depending on the availability of growth factors, which can salvage the bone marrow impact of radiation by granulopoiesis/lymphopoicsis induction. Drugs likc filgrastim (G-CSF):2.5–5 ug/kg/day subcutaneously daily/pegfilgrastim (PegG-CSF) 6 mg once daily subcutaneously/sargramostim (GM-CSF) 5–10 ug/kg/day subcutaneously may be helpful [10].
(b) Broad spectrum antibiotics + anti-aerobics + antiviral + antifungal drug combinations may be tried judiciously depending on the condition of the patient; high risk patients may be shifted to special ITU?/special rooms with provision for monolaminar flow of air/O_2 support.
(c) Allogenic stem cell transplant with HLA matching.
(d) Cord blood stem cell transplant with or without myeloablation or minimal myeloablation.
(e) Cord blood, which is a rich source of hematopoietic, mesenchymal, angiogenesis induction stem cells apart from embryonal stem cells and unidentified stem cells and supportive cytokines and growth factors, is a nature made cocktail which has the potential to be extremely effective in situations of bone marrow refractions or failure triggering situations.

(f) To understand the effectiveness in cases of radiation injury, it may be noted that bone marrow transplant requires stem cells with perfect HLA match preferably 8/8 for good results, like matching of HLA-A-B-C and also DRB1; even HLA factors like DQ/DP may have a possible role in better engraftment; but in case of cord blood, even a mismatched transplant is feasible. Further, graft-versus-host (GVH) reaction is severe at times with bone marrow transplant, but with cord blood, the GVH response is often mild (if at all) after the match. Donation transplant coordination after the match is not required for cord blood but is essential and complex with stem cells collected from bone marrow. Cord blood donation causes no risk to the mother.

Resuscitation and Preventing the Bone Marrow Failure

Blood is an essential component of any shock management, whatever might be the trigger of the shock /shock-like syndrome that ensues, be it disaster due to natural causes or a man-made disaster. Cord blood, because of its rich mix of fetal and adult hemoglobin, high platelet and WBC counts, and plasma filled with cytokine and growth factors, as well as its hypo-antigenic nature and altered metabolic profile, has all the potentials of a real and safe alternative to adult blood transfusion. In the human system, earlier investigators used cord whole blood in case of anemic patients of different etiologies, viz., malarial anemia [11], geriatric anemia [12–14], thalassemia [15, 16], HIV [17], and leprosy [18], only to name a few. Its emergency use potential has been suggested whenever there is a true need of blood, i.e., in emergency, as seen in Ireland [19], or an accidental collateral victim of allied bombing in Iraq war [20] where there is an emergency need for a true blood substitute.

Scientists have searched meticulously all available sources for a blood substitute; there was great hope when *Nature* published an early news that scientists have found a blood substitute in the sea worm, which has the potentialities to replace the RBC functions [21]. However, cord blood was never been looked into seriously as a blood substitute. This blood, which is now showing its potentials as a blood substitute, was earlier discarded as waste.

Prof. Andrew Burd, an important contributor in the field, has roughly estimated the total wastage of the cord blood per year [22]:

Umbilical cord blood	Volume (cesarean section)	1,560,000 l
	Volume (vaginal delivery)	7,225,000 l
	Total volume	8,785,000 l

Cord blood produce anti-inflammatory cytokines like IL1 antagonist; in addition, it lacks the expression of proinflammatory cytokinesIL1, TNF alpha, IL6, IL12, and other macrophage inflammatory protein. There overall effect is the suppression of inflammation. These activated macrophages eventually go to the cord blood to give it its unique characteristics [23]. In fine, placental umbilical cord, whole blood is a unique cocktail made by nature which has enormous potentialities in transfusion, transplantation, and molecular therapy impact of its cytokine content as a whole.

Conclusion

Strategies to treat acute radiation syndrome (ARS) resulting from large-scale radiological or nuclear accidents remain a challenge for the survivor. The main impact of radiation is on the bone marrow of the victim. HLA-matched standardized allogenic transplant is difficult to translate in reality in an emergency situation. On an average, the adult bone marrow cellular load for the transplant range from 1.64 to 8.6 × 10 to the power 9 viable nucleated bone marrow cells. The role of adult bone marrow stem cell therapy for the treatment of a radiation victim is very controversial because of nonavailability of critical treatment data on actual victims.

When dealing with mass casualties in nuclear terrorism, or nuclear accidents, it is noticed that dehydration claims large numbers of victims due to radiation-induced vomiting. Large numbers of medical or paramedical personnel would be necessary to provide the basic needs of fluid and nutrition for the injured as well as medications. However, regardless of the final decision on the treatment, radiation victims basically must have their destroyed marrow replaced where cord blood can play a very positive life-saving role [24]. The treatment for nuclear radiation has been a topic of debate since the first nuclear accident was reported in 1954. Benjamin K Sovacool [25] reported the occurrence of 99 cumulative accidents globally in nuclear power plants till date. While 57 % of accidents occurred in the USA, other notable accidents are the Chernobyl disaster (1986), the Fukushima Daiichi nuclear disaster (2011), Three Mile Island accident (1979), and the SL-I accident (1961). However, pronuclear advocates claim that apart from the Chernobyl disaster, there were no human casualties due to exposure to radiation in any of the other accidents [26], but eternal vigilance against nuclear terrorism should not be relaxed as it could be a real threat for the world.

References

1. Mathe G et al (1959) Transfusions and grafts of homologous bone marrow in humans after accidental high dosage irradiation. Rev Fr Etud Clin Biol 4(3):226–238
2. Ende N, Lu S, Ende M, Giuliani D, Ricafort RJ, Alcid MG, Deladisma MD, Bagtas-Ricafort L (1996) Potential effectiveness of stored cord blood (non-frozen) for emergency use. J Emerg Med 14(6): 673–677
3. Ende N, Ponzio NM, Giuliani D, Bagga PS, Godyn J, Ende M, Athwal RS (1995) The effect of human cord blood on SJL/J mice after chemoablation and irradiation and its possible clinical significance. Immunol Invest 24(6):999–1012
4. Kovalenko OA, Azzam EI, Ende N (2013) Human umbilical-cord-blood mononucleated cells enhance the survival of lethally irradiated mice: dosage and the window of time. J Radiat Res 54(6): 1010–1014
5. Mickelson AB, ed. (2012) Medical Consequences of Radiological and Nuclear Weapons, Textbook of Military Medicine series. Office of the Surgeon General, Department of the Army, United States of America, Falls Church, VA, and Borden Institute, Fort Detrick, Maryland, USA
6. Gorin NC, Fliedner TM, Gourmelon P et al (2006) Consensus conference on European preparedness for haematological and other medical management of mass radiation accidents. Ann Hematol 85(10): 671–679
7. American society of clinical Oncology (1994) Recommendations for use of haematopoietic colony stimulating factors: evidence based clinical practice guidelines. J Clin Oncol 12(11):2471–2508
8. Fliedner TM (2006) Nuclear terrorism: the role of haematology in coping with its health consequences. Curr Opin Hematol 13(6): 436–444
9. Singh VK, Shafran RL, Jackson WE 3rd, Seed TM, Kumar KS (2006) Induction of cytokines by radioprotective tocopherol analogs. Exp Mol Pathol 81(1): 55–61
10. SV K, Belyi DA et al (2011) Hematopoietic cell infusion for the treatment of nuclear disaster victims: new data for the Chernobyl accident. Int J Radiat Biol 87(8):846–850
11. Clarke T (2002) Newborns might help malaria kids—blood from umbilical cords could treat anemia caused by tropical disease. Available at: http://www.nature.com/nsu/021111/021111-11html
12. Bhattacharya N (2004) Placental umbilical cord whole blood transfusion. J Am Coll Surg 199(2): 347–348
13. Bhattacharya N, Bandopadhyay T, Bhattacharya M, Bhattacharya S. Do not discard 99.99 % of the human placental umbilical cord blood for the sake of stem cells only. Available at: http://bmj.com/cgi/eletters/323/7304/60#16874. Accessed 5 Oct 2001
14. Bhattacharya N, Bandyopadhyay T, Bhattacharya M, Bhattacharya S. Immunization and fetal cell /tissue transplant: a new strategy for geriatric treatment. Available at: http://bmj.com/cgi/eletters/323/7320/1025/b#21055. Accessed 6 Apr 2002
15. Bhattacharya N (2008) Study of the utility of placental cord blood in meeting the transfusion needs of beta-thalassaemic patients. Reg Health Forum 12(2):16–27
16. Bhattacharya N. The safe use of placental umbilical cord whole blood transfusion in patients suffering with anemia and thalassemia in under-resourced regions of the world. Available at: http://bmj.com/cgi/eletters/321/7269/1117#62372. Accessed 9 June 2004
17. Bhattacharya N. Umbilical cord whole blood transfusion in HIV patients with anemia and emaciation. Available at: http://bmj.com/cgi/eletters/327/7414/562-a#59738. Accessed 17 May 2004
18. Bhattacharya N. A preliminary study report on placental umbilical cord blood transfusion in victims of anemia with leprosy in under-resourced regions of the world. Available at: http://bmj.comcgi/eletters/328/7454/1447/=63828. Accessed 22 June 2004

19. Bhattacharya N. Umbilical cord whole blood transfusion: a suggested strategy to combat blood scarcity in Ireland. Available at: http://bmj.com/cgi/eletters/324/7330/134/c#19096. Accessed 27 Jan 2002
20. Bhattacharya N. Utilization of a genuine blood substitute: a suggestion to the medical fraternity in Iraqi hospital. Available at: http://bmj.com/cgi/eletters/326/7391/675#30850. Accessed 30 Mar 2003
21. Hoag H. Blood substitute from worm shows promise—hemoglobin from sea creature could replace red cells. Available at: http://www.nature.com/nsu/030602/030602-7html. Accessed 4 June 2003
22. Burd A et al (2011) A massive wastage of global resource. In: Bhattacharya N, Stubblefield P (eds) Regenerative medicine using pregnancy-specific biological substances. Springer, London/New York, pp 1–7
23. Chang M-DY, Polar JW, Khalile H et al (1993) Mouse placental macrophages have a decreased ability to present antigen. Proc Natl Acad Sci U S A 90: 462–466
24. Ende N et al (2011) Emergency use of human cord blood. In: Bhattacharya N, Stubblefield P (eds) Regenerative medicine using pregnancy-specific biological substances. Springer, London/New York, pp 85–91
25. Sovacool BK (2010) A critical evaluation of nuclear power and renewable electricity in Asia. J Contemp Asia 40(3):393–400
26. Arm ST (2010) Nuclear energy a vital component of our energy future. In: Chemical engineering progress. American Institute of Chemical Engineers, New York, pp 27–34, ISSN0360-7275

Mesenchymal Stem Cell Treatment Option to Manage Autoimmune Disorders: A Technically Feasible Integration

Subadra Dravida and R. Indarapu

Introduction

The T and B lymphocytes of white blood cells are "soldiers" of the immune system, which originate in the bone marrow. When a healthy individual is infected, the body reacts by activating a variety of immune cells sometimes, while at other times, the antigens bind to the major histocompatibility complex (MHC) proteins, also known as human leukocyte antigen (HLA) molecules, on the surface of the antigen-presenting cells. CD4 helper T cells and antigen-specific B cells are the other components of the immune system participating and adapting to the immune responses.

An autoimmune disease develops when the immune system, which is supposed to defend the body against disease, decides the healthy cells as foreign and attacks, annihilating them. Depending on the type of manifestation, an autoimmune disease can affect one or many different types of body tissues sometimes causing abnormal organ growthand changes in organ function. Autoimmune disease affects up to 50 million Americans [1], according to the American Autoimmune Related Diseases Association (AARDA), while there is a lack of current prevalence data from India. Additionally, under-projection and underestimation of prevalence occurs.

Mesenchymal stromal cells as bone-forming cells in the bone marrow [2], multipotent in nature, are often named as mesenchymal stem cells (MSCs). MSCs are immune privileged since they exhibit low levels of major histocompatibility (MHC) class I molecules, rarely express cell surface MHC class II, and do not express co-stimulatory molecules (CD40, CD40L, CD80, CD86) and they escape T cell recognition [3]. Their effects on immunocompetent cells are not MHC restricted, permitting allogenic MSCs to be used with no need to match with host human leukocyte antigens (HLAs). Unlike blood stem cells originating from bone marrow, MSCs can be isolated from a variety of other sources including placenta, umbilical cord blood/tissue, adipose tissue, teeth and menstrual fluid, and urine [4]. The discovery that MSCs were found to escape T-cell recognition, suppress T-cell response to mitogens, and also prolong skin graft survival in baboons by Bartholomew et al. in 2002 [5] indicated new and interesting features of these progenitor cells providing the basis of premise as "therapeutic off-the-shelf tool" in cell therapies. Their ability to differentiate into traditional mesodermal tissues led to their prominence in the regenerative medicine.

S. Dravida (✉)
Tran-Scell Biologics Pvt. Ltd., 100/1, Ravi Chambers, Madhapur, Jubilee Hills, Hyderabad 500033, India
e-mail: suba.dravida@gmail.com

R. Indarapu
Curatum Cell Sciences Pvt. Ltd., Sowbhagyapuram Road, Hyderabad, India

I. Somasundaram (ed.), *Stem Cell Therapy for Organ Failure*,
DOI 10.1007/978-81-322-2110-4_13, © Springer India 2014

MSCs as Cellular Therapeutic Tools in Autoimmune and Inflammatory Disease

Autologous or allogenic MSCs have been tested in a vast and varied immune-mediated disease models like a model of multiple sclerosis—diabetic NOD/scid mice [6]—lupus murine models [7–10], and collagen-induced arthritis model [11, 12], establishing preclinical efficacy. The mechanisms by which MSCs exert immunomodulatory effects are speculated to involve multiple pathways. It has been shown that MSCs suppressive effects require a previous priming step that occurs in the presence of an inflammatory environment and is mediated by the secretion of specific cytokines [13, 14]. IFN-γ, alone or together with tumor necrosis factor (TNF)-α, IL-1α, or IL-1β, are required to trigger the expression by MSCs of high levels of soluble factors involved in immunosuppression such as IDO, HGF, TGF-β, and NO [15, 16]. MSCs have also been shown to induce the generation of functional Tregs both in vitro and in vivo [17, 18]. In MLR/lpr mice, transplantation of MSCs from many sources (bone marrow, umbilical cord, or exfoliated deciduous teeth) can restore Treg cells and induce a significant reduction in Th17 levels, consequently upregulating the ratio of Treg/Th17 cells [8]. It has also been shown that MSCs also generate functional active CD4+ CD25+ Foxp3+ T-regulatory cells during the in vitro differentiation phase of Th1 and Th17 cells [19].

Available Data on Clinical Trials

In 2014, 44 MSC registered clinical trials were found on the http://clinicaltrials.gov/ in autoimmune diseases, while 32 registrations fell under the open category. Among the regions that have registered in MSC trials in autoimmune disorders, East Asia scores the maximum number of studies although the statistics in prevalence vary.

Rheumatoid arthritis (RA) is a chronic systemic disease, which is characterized by chronic inflammation in the synovial tissue. RA results in the destruction of cartilage, bone and ligaments, and joint deformity. The underlying hypothesis is that umbilical cord-derived mesenchymal stem cell (UC-MSCs) has anti-inflammatory effects and thus potentially alleviates the progression of rheumatoid arthritis. Preclinical data suggested that in RA model, MSCs inhibited the Th17 CD4 T-cell subset and induced an anti-inflammatory monocyte type, with the subsequent simultaneous reduction in the key cytokines, IL-17, IL-6, and TNF-alpha [20]. The randomized, double-blind placebo-controlled dose escalation study evaluated the safety, tolerability, and effectiveness of a single intravenous infusion of two MSC dose levels in 48 patients (www.mesoblast.com).

Type 1 diabetes (T1D) is a more severe form and insulin-dependent type. It is sometimes called "juvenile" diabetes, because T1D usually develops in children and teenagers, although it can develop at any age. The immune system mistakenly sees the insulin-producing cells in the pancreas as foreign and destroys them. There are 10 different clinical trials registered with global clinical trials registry wherein the biological intervention proposed was MSCs either combined with mononuclear cells or co-transplanted with islet cells. Owing to the MSCs immunomodulatory properties, MSC therapy was strategized for T1D in humans with strong animal models tested [21]. A phase II, multicentric, randomized, double-blind, placebo-controlled study to evaluate the safety and efficacy of MSCs for the treatment of recently diagnosed T1D started in 2008 had accumulated interesting positive follow-up data on C-peptide responses along with changes in levels of IA-2 autoantibodies and progressing to phase III in the USA (Clinical Trial Identifier: NCT00690066).

Psoriasis (P) is a chronic skin problem that causes skin cells to grow too quickly, resulting in thick, white, silvery, or red patches of skin. Psoriasis occurs when the immune system overreacts, causing inflammation and flaking of skin. The causative factors flaring up include cold and dry climate, infections, stress, dry skin, and taking certain medicines. Fadi Braiteh reported a case report of complete remission of P after hematopoietic stem cell transplantation and not MSCs in 2008.

Graft-versus-host disease (GVHD) is a common complication following an allogenic tissue transplant. It is commonly associated with bone marrow transplants and tissue grafts. GVHD is the most calculated application for MSCs, where the transplanted hematopoietic stem cells comprising of donor T cells attack [22] an immune-compromised recipient. Case report in 2004 and a later phase II clinical study involving 55 steroid-resistant patients (25 children and 30 adults) with severe acute disease reported amazing response, with improvement of liver and intestinal function with no side effects. Osiris Therapeutics Incorporation's Prochymal™ (bone marrow-derived MSCs) was approved by Canadian health regulators in 2012 to treat acute GVHD in children, making it the first stem cell drug to be approved for a systemic disease anywhere in the world. Treatment with Prochymal™ resulted in a statistically significant improvement in survival when compared with a historical control population of pediatric patients with refractory GvHD.

Crohn's disease (CD) is an inflammatory bowel disease (IBD) with inflammation of the lining of digestive tract that can lead to abdominal pain, severe diarrhea, and even malnutrition. Phase 1 clinical trial using autologous adipose-derived MSCs [23] reported no adverse events, while phase II multicentric randomized controlled trial [24] on 49 patients with complex perianal fistulas published the efficacy of local injections on healing fistulas (6/8) with no adverse effects. Onken et al. and Duijvestein et al. published mixed responses of their clinical activity with no significant efficacy observed in a small group of patients treated with allogenic bone marrow-derived MSCs (BM-MSC) [25, 26].

Multiple sclerosis (MS) is another autoimmune inflammatory demyelinating disease of the central nervous system associated with major disability. Mohyeddin Bonab et al. and Karussis et al. reported no adverse events with autologous MSC treatment of MS. Further phase I/II studies involving 10–15 patients reported stability during follow-up (6–28 months) [27]. Karussis et al. published increased regulatory T cells with activation markers on dendritic cells upon MSC transplantation [28]. Freedmann et al. reported culture of the cells and the MS treatment protocols based on international transplant guidelines [29]. Connick et al. reported a study including 10 patients with MS treated with an intravenous infusion of autologous MSCs [30].

Systemic lupus erythematosus (SLE) is a systemic autoimmune connective tissue disease that can affect any part of the body. SLE often harms the heart, joints, skin, lungs, blood vessels, liver, kidneys, and nervous system. Genetic and environmental factors produce autoantibodies that mistakenly attack and destroy the body's own healthy cells and tissue. These autoantibodies also trigger inflammation in SLE condition which can lead to organ damage. It is both a type II and a type III hypersensitivity reaction in which bound antibody-antigen pairs (immune complexes) precipitate and cause further immune responses in SLE condition. There are six registered global clinical trials on MSCs as therapeutic cells to treat SLE. Sun et al. reported the use of umbilical cord-derived MSCs (UC-MSCs) in severe lupus patients ($n = 16$). Five of 15 renal cases had histological confirmation of proliferative nephritis, 11 were preconditioned with cyclophosphamide prior to MSC infusion. Follow-up was for 8–25 months wherein significant improvement was verified for SLEDAI score, serum albumin, 24 h urinary protein, serum creatinine, serum complement, and anti-dsDNA antibodies [31, 32].

Conclusion

Repository of meaningful information on clinical trials using MSCs to address or treat autoimmune disorders links the biology of MSCs to beneficial immunomodulation. Even if the in vivo role of endogenous MSCs remains exploratory, binding the therapeutic effects of ex vivo expanded MSCs for immune regulation seems to have a significant clinical potential. Long-term safety concerns and sustained clinical outcome remain to be updated and not an issue as the proposal to treat autoimmune diseases with MSCs is a novel and nascent theory. The single or repeated MSC therapy to cure the diseases needs to be explored in controlled and labeled clinical trials given the unknown interactions with host cells [33]. Defining regulatory and technical conditions would allow the development of better clinical parameters to completely manage the autoimmunity.

References

1. Autoimmune Disease Fact Sheet (n.d.) womenshealth.gov. Retrieved 10 July 2012 from http://womenshealth.gov/publications/our-publications/fact-sheet/autoimmune-diseases.cfm
2. Fridenshteĭn AJ, Piatetskiĭ-Shapiro II, Petrakova KV (1966) Osteogenesis in transplants of bone marrow cells. J Embryol Exp Morphol 16:381–390
3. Chamberlain G, Fox J, Ashton B, Middleton J (2007) Concise review: mesenchymal stem cells: their phenotype, differentiation capacity, immunological features, and potential for homing. Stem Cells 25:2739–2749
4. Hass R, Kasper C, Bohm S, Jacobs R (2011) Different populations and sources of human mesenchymal stem cells (MSC): a comparison of adult and neonatal tissue-derived MSC. Cell Commun Signal 9:12
5. Bartholomew A, Sturgeon C, Siatskas M, Ferrer K, Mcintosh K, Patil S, Hardy W, Devine S, Ucker D, Deans R, Moseley A, Hoffman R (2002) Mesenchymal stem cells suppress lymphocyte proliferation in vitro and prolong skin graft survival in vivo. Exp Hematol 30:42–48
6. Lee RH, Seo MJ, Reger RL, Spees JL, Pulin AA, Olson SD, Prockop DJ (2006) Multipotent stromal cells from human marrow home to and promote repair of pancreatic islets and renal glomeruli in diabetic NOD/scid mice. Proc Natl Acad Sci U S A 103:17438–17443
7. Sun L (2010) Allogenic mesenchymal stem cells transplantation in refractory systemic lupus erythematosus: a pilot clinical study. Ann Rheum Dis 69:1423–1429
8. Gu Z, Akiyama K, Ma X, Zhang H, Feng X, Yao G, Hou Y, Lu L, Gilkeson GS, Silver RM, Zeng X, Shi S, Sun L (2010) Transplantation of umbilical cord mesenchymal stem cells alleviates lupus nephritis in MRL/lpr mice. Lupus 19:1502–1514
9. Chang JW, Hung SP, Wu HH, Wu WM, Yang AH, Tsai HL, Yang LY, Lee OK (2011) Therapeutic effects of umbilical cord blood-derived mesenchymal stem cell transplantation in experimental lupus nephritis. Cell Transplant 20(2):245–257
10. Schena F, Gambini C, Gregorio A, Mosconi M, Reverberi D, Gattorno M, Casazza S, Uccelli A, Moretta L, Martini A, Traggiai E (2010) Interferon γ-dependent inhibition of B cell activation by bone marrow-derived mesenchymal stem cells in a murine model of systemic lupus erythematosus. Arthritis Rheum 62:2776–2786
11. Augello A, Tasso R, Negrini SM, Cancedda R, Pennesi G (2007) Cell therapy using allogeneic bone marrow mesenchymal stem cells prevents tissue damage in collagen-induced arthritis. Arthritis Rheum 56:1175–1186
12. González MA, González-Rey E, Rico L, Büscher D, Delgado M (2009) Treatment of experimental arthritis by inducing immune tolerance with human adipose-derived mesenchymal stem cells. Arthritis Rheum 60:1006–1019
13. Jones S, Horwood N, Cope A, Dazzi F (2007) The antiproliferative effect of mesenchymal stem cells is a fundamental property shared by all stromal cells. J Immunol 179:2824–2831
14. Jorgensen C (2010) Mesenchymal stem cells immunosuppressive properties: is it specific to bone marrow-derived cells? Stem Cell Res Ther 1(2):15
15. Aggarwal S, Pittenger MF (2005) Human mesenchymal stem cells modulate allogenic immune cell responses. Blood 105:1815–1822
16. Krampera M, Cosmi L, Angeli R, Pasini A, Liotta F, Andreini A, Santarlasci V, Mazzinghi B, Pizzolo G, Vinante F, Romagnani P, Maggi E, Romagnani S, Annunziato F (2006) Role for interferon-gamma in the immunomodulatory activity of human bone marrow mesenchymal stem cells. Stem Cells 24:386–398
17. González-Rey E, González MA, Varela N, O'valle F, Hernández-Cortés P, Rico L, Büscher D, Delgado M (2010) Human adipose-derived mesenchymal stem cells reduce inflammatory and T cell responses and induce regulatory T cells in vitro in rheumatoid arthritis. Ann Rheum Dis 69:241–248
18. Prevosto C, Zancolli M, Canevali P, Zocchi MR, Poggi A (2007) Generation of CD4+ or CD8+ regulatory T cells upon mesenchymal stem cell-lymphocyte interaction. Haematologica 92:881–888
19. Luz-Crawford P, Djouad F, Fernández X, Contreras R, Kurte M, Apablaza F, Jorgensen C, Figueroa F, Carrión F (2012) Mouse MSCs induce a CD4CD25Cfoxp3 Treg subpopulation during the differentiation process of Th1 and Th17 cells. ISSCR 10th annual meeting, Japan, 13–16 June 2012 (abstract accepted)
20. Newman RE, Yoo D, LeRoux MA, Danilkovitch-Miagkova A (2009) Treatment of inflammatory diseases with mesenchymal stem cells. Inflamm Allergy Drug Targets 8:110–123
21. Abdi R, Fiorina P, Adra CN, Atkinson M, Sayegh MH (2008) Immunomodulation by mesenchymal stem cells: a potential therapeutic strategy for type 1 diabetes. Diabetes 57(7):1759–1767. doi:10.2337/db08-0180, 1939-327X
22. English K, French A, Wood KJ (2010) Mesenchymal stromal cells: facilitators of successful transplantation? Cell Stem Cell 7:431–442
23. García-Olmo D, García-Arranz M, Herreros D, Pascual I, Peiro C, Rodríguez-Montes JA (2005) A phase

I clinical trial of the treatment of Crohn's fistula by adipose mesenchymal stem cell transplantation. Dis Colon Rectum 48:1416–1423
24. García-Olmo D, Herreros D, Pascual I, Pascual JA, Del-Valle E, Zorrilla J, De-La-Quintana P, García-Arranz M, Pascual M (2009) Expanded adipose-derived stem cells for the treatment of complex perianal fistula: a phase II clinical trial. Dis Colon Rectum 52: 79–86
25. Onken J, Gallup D, Hanson J, Pandak M, Custer L (2006) Successful outpatient treatment of refractory Crohn's disease using adult mesenchymal stem cells. American College of Gastroenterology Conference, Las Vegas, Abstract 121
26. Duijvestein M, Vos AC, Roelofs H, Wildenberg ME, Wendrich BB, Verspaget HW, Kooy-Winkelaar EM, Koning F, Zwaginga JJ, Fidder HH, Verhaar AP, Fibbe WE, Van Den Brink GR, Hommes DW (2010) Autologous bone marrow-derived mesenchymal stromal cell treatment for refractory luminal Crohn's disease: results of a phase I study. Gut 59(12):1662–1669
27. Mohyeddin Bonab M, Yazdanbakhsh S, Lotfi J (2007) Does mesenchymal stem cell therapy help multiple sclerosis patients? Report of a pilot study. Iran J Immunol 4:50–57
28. Karussis D, Karageorgiou C, Vaknin-Dembinsky A (2010) Safety and immunological effects of mesenchymal stem cell transplantation in patients with multiple sclerosis and amyotrophic lateral sclerosis. Arch Neurol 67: 1187–1194
29. Freedman MS, Bar-Or A, Atkins HL, Karussis D, Frassoni F, Lazarus H, Scolding N, Slavin S, Le Blanc K, Uccelli A, MSCT Study Group (2010) The therapeutic potential of mesenchymal stem cell transplantation as a treatment for multiple sclerosis: consensus report of the International MSCT Study Group. Mult Scler 16(4):503–510
30. Connick P, Kolappan M, Crawley C, Webber DJ, Patani R, Michell AW, Du MQ, Luan SL, Altmann DR, Thompson AJ, Compston A, Scott MA, Miller DH, Chandran S (2012) Autologous mesenchymal stem cells for the treatment of secondary progressive multiple sclerosis: an open-label phase 2a proof-of-concept study. Lancet Neurol 11(2):150–156
31. Sun L, Wang D, Liang J, Zhang H, Feng X, Wang H, Hua B, Liu B, Ye S, Hu X, Xu W, Zeng X, Hou Y, Gilkeson GS, Silver RM, Lu L, Shi S (2010) Umbilical cord mesenchymal stem cell transplantation in severe and refractory systemic lupus erythematosus. Arthritis Rheum 62:2467–2475
32. Spaeth EL, Dembinski JL, Sasser AK, Watson K, Klopp A, Hall B, Andreeff M, Marini F (2009) Mesenchymal stem cell transition to tumor-associated fibroblasts contributes to fibrovascular network expansion and tumor progression. PLoS One 4:e4992
33. van Velthoven Cindy TJ, Kavelaars A, van Bel F, Heijnen CJ (2010) Repeated mesenchymal stem cell treatment after neonatal hypoxia–ischemia has distinct effects on formation and maturation of new neurons and oligodendrocytes leading to restoration of damage, corticospinal motor tract activity, and sensorimotor function. J Neurosci 30(28):9603–9611

Part VI

Stem Cells and Organ Systems: Gastrointestinal System

Overview of Pancreatic Replacement of β-Cells from Various Cell Sources

Subhanwita Sarkar Dey, Noriko Yoshida, and Kouichi Hasegawa

Diabetes Types, Cause, and Current Therapy

According to the estimation by the World Health Organization (WHO), nearly 347 million people all over the world suffer from diabetes [1], and the number is likely to be increased to 552 million by 2030 [2]. Diabetes is a disease caused by high blood sugar. Mainly two types of diabetes have been categorized. Type I diabetes is caused by an autoimmune attack on pancreatic β-cells, which produce insulin and control blood sugar level, and a consequent of β-cell deficiency [3]. In type I diabetes, the onset of symptoms has been found to occur when the β-cell mass falls below 20 % of the normal range [4]. Type II diabetes is the more predominant form which shows metabolic disorders, "peripheral insulin resistance" leading to β-cell failure [5]. In type II diabetes, the β-cell mass decreases to 40–60 % of the normal level, and external insulin administration is not sufficient for normal functioning of islet cells [6, 7, 4]. Therefore, in order to increase the number of islet cells, "replacement therapies" are being fostered for producing insulin from different sources [8]. Currently, islet transfer from donated tissues is considered as one of the options for increasing the number of islet cells; however, the donation number is very small [9]. Although islet transplantation research has been progressed with evident success, the longevity of this procedure still remains unclear [9]. Recipients staying euglycemic without the use of any exogenous insulin more than four years post-islet transplant has been very rare [9]. The concern still remains having insulin-producing foreign cells within the hepatic parenchyma, with portal pressures resulting from islet infusion. The second concern is regarding the fact that the islet recipient can be sensitized against donor tissue types, making it a narrower chance to find a donor for second transplant [9]. Therefore, to escape the risk of pancreatitis associated with intrapancreatic grafting, non-pancreatic systems like the liver portal system, omentum, are presently used for transplantation

S. Sarkar Dey
Institute for Stem Cell Biology and Regenerative Medicine (inStem), National Centre for Biological Sciences (NCBS), Tata Institute of Fundamental Research (TIFR), GKVK Campus, Bellary Road, Bangalore 560065, India
e-mail: subhanwitas@instem.res.in

N. Yoshida
Institute for Integrated Cell-Material Sciences (iCeMS), Kyoto University, Yoshida-Ushinomiya-cho, Sakyo-ku, Kyoto 606-8501, Japan
e-mail: nyoshida@icems.kyoto-u.ac.jp

K. Hasegawa (✉)
Institute for Stem Cell Biology and Regenerative Medicine (inStem), National Centre for Biological Sciences (NCBS), Tata Institute of Fundamental Research (TIFR), GKVK Campus, Bellary Road, Bangalore 560065, India

Institute for Integrated Cell-Material Sciences (iCeMS), Kyoto University, Yoshida-Ushinomiya-cho, Sakyo-ku, Kyoto 606-8501, Japan
e-mail: kouichih@instem.res.in; khasegawa@icems.kyoto-u.ac.jp

I. Somasundaram (ed.), *Stem Cell Therapy for Organ Failure*,
DOI 10.1007/978-81-322-2110-4_14, © Springer India 2014

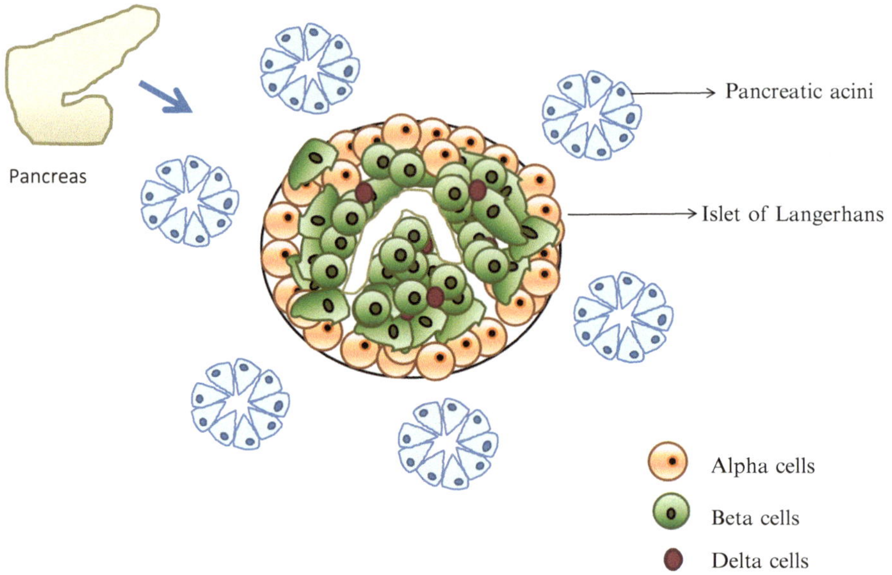

Fig. 1 A scheme of cell types in typical pancreas

of β-cells [4, 10]. However, ectopically grafted β-cells have two problems: one is decrease in their functionality after a year of engraftment and second is the need for lifelong immunosuppressive therapy. The solution to the latter problem would be to "induce immune tolerance" [4]. The narrow success has encouraged alternative regenerative therapy to spring up, where precursor/stem cells can be used to produce insulin-producing cell, slowly developing to be a more efficient substitute.

The basic question still remains regarding the orchestrated action from a pluri- or multipotent stem cells into differentiated states, characteristic for pancreatic development [11]. All five different pancreatic endocrine lineages as well as the exocrine, acinar, and ductal cells (Fig. 1) are derived from a pool of early pancreatic progenitor cells [12]. Ultimately, cell differentiation occurs by "well-orchestrated gene expression patterns" controlled by specific combinations of transcription factors [8]. Various factors like Sox9, Ngn3, and Pdx1 have been shown to be crucial in differentiation of beta cell lineage in mice (Fig. 2). Endoderm progenitors take pancreatic fate on activation of Pdx1 and Ptf1a in mouse [13, 8]. Pancreatic progenitors give rise to endocrine ductal and acinar progenitors [8]. Endocrine progenitors

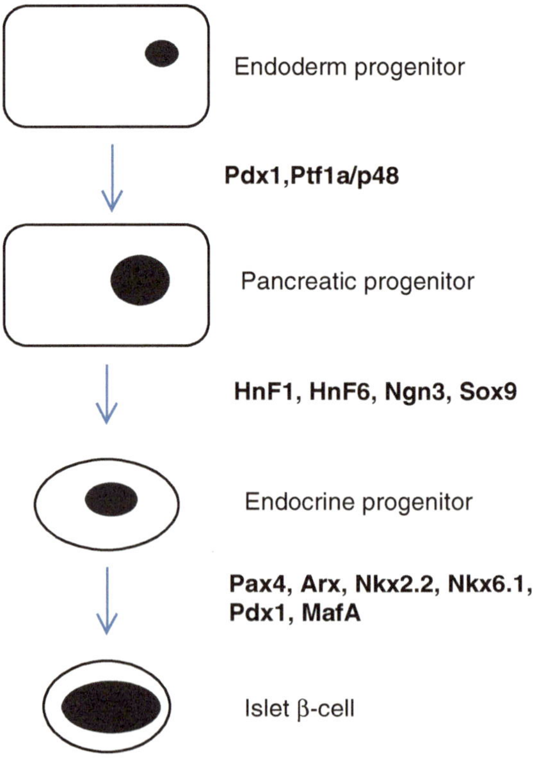

Fig. 2 A scheme of various factors involved in mouse β-cell differentiation

Table 1 Factors involved in generating IPCs

Factors to generate IPC	References
Activin A	[19, 20, 47]
Betacellulin	[64]
Basic fibroblast growth factor (bFGF or FGF2)	[18–20]
Epidermal growth factor (EGF)	[65]
Exendin-4	[20]
Gastrin	[66]
Glucagon-like peptide	[20]
Glucose	[18]
Hepatocyte growth factor (HGF)	[20]
Insulin-like growth factor	[20]
Keratinocyte growth factor (KGF or FGF7)	[18]
Nicotinamide	[18]
Retinoic acid	[18]

are then differentiated into α, β, δ, PP, and ξ cells. The identification of Neurogenin3 (Ngn3), a bHLH transcription factor, is very important for endocrine development. Ngn3-expressing cells (Ngn3+) function as endocrine precursor cells and produce all five hormone-secreting pancreatic cells [13]. Pax4 and Arx are both expressed in the mouse embryonic pancreas [8]. During the later developmental stages, Pax4is restricted to β- and δ-cells but not to Arx(ε- and ξ-cells) [14, 15]. Apart from the factors discussed above, various other external factors used for differentiating beta cells from various sources are summarized in Table 1. We have discussed many of these extrinsic factors in the differentiation protocol later in this chapter.

Insulin-Producing Cell Sources

There are various sources from which one could produce insulin-producing cells (IPCs). Figure 3 illustrates an overview of all cell-type differentiation to IPC. Cell sources from which islet cells are differentiated are human pluripotent stem cells, mesenchymal stem cells, transdifferentiation from various non β-cell sources, and adult pancreatic progenitor cells. This section discusses each of these cell sources used in recent applications, citing advantages and disadvantages for each of them.

Human Pluripotent Stem Cells

Human pluripotent stem cells (hPSCs) include human embryonic stem cells (hESCs) and human-induced pluripotent stem cells (hiPSCs) and have the capacity to differentiate almost all cell types in the body including IPCs. Both hESCs and hiPSCs are good source for the regenerative medicine in diabetes and generate "glucose-responsive insulin-secreting cells in vitro" [16–18]. hiPSC are ethically more acceptable than hESCs, especially for generating autologous stem cells and their derivatives [4]. So far, there have been several approaches identified for the differentiation of hPSCs into IPCs.

One such approach for differentiating hPSCs into IPCs is based on spontaneous differentiation. The selection of nestin-positive progenitor cells through embryoid body (EB) formation is one example, while acquiring pancreatic progenitor cells via definitive endoderm is another example of such approach [18]. The nestin-positive progenitor selection strategy relies on initial EB

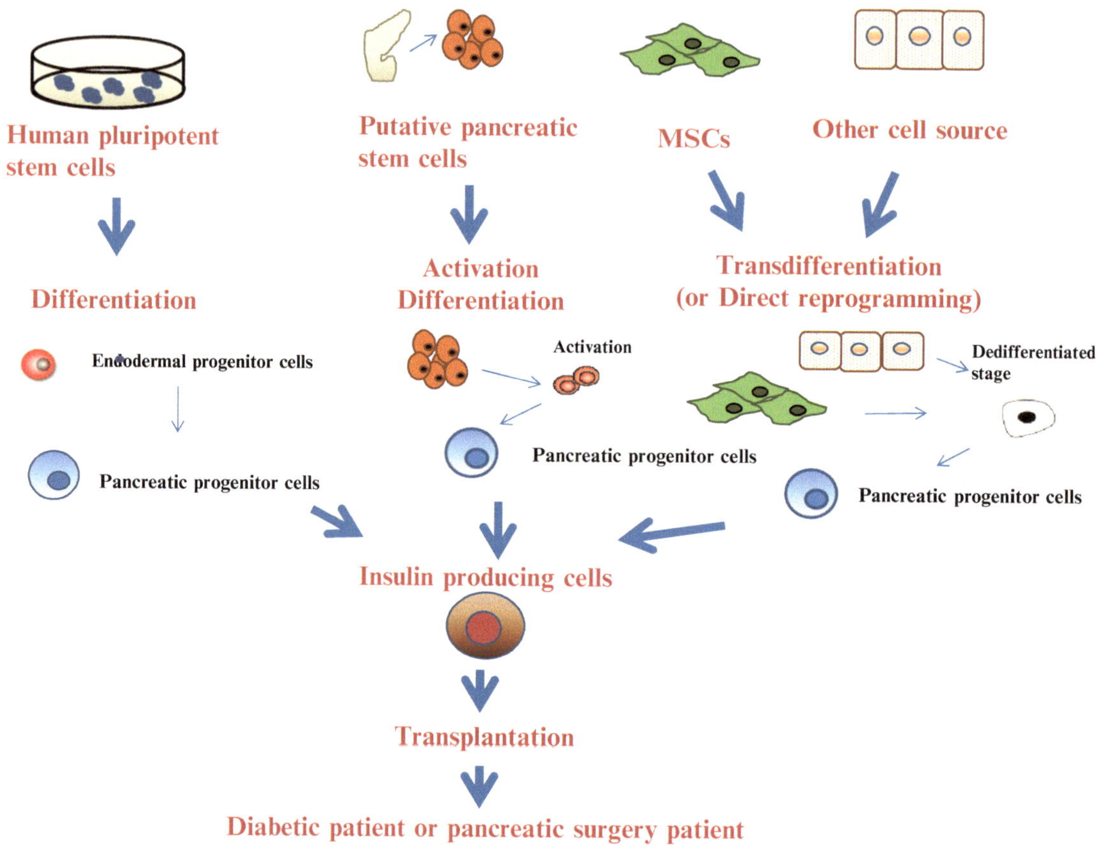

Fig. 3 Overview of all cell-type differentiation to insulin-producing cell

formation. EB further differentiates to all three germinal layers including endoderm (SOX9+), mesoderm (FLK1+), and ectoderm (Nestin+). Sequential treatment of EBs with a cocktail of growth factors enforces a lineage commitment pathway that initially forms nestin-positive cells, which subsequently differentiated into endocrine progenitors and ultimately into IPCs. The IPCs in the suspension culture aggregate into islet-like clusters. For the other protocol, the hESCs are first dissociated into small clumps and treated with Wortmannin, N2/B27 [18]. Further, the cells are cultured in IMDM/F12 medium with retinoic acid and fibroblast growth factor 7 (FGF7, also known as KGF) to produce IPC [18]. An alternate approach involves culture and plateEBs in insulin-transferrin-selenium-fibronectin medium, followed by medium supplemented with N2/B27 and basic fibroblast growth factor (bFGF, also known as FGF2) [18]. The next step involves lowering glucose concentration in the medium, followed by the withdrawal of bFGF and addition of nicotinamide [18]. Dissociating the cells and growing them in suspension result in the formation of clusters which exhibit higher insulin secretion and possess longer durability than cells grown as monolayers [18].

The other approach of differentiating hPSCs into IPCs are based on pancreatic development. Figure 4 depicts typical differentiation of hPSCs via definitive endoderm to IPC. hPSCs are differentiated to mesendoderm and definitive endoderm, which is further differentiated to pancreatic progenitor cells and finally to IPCs [4]. A protocol published very recently showed generation of insulin-producing cells from hPSCs using small molecules. In this protocol, treatment with activin A and a GSK3β (Glycogen synthase kinase 3) inhibitor CHIR99021 (an activator of Wnt signaling) enhances endodermal

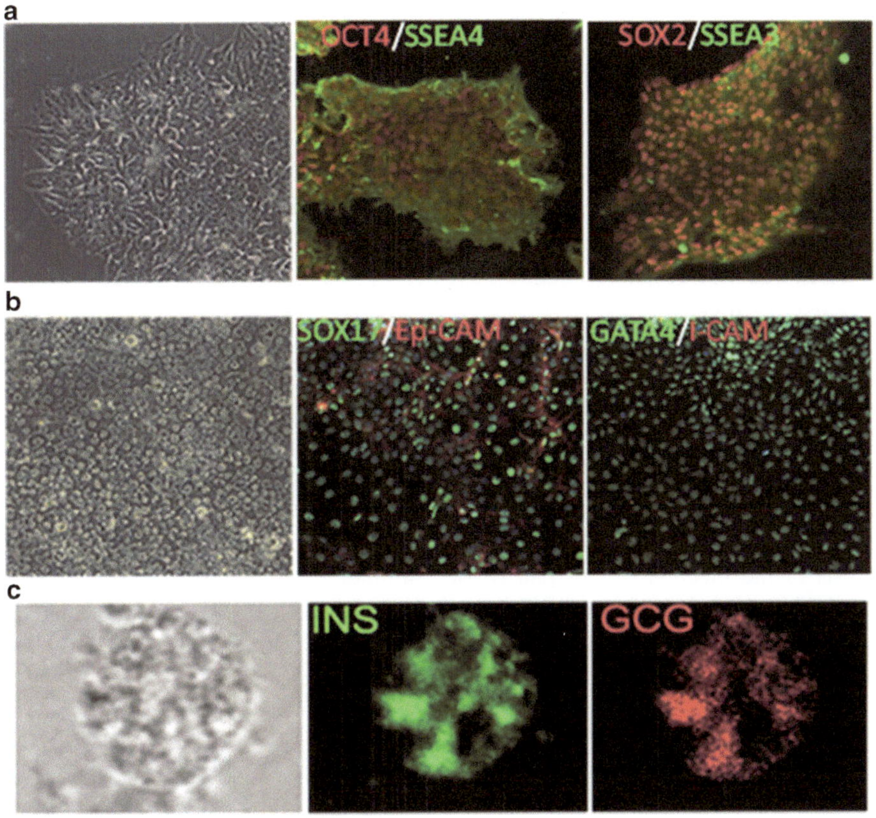

Fig. 4 An example of insulin-producing cells differentiated from hPSCs. (**a**) Undifferentiated hiPSCs. Phase contrast (*left panel*), immunofluorescence of undifferentiated marker OCT4 (red signals in *center panel*), SSEA4 (green signals in *center panel*), SOX2 (red signals in *right panel*), and SSEA3 (green signals in *right panel*).(**b**) Definitive endoderm cells differentiated from hiPSCs. Phase contrast (*left panel*), immunofluorescence of definitive endoderm marker SOX17 (green signals in *center panel*) and GATA4 (green signals in *right panel*), epithelial marker Ep-CAM (red signals in *center panel*), and hepatocyte marker I-CAM (red signals in *right panel*, undetected). (**c**) Pancreatic islet cells differentiated from hPSCs (Photos provided from Dr. Nan Sook Lee and Dr. Robert H Chow, University of Southern California). Phase contrast (*left panel*), immunofluorescence of β-cell marker insulin (green signals in *center panel*) and α-cell marker glucagon (red signals in *right panel*)

differentiation. This has been followed by the treatment of retinoic acid, a bone morphogenic protein (BMP) inhibitor, and a transforming growth factor-β (TGFβ) inhibitor to induce efficient differentiation of pancreatic progenitor cells (PDX1+ and NGN3+) from definitive endoderm [19]. Mixture of forskolin, dexamethasone, and a TGFβ inhibitor has been found to induce the differentiation of IPCs with 10% of the cells becoming insulin positive [19]. Various research groups have used activin to increase IPC content. Activin along with Wnt3a or CHIR99021 increases endodermal differentiation in vitro, mimicking the coordinated expression and action of both activin/nodal and Wnt during primitive streak formation [19, 20]. Kunisada et al. also compared Wnt3a and CHIR99021 in the same condition and showed CHIR99021 as more efficient than Wnt3a in inducing SOX17- and FOXA2-positive endodermal cells [19]. Various factors regulate pancreatic differentiation during embryonic development such as FGF and BMP pathways such as bFGF, FGF7, FGF10, and Noggin (an endogenous protein BMP inhibitor). Incretins and growth factors such as insulin-like growth factor-1 (IGF-1), hepatocyte growth factor

(HGF), and glucagon-like peptide-1 (GLP-1)/exendin-4 (a peptide analog of GLP-1) have also been used to facilitate differentiation [20].

Jiang et al. reported another simple approach involving serum-free medium which could induce IPCs from hESCs [21]. Sodium butyrate and activin have been used to generate definitive endoderm co-expressing CXCR4 and SOX17 and CXCR4 and FOXA2. IGF and nicotinamide have also been used at the final stages [21].

These protocols for differentiation of hPSCs into IPCs are promising; however, these methods are laborious, expensive, as well as time consuming [18]. Although these protocols can successfully be used to differentiate hPSCs into IPCs, it is still debated which protocol is better suited for future use. We strongly feel that a greater in-depth analysis of the mechanisms of pancreatic islet β-cell development will be necessary for further development of efficient protocols in IPC differentiation from hPSCs. Regarding enhancing efficiency of IPC differentiation, quality or varying characteristics of hPSCs are a matter of concern. Recently, Wen et al. generated a panel of hPSCs and examined the efficiency of differentiation among the hPSC lines [22]. The authors found most hPSC lines were induced into PDX1-positive progenitor cells expressing PDX1, MAFA, GLUT2, and insulin at the final stages of differentiation. However, the efficiency of differentiation and the expression of each differentiation step markers, such as SOX17, SOX9, PDX1, and insulin, were observed on different days of differentiation between hPSC lines [23].

Although a lot of research has been focused on using hPSCs showing promising results about the formation of IPC, one must not forget to survey the disadvantages of using hPSCs, especially in treating diabetes [4]. One of the problems with hPSC transplantation is the risk of formation of teratomas due to contamination of undifferentiated hPSC [16, 24]. Another disadvantage in hiPSCs is the DNA-based reprogramming techniques that could lead to insertional mutagenesis or oncogenic reprogramming factors resulting in tumor formation. Non-integrative techniques are preferred for iPSC clinical application. However, eliminating DNA insertion alone for generating iPSC may not solve the other problems associated with genetic instabilities and abnormalities acquired during both somatic differentiation and reprogramming [4]. Moreover, ethical and safety issues regarding usage of hPSCs still remain as the biggest challenge and must be resolved before clinical trials.

Mesenchymal Stem Cells

Mesenchymal stem cells (MSCs), a group of stem cells, are being used for improving islet transplantation. Very recently, Wu et al. found that co-transplantation of human bone marrow-derived mesenchymal stem cells (hBMSCs) could prevent immune rejection and improve human islet transplantation [25]. However, the capacity of hBMSCs to differentiate to IPCs still remains controversial. Few groups have tried establishing this pathway of differentiation. Xie et al. used the traditional two-step induction to differentiate MSCs into IPCs [26]. They showed that injured pancreatic tissue extract could effectively promote the transdifferentiation efficiency and maturity of IPCs by the traditional induction [26]. On the other hand, Gershengorn et al. successfully showed that fibroblast-like cells derived from adult human islets donated postmortem proliferate readily in vitro [27]. These mesenchymal-type cells, which exhibit no hormone expression, can then be induced to differentiate into hormone-expressing islet-like cell aggregates, which reestablish the epithelial character typical of islet cells [27]. It has also been shown that human pancreatic islet cells migrate to form mesenchymal populations in growth-promoting conditions and further human islet-derived mesenchymal cells differentiate into hormone-producing cell aggregates [28]. Proliferating populations of mesenchymal-like islet-derived progenitor cells have shown great plasticity in serum-depleted condition, and Dalvi et al. believe that "these mesenchymal cells generated in vitro may be better islet progenitors" [28]. Bone marrow cells from male mice (which express Cre-Lox tagged with EGFP (Green fluorescent protein) when the insulin gene is actively transcribed) have been used and transplanted into lethally irradiated recipient female mice; these GFP positive cells were found in the islets [29]. Unfortunately, contradictions have been found in this data [30].

Adult Pancreatic Progenitor Cells

Stem/progenitor cells are undifferentiated long-lived cells and have the capacity to retain their stem cell identity by self-renewal and undergo multilineage differentiation in various tissue/organ [31]. Several approaches have been undertaken to identify a population of resident human pancreatic stem cells. Although some progenitor cell markers can be histochemically identified in the adult pancreas, especially after severe injury, evidences that these progenitor/precursor cells differentiated from a stem cell population have been difficult to obtain [32, 33]. in vivo analyses indicate that no population of cells resides in the human adult pancreas that proliferates at the high rates characteristic of stem cells in gut or skin [33, 34]. However, neither study could exclude the possibility of existence of dedicated stem/progenitor cells with quiescent or a slower mitotic rate. If such stem cells could be isolated and expanded, they might be more amenable to differentiation into β-cells than either undifferentiated hESCs or hiPSCs. Identification of stem and/or progenitor cells in the human adult pancreas has been an area of intense investigation in the past decades, but the results remain controversial. In rodents, there are pancreatic stem cells that can regenerate the pancreas, and several studies have been done on pancreatic progenitor cell nature in various culture conditions. However, live cell isolation of the progenitor cells from adult human pancreas and characterization of the nature of these cells remain largely unknown.

In mouse, three independent studies of the transcription factors hepatocyte nuclear factor 1β (Hnf-1β) [35] and Sox-9 [12, 36] in embryonic mice, normal adult mice, and mice after pancreatic duct ligation have been performed to demonstrate pancreatic progenitor cell location and activities. A fourth study has been done on mucin-1 as a duct-cell tracer in the fetal and postnatal mouse pancreas [37]. Tamoxifen-inducible Cre-mediated lineage tracing experiments have been conducted to test whether duct cells are a source of progenitor, using Cre driven by the carbonic anhydrase-II promoter [38], the Hnf1b promoter [35], or the Muc1 (mucin-1) promoter [37]. CreER experiment showed that certain percentage of the β-cells in the ligated portion of the pancreas, whereas for both Hnf1b-CreERT2 and Muc1-CreERT2 mice, the results concluded that duct cells do not convert into islet cells in the uninjured pancreas after birth [38, 35, 37]. Xu et al. using injured mouse, have shown that the progenitor population activated during injury expresses the gene for Ngn3 [39]. The evidence that a "fraction of β-cells are produced from alpha cells [40]" expressing the glucagon promoter under extreme injury conditions, further suggests that β-cell progenitors could have an islet origin.

Some other studies in mouse have demonstrated genetic markers of the pancreatic progenitors. A lineage tracing experiment using Ins2 (insulin-2) promoter CreER transgenic mouse line has provided evidence that new β-cells are primarily derived from cells with at least some Ins2 promoter activity [33, 41]. Ins2 promoter activity is not fully restricted to β-cells, unlike the Ins1 promoter that is activated later in development and marks a majority of mature β-cells [42, 43]. Ngn3 might be a good marker for endocrine progenitor cells. Ngn3 is only expressed at a very low level in differentiated endocrine cells, but is a characteristic of endocrine differentiation from embryonic progenitor cells [4]. Increase in Ngn3 expression in the pancreata of adult mice post-pancreatic duct ligation has confirmed there activation of "dormant" endocrine progenitors [37]. Genetic lineage tracing has revealed that these cells do not further differentiate into β-cells [12]. Some evidences suggest that adult putative stem cells can give rise to populations with the characteristics of islet cells and other lineages. This indicates their multipotent nature and is also characterized by high levels of aldehyde dehydrogenase activity, enabling their

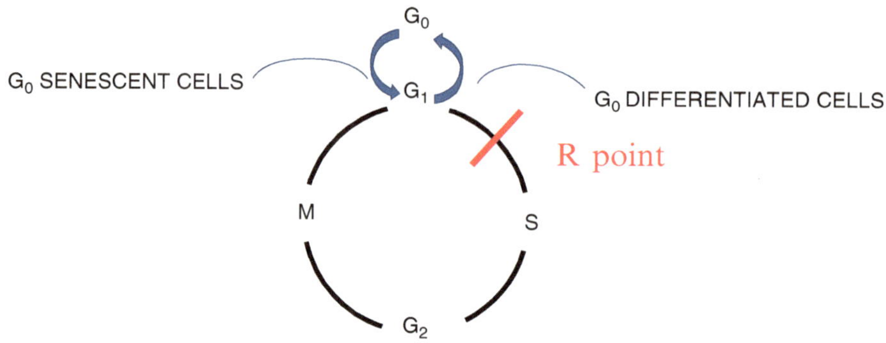

Fig. 5 A scheme of molecular switch of adult progenitor cells of quiescent and senescent stages

isolation by flow cytometry using a fluorogenic enzyme substrate; however, the sorting number is ultimately low [44]. These cells proliferate and form cellular aggregates that display a capacity for endocrine and exocrine differentiation [45]. However, genetic lineage does not verify this result. There are several other markers such as Sox9 in mouse; however, there is no validation of such markers for human pancreatic progenitors at the moment.

The other studies on mouse pancreatic progenitors have demonstrated activators of the cells. Somatic stem cells are able to enter reversible (quiescent) or irreversible (senescent and differentiated) G0 states from the G1 phase of the cell cycle before the restriction point (R point), and once these cells reach the R point, they are committed to the next round of the cell cycle (Fig. 5) [46]. Subpopulations of stem cells reside in the quiescent state and enter the cell cycle when they become activated in response to appropriate extrinsic signals [46]. Evidence suggests that many of the genes essential for embryonic development of the islet are maintained at low levels in the environment of adult β-cell [33]. Several groups have carried out studies to determine whether soluble growth factors, known to regulate pancreatic β-cell development, play a role in maintaining adult β-cell identity. Emerging evidences suggest a prominent role for the members of the TGFβ superfamily. Using a factorial design high-content screening approach, Szabat et al. have recently found that activin A and follistatin have reciprocal effects on the maturity of adult β-cells [47]. On other hand, Dioum et al. have identified a family of neurogenic 3,5-disubstituted isoxazoles (ISX) that increases the expression of neurogenic differentiation 1 (NEUROD1, also known as BETA2) in human cadaveric islets; this transcription factor functions in neuronal and pancreatic β-cell differentiation and is essential for insulin gene transcription [48]. Within the first few hours of exposure, ISX has been found to cause biphasic activation of ERK1/2 and increase bulk histone acetylation [48]. The authors also found that although there was little effect on histone deacetylase activity, Isx increased histone acetyl transferase activity in nuclear extracts [48]. The Notch signaling pathway is a crucial regulator of progenitor dynamics in the mammalian pancreas [49]. Genetic deletion of Notch signaling components in the pancreatic domain results in differentiation of presumptive Notch responsive progenitors into endocrine cells including β-cells [49]. High levels of Notch signaling correlate with quiescence/slow cycling, whereas lower levels are associated with progenitor amplification; in addition, the study also revealed that endocrine differentiation requires strong Notch signaling downregulation. Therefore, playing with the signaling cascade which regulates endocrine differentiation (e.g., cell cycle, notch regulation) could be a possible mechanism of islet cell differentiation. Apart from differentiation of stem cells, there are signals that can induce self-proliferation of β-cells, a potential target to produce IPCs. Recently, a group from Harvard stem cell unit has shown that transient expression of endogenous betatrophin (secretory in nature) in mouse liver significantly and specifically promotes pancreatic β-cells proliferation, expands islet cell mass, and improves glucose tolerance in mouse [5].

All of these studies have been performed in mouse. Unlike mouse, human cannot regenerate pancreas. Rare multipotent progenitor cells have been isolated from human pancreata, where the progenitor cells were presumably residing in islets [50]. The study also revealed that these cells could proliferate and generate progeny with pancreatic and neural features in vitro. Genetic lineage tracing of the same cells in mouse revealed that these progenitor cells were derived from cells that expressed insulin. However, the existence and use of adult stem cells in human pancreatic islet regeneration still remains unsure. Therefore, further study is required first to identify biomarkers, next to determine their differentiation capacities, and, finally, to understand how do they control diabetes.

Transdifferentiation of Islet Cells from Non-β-Cell Source

The concept of "transdifferentiation" or "direct reprogramming" has been extensively used in this field. In transdifferentiation, using ex vivo or in vivo approaches, a "detour method" can be applied to convert one type of differentiated cells into another phenotype "without complete dedifferentiation" [51]. The key transcription factors involved in pancreas are PDX1, NGN3, and MafA. Using these factors, transdifferentiation of exocrine acinar cells into endocrine β-cells has been demonstrated [52]. Pancreatic α-cells are more abundant than β-cells and have been a target for a while. Courtney et al. showed a spontaneous transition of α-cell to β-cell type [53]. The authors demonstrated that pancreatic α-cells can also be converted into β-cells by inactivation of Arx gene in pancreatic α-cells [53].

The cell types which share the same origin of pancreatic β-cell might be a good potential target to transdifferentiate and produce insulin-producing cells. Also upregulated genes involved in β-cell development/differentiation would be potential targets to induce reprogramming. Although the knowledge of human pancreatic development is strictly limited, β-cell phenotype characteristic is conserved across species. However, understanding of islet cell development comes from developmental studies in mice which have significant differences (e.g., the relative roles of glucose transport and phosphorylation as a part of glucose sensing [54], responsiveness to glucokinase activators [55], glucose-induced desensitization [56], responses to galanin [57] and melatonin [58], the roles of Pax4 [59] and p57Kip2 [60] in β-cell proliferation, and, central to β-cell function, regulation of the insulin promoter [61]) from humans [62]. Studies on human system have relied on histological samples or in vitro, in vivo studies using human fetal pancreata. Studies have been conducted on fetal pancreata mainly at 18–24 gestational weeks [8]. At this late stage of pancreatic development, the human pancreas is already rich in endocrine cells; therefore, the analysis of islet differentiation under these conditions might be somewhat limited [8, 67]. In summary, although rodent and human pancreas development may share similarities, further studies in human pancreas using human stem cells would be necessary for further development of transdifferentiation and/or direct reprogramming of β-cells.

Clinical Trials

Among all sources of IPC formation for cell transplantation purposes, strategies based on hPSCs seem to have the highest translational potential; however, the safety issues are of major concern. Formation of teratoma and allorejections are the major issues.

Although the use of iPSC is not as risky in immune rejection, the threats are still associated with DNA transfer/mutations in iPSC generation. In addition, ethical issues are also involved if autologous cells are used before it comes to clinical trial. The main challenges associated with this approach are as follows: (i) checking on regulation of expression by the insulin gene to make sufficient amount of insulin, (ii) prevention of immune response to these genetically modified cells, and (iii) autoimmunity that previously resulted in killing of patient's insulin-producing β-cells (http://www.asgct.org/). Preclinical studies to overcome the patient's aberrant immune response to pancreatic β-ells by genetically incorporating immune suppressive proteins (cytokines, such as TGFβ and interleukin 10 (IL-10)) into the pancreas may suppress autoimmune response allowing the patient's own insulin-secreting β-cells to survive (http://www.asgct.org/).

Currently, there are more than 20 ongoing human clinical trials utilizing one of the following cell sources (allogeneic pancreatic β-cells, mesenchymal stem cells, cord blood stem cells, or autologous adipose-derived stem cells) to transfer or generate pancreatic β-cells in patients with type I diabetes (http://www.asgct.org/.) For the hPSC-derived IPCs to be used for clinical use in diabetes therapy, a differentiation protocol with significant detail and improvement needs to be developed [18]. Several research groups are trying to establish protocols; however, none of the protocols has shown robust insulin production from hPSC till date. The safety usage and protecting these cells from the immune system are the main challenges yet to resolve. To improve cell therapy trials, researchers either added specific nutrients (vitamin D, omega 3 fatty acids which are usually low in diabetics) along with the immunosuppressant or deleted the β-cell targeted autoimmune cells (for type 1)(http://www.asgct.org/). Collectively, these novel approaches represent the advent of a new and promising era in type I diabetes research.

The existence and usage of adult pancreatic stem cells in differentiation and/or transdifferentiation of IPC still need to be explored. Recently, gene therapy researchers have identified three genes (Ngn3, Pdx1, and Mafa) that can convert adult pancreatic exocrine cells into pancreatic β-cells [52]. In preclinical animal models, these reprogrammed exocrine cells are found to be identical to β-cells in size, shape, and expression of essential genes and result in reduced blood sugar levels. Other cells that have been shown to be susceptible to reprogramming to become beta-like cells include pancreatic α-cells [53], pancreatic ductal cells [64], and adult stem cells in the liver [65]. However, similar researches are needed to be explored in humans.

Conclusions and Hypothesis

There are several potential cell sources that can be differentiated/reprogrammed into pancreatic β-cells. However, to find out most efficient ways to cure diabetes, there are several issues to be resolved and improved, for example, the efficient methods of producing pancreatic IPCs, evaluation of these cells, transplantation methods including safety, and long-term observation of the function of the transplanted cells. We propose here that using progenitor markers and live cell isolation could lead us to have an in-depth characterization of adult pancreatic progenitor cells, which would further help us to differentiate islet cells. Figure 6 illustrates a schematic representation of in vivo generation of insulin-producing cells. It has recently been seen that Sox9-positive cells can differentiate all cell types in pancreas and regenerate whole pancreas in mouse [36]. Sox9 is expressed through the biliary and pancreatic ductal epithelium connected to intestinal stem cell zone [36]. This study also suggests that the fundamental function of embryonic Sox9 might maintain cells in an undifferentiated state in the adult, and Sox9 marks the precursor cell population during physiological cell replacement and/or during regenerative process after injury. In human pancreas, there exists a small population of SOX9-positive cells too [66]. Therefore, we hypothesized that SOX9 is a progenitor marker for isolation of pancreatic progenitor marker in human also. Literature study has revealed that FACS-sorted centroacinar/terminal ductular cells are uniquely able to form self-renewing "pancreatospheres" in suspension culture [45], even when plated at clonal density in mouse; however, similar self-renewing processes have not been observed in human. In addition, SOX9 is also a transcription factor located in nuclei, and so far there is no way to isolate living SOX9-positive cells from human pancreas. Therefore, biomarkers are needed to isolate live progenitor cells

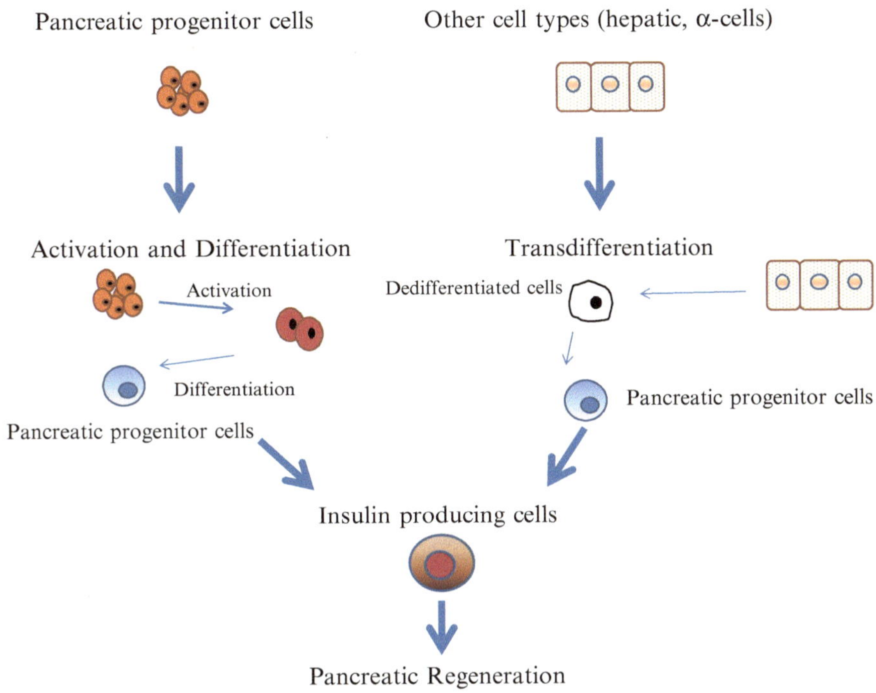

Fig. 6 Overview of in vivo generation of insulin-producing cells

from adult human tissues for uniquely identifying the progenitor cells within the pancreatic cell population. Using this technique, one may isolate pancreatic progenitor cells and differentiate cells with known factors or use small molecules by high throughput screening. We have recently found a monoclonal antibody that recognizes the surface of SOX9-positive cells in pancreas [66] and have started isolation of these cells from adult pancreas.

We believe that the basic studies of human pancreas development, differentiation of β-cells, and nature of adult pancreatic progenitor cells will be important for further development of regenerative medicine in human pancreas leading to cure diabetes and/or pancreatic cell loss including pancreatic cancer surgery in future.

References

1. World Health Organization (WHO) (2013) "Diabetes" Media Centre, Fact Sheet No. 312. http://www.who.int/mediacentre/factsheets/fs312/en/
2. International Diabetes Federation (2013) IDF diabetes atlas, 6th edn. International Diabetes Federation, Brussels. http://www.idf.org/diabetesatlas
3. Yoon J, Jun H (2005) Autoimmune destruction of pancreatic b cells. Am J Ther 591:580–591
4. Bouwens L, Houbracken I, Mfopou JK (2013) The use of stem cells for pancreatic regeneration in diabetes mellitus. Nat Rev Endocrinol 9:598–606
5. Yi P, Park J-S, Melton DA (2013) Betatrophin: a hormone that controls pancreatic β cell proliferation. Cell 153:747–758
6. Butler AE et al (2007) Modestly increased beta cell apoptosis but no increased beta cell replication in recent-onset type 1 diabetic patients who died of diabetic ketoacidosis. Diabetologia 50:2323–2331
7. Rahier J, Guiot Y, Goebbels RM, Sempoux C, Henquin JC (2008) Pancreatic beta-cell mass in European subjects with type 2 diabetes. Diabetes Obes Metab 10 Suppl 4:32–42
8. Rojas A, Khoo A, Tejedo JR, Bedoya FJ, Martín F (2010) Islet cell development. Adv Exp Med Biol 654:59–75
9. Naftanel MA, Harlan DM (2004) Pancreatic islet transplantation. PLoS Med 1:e58; quiz e75
10. Jacobs-Tulleneers-Thevissen D et al (2010) Human islet cell implants in a nude rat model of diabetes survive better in omentum than in liver with a positive influence of beta cell number and purity. Diabetologia 53:1690–1699

11. Shih HP et al (2012) A Notch-dependent molecular circuitry initiates pancreatic endocrine and ductal cell differentiation. Development 139:2488–2499
12. Kopp JL et al (2011) Sox9+ ductal cells are multipotent progenitors throughout development but do not produce new endocrine cells in the normal or injured adult pancreas. Development 138:653–665
13. Gu G, Dubauskaite J, Melton DA (2002) Direct evidence for the pancreatic lineage: NGN3+ cells are islet progenitors and are distinct from duct progenitors. Development 129:2447–2457
14. Sosa-pineda B (2004) Molecules and the gene Pax4 is an essential regulator of pancreatic β-cell. Development 18:289–294
15. Collombat P et al (2003) Opposing actions of Arx and Pax4 in endocrine pancreas development. Genes Dev 17:2591–2603
16. Kroon E et al (2008) Pancreatic endoderm derived from human embryonic stem cells generates glucose-responsive insulin-secreting cells *in vivo*. Nat Biotechnol 26:443–452
17. Assady S et al (2001) Insulin production by human embryonic stem cells. Diabetes 50:1691–1697
18. Wei R et al (2013) Insulin-producing cells derived from human embryonic stem cells: comparison of definitive endoderm- and nestin-positive progenitor-based differentiation strategies. PLoS One 8:e72513
19. Kunisada Y, Tsubooka-Yamazoe N, Shoji M, Hosoya M (2012) Small molecules induce efficient differentiation into insulin-producing cells from human induced pluripotent stem cells. Stem Cell Res 8: 274–284
20. Hosoya M (2012) Preparation of pancreatic β-cells from human iPS cells with small molecules. Islets 4:249–252
21. Jiang J et al (2007) Generation of insulin-producing islet-like clusters from human embryonic stem cells. Stem Cells 25:1940–1953
22. Wen Y, Chen B, Ildstad ST (2011) Stem cell-based strategies for the treatment of type 1 diabetes mellitus. Expert Opin Biol Ther 11:41–53
23. Zhang D et al (2009) Highly efficient differentiation of human ES cells and iPS cells into mature pancreatic insulin-producing cells. Cell Res 19:429–438
24. Sui L, Mfopou JK, Chen B, Sermon K, Bouwens L (2013) Transplantation of human embryonic stem cell-derived pancreatic endoderm reveals a site-specific survival, growth, and differentiation. Cell Transplant 22:821–830
25. Wu H, Wen D, Mahato RI (2013) Third-party mesenchymal stem cells improved human islet transplantation in a humanized diabetic mouse model. Mol Ther 21:1778–1786
26. Xie H et al (2013) Role of injured pancreatic extract promotes bone marrow-derived mesenchymal stem cells efficiently differentiate into insulin-producing cells. PLoS One 8:e76056
27. Gershengorn MC et al (2004) Epithelial-to-mesenchymal transition generates proliferative human islet precursor cells. Science 306:2261–2264
28. Dalvi MP, Umrani MR, Joglekar MV, Hardikar AA (2009) Human pancreatic islet progenitor cells demonstrate phenotypic plasticity *in vitro*. J Biosci 34:523–528
29. Ianus A, Holz GG, Theise ND, Hussain MA (2003) *In vivo* derivation of glucose-competent pancreatic endocrine cells from bone marrow without evidence of cell fusion. J Clin Invest 111:843–850
30. Taneera J et al (2006) Failure of transplanted bone marrow cells to adopt a pancreatic beta-cell fate. Diabetes 55:290–296
31. Weissman IL (2000) Stem cells: units of development, units of regeneration, and units in evolution. Cell 100:157–168
32. Ku HT (2008) Minireview: pancreatic progenitor cells–recent studies. Endocrinology 149:4312–4316
33. Szabat M et al (2012) Maintenance of β-cell maturity and plasticity in the adult pancreas: developmental biology concepts in adult physiology. Diabetes 61:1365–1371
34. Teta M, Rankin MM, Long SY, Stein GM, Kushner JA (2007) Growth and regeneration of adult beta cells does not involve specialized progenitors. Dev Cell 12:817–826
35. Solar M et al (2009) Pancreatic exocrine duct cells give rise to insulin-producing beta cells during embryogenesis but not after birth. Dev Cell 17: 849–860
36. Furuyama K et al (2011) Continuous cell supply from a Sox9-expressing progenitor zone in adult liver, exocrine pancreas and intestine. Nat Genet 43: 34–41
37. Kopinke D, Murtaugh LC (2010) Exocrine-to-endocrine differentiation is detectable only prior to birth in the uninjured mouse pancreas. BMC Dev Biol 10:38
38. Inada A et al (2008) Carbonic anhydrase II-positive pancreatic cells are progenitors for both endocrine and exocrine pancreas after birth. Proc Natl Acad Sci U S A 105:19915–19919
39. Xu X et al (2008) Beta cells can be generated from endogenous progenitors in injured adult mouse pancreas. Cell 132:197–207
40. Thorel F et al (2010) Conversion of adult pancreatic alpha-cells to beta-cells after extreme beta-cell loss. Nature 464:1149–1154
41. Dor Y, Brown J, Martinez OI, Melton DA (2004) Adult pancreatic beta-cells are formed by self-duplication rather than stem-cell differentiation. Nature 429:41–46
42. Deltour L et al (1993) Differential expression of the two nonallelic proinsulin genes in the developing mouse embryo. Proc Natl Acad Sci 90:527–531
43. Szabat M, Luciani DS, Piret JM, Johnson JD (2009) Maturation of adult beta-cells revealed using a Pdx1/insulin dual-reporter lentivirus. Endocrinology 150:1627–1635
44. Erickson RP, Grimes J, Venta PJ, Tashian RE (1995) Expression of carbonic anhydrase II (CA II)

promoter-reporter fusion genes in multiple tissues of transgenic mice does not replicate normal patterns of expression indicating complexity of CA II regulation *in vivo*. Biochem Genet 33: 421–437
45. Rovira M et al (2010) Isolation and characterization of centroacinar/terminal ductal progenitor cells in adult mouse pancreas. Proc Natl Acad Sci U S A 107:75–80
46. Cheung TH, Rando TA (2013) Molecular regulation of stem cell quiescence. Nat Rev Mol Cell Biol 14:329–340
47. Szabat M, Johnson JD, Piret JM (2010) Reciprocal modulation of adult beta cell maturity by activin A and follistatin. Diabetologia 53:1680–1689
48. Dioum EM et al (2011) A small molecule differentiation inducer increases insulin production by pancreatic β cells. Proc Natl Acad Sci U S A 108: 20713–20718
49. Ninov N, Borius M, Stainier DYR (2012) Different levels of Notch signaling regulate quiescence, renewal and differentiation in pancreatic endocrine progenitors. Development 139:1557–1567
50. Seaberg RM et al (2004) Clonal identification of multipotent precursors from adult mouse pancreas that generate neural and pancreatic lineages. Nat Biotechnol 22:1115–1124
51. Nicholas CR, Kriegstein AR (2010) Regenerative medicine: cell reprogramming gets direct. Nature 463:1031–1032
52. Zhou Q, Brown J, Kanarek A, Rajagopal J, Melton DA (2008) *In vivo* reprogramming of adult pancreatic exocrine cells to beta-cells. Nature 455:627–632
53. Courtney M et al (2013) The inactivation of Arx in pancreatic α-cells triggers their neogenesis and conversion into functional β-like cells. PLoS Genet 9:e1003934
54. Schuit FC (1997) Is GLUT2 required for glucose sensing? Diabetologia 40:104–111
55. Johnson D et al (2007) Glucokinase activators: molecular tools for studying the physiology of insulin-secreting cells. Biochem Soc Trans 35: 1208–1210
56. Zawalich WS, Bonnet-Eymard M, Zawalich KC (1998) Glucose-induced desensitization of the pancreatic beta-cell is species dependent. Am J Physiol 275:E917–E924
57. McDonald TJ et al (1994) Canine, human, and rat plasma insulin responses to galanin administration: species response differences. Am J Physiol 266:E612–E617
58. Ramracheya RD et al (2008) Function and expression of melatonin receptors on human pancreatic islets. J Pineal Res 44:273–279
59. Brun T et al (2008) The diabetes-linked transcription factor Pax4 is expressed in human pancreatic islets and is activated by mitogens and GLP-1. Hum Mol Genet 17:478–489
60. Potikha T, Kassem S, Haber EP, Ariel I, Glaser B (2005) p57Kip2 (cdkn1c): sequence, splice variants and unique temporal and spatial expression pattern in the rat pancreas. Lab Invest 85:364–375
61. Hay CW, Docherty K (2006) Comparative analysis of insulin gene promoters: implications for diabetes research. Diabetes 55:3201–3213
62. Piper Hanley K et al (2010) *In vitro* expression of NGN3 identifies RAB3B as the predominant Ras-associated GTP-binding protein 3 family member in human islets. J Endocrinol 207: 151–161
63. Al-Hasani K et al (2013) Adult duct-lining cells can reprogram into β-like cells able to counter repeated cycles of toxin-induced diabetes. Dev Cell 26: 86–100
64. Banga A, Akinci E, Greder LV, Dutton JR, Slack JM (2012) In vivo reprogramming of Sox9+ cells in the liver to insulin-secreting ducts. Proc Natl Acad Sci U S A 109:15336–15341
65. Stamp LA et al (2012) The GCTM-5 epitope associated with the mucin-like glycoprotein FCGBP marks progenitor cells in tissues of endodermal origin. Stem Cells 30:1999–2009
66. Demeterco C, Beattie GM, Dib SA, Lopez AD, Hayek A (2000) A role for activin A and betacellulin in human fetal pancreatic cell differentiation and growth. J Clin Endocrinol Metab 85:3892–3897

Feto-Maternal Mesenchymal Stem Cell Transplantation for Treating Diabetes

Ramesh R. Bhonde and Vijayalakshmi Venkatesan

Introduction

In recent years, stem cells have reached to greater heights due to its therapeutic potential in treating degenerative diseases. The long-term objective of regenerative medicine or stem cell therapy is to treat patients with their own stem cells. A better understanding of stem cell biology would almost certainly allow for the establishment of efficient and reliable cell transplantation experimental programs in the clinic. These stem cells can be isolated from various sources such as the bone marrow, adipose tissue, pancreas, liver, placenta, umbilical cord blood, Wharton's jelly, and so on. However, the present investigation demands a readily available, abundant, reliable noninvasive, and efficient source of stem cell that could be a valuable tool in regenerative medicine. Such sources that could be explored in the aforesaid category are the feto-maternal mesenchymal stem cells (biological trash) such as placenta, cord blood, and cord tissue. These could be transformed into a treasure in treatment of multitude of diseases, especially in the treatment of diabetes.

This chapter reviews the potential use of feto-maternal MSCs to yield abundant glucose-responsive insulin-producing cells as ideal candidates for treating diabetes mellitus. It has been hypothesized that these undifferentiated feto-maternal mesenchymal stem cells may be used to replenish the islet mass in diabetic patients, thus making MSC transplantation (a form of cell therapy that has already proven effective at clinically restoring normoglycemia) available to millions of prospective diabetic patients.

Significance of Feto-Maternal Source

The most principal redundant source under the scan of cell therapists is the fetal placenta and the umbilical cord. Details into its components and development, maintenance, as well as physiologic functions such as immunotolerance and regenerative capacity of these extra-embryonic tissue-derived stem cells had created great deal of excitement to contemporary researchers [1]. There are multitude supportive evidences that make placenta a better source of stem cells; these include its high plasticity as evidenced by its early stage embryological development as well as its role in feto-maternal tolerance [2]. In addition, the noninvasive nature of this pregnancy discard tissue makes it an attractive source for stem cells.

R.R. Bhonde (✉)
School of Regenerative Medicine, Manipal Institute of Regenerative Medicine, Manipal University, GKVK Post, Bellary Road, Allalasandra, Near Hotel Royal Orchid, Yelahanka, Bangalore 560065, India
e-mail: rr.bhonde@manipal.edu

V. Venkatesan
Department of Stem Cell Research, National Institute of Nutrition (ICMR), Secunderabad, Hyderabad 500007, India
e-mail: v.venkateshan@gmail.com

I. Somasundaram (ed.), *Stem Cell Therapy for Organ Failure*,
DOI 10.1007/978-81-322-2110-4_15, © Springer India 2014

The fetal component of term placenta amnion contains amniotic epithelium and amniotic mesenchyme whereas the chorion composed of chorionic mesenchyme and chorionic trophoblast.

The plasticity of amniotic membrane is evidenced by the presence of amniotic mesenchyme as well as the embryonic nature of amniotic epithelium. Chorion part of placental tissue had undergone meticulous scrutiny to identify mesenchymal stromal cells. Several reports illuminate the higher yield coupled with extensive proliferative capacities for chorion-derived stromal cells [3, 4]. Besides, their suitability in cellular therapeutics is better illustrated by their fibroblastic morphology, plastic adherence, surface antigenic expression, as well as multidifferentiation capacity [4–7]. The placenta is involved in maintaining fetal tolerance and contains cells that display immunomodulatory properties [2]. This feature coupled with its redundancy and other advantages makes it a suitable candidate for cellular therapies and transplantation biologists.

Similar to the placental source of redundant tissue, umbilical cord matrix tissue that connects the placenta and the fetus during pregnancy, which is discarded after delivery, is yet another potent source of stem cells. The advantages of the redundant umbilical cord tissue are similar as that of placenta in redundancy, availability, reliability, and efficiency [3, 8]. It can provide an inexhaustible source of stem cells for therapy without involving any invasive procedures or ethical concerns. The umbilical cord is a fetus-derived tissue, and the WJ-derived MSCs share some properties unique to fetal-derived MSCs, like having faster proliferation and greater ex vivo expansion capabilities than adult MSCs. In addition, the beneficial properties of umbilical cord-derived MSCs, for instance, multitude differentiation potential along with angiogenic potency, make them a significant source [9, 10]. To take these research pursuits to cell-based therapy, it is of vital importance to study the innate property of stem cells to generate clinical quality of MSCs. It is of prime importance to look into the credentials of homing in tissue-specific stem cells, which will pave way for clinical translation. Hence, the ability of these cells to migrate, extravagate, and ultimately home to the site of injury has also been demonstrated. Reckoning on its clinical application, several preclinical animal models of human disease such as neurodegenerative disease, cancer, diabetes, and heart disease have been reported [11–14].

Recently, the placenta and umbilical cord stem cells were reported to exhibit some of the characteristics of pluripotent properties including expression of stem cell markers such as c-kit, Thy-1, oct-4, SOX2, Nanog, hTERT, SSEA-1, SSEA-3, SSEA-4, TRA-1-60, and TRA-1-81 [4, 6, 15–17]. Besides, these isolated hPDMSCs and hucMSCs exhibited mesenchymal proteins stro1, vimentin, nestin, and surface markers CD90, CD105, CD73, CD44, CD29 and were negative for CD45, CD34, CD31 representing its homogeneity of MSC [16–18]. Besides, its long-term in vitro expansion has also been demonstrated, and no abnormalities were identified irrespective of extensive culturing [6, 19]. They also possessed differentiation capacity into all three germ layers such as ectoderm, mesoderm, and endoderm. These functional placental-/umbilical cord-derived stem cells can be isolated from either fresh or frozen term human placentas; this implies that if each individual's placenta/cord is stored at birth instead of thrown away, these cells can be harvested in the future if therapeutic need arises. This potential represents a major breakthrough in the stem cell field both for its autologous and allogenic transplantation studies. Thus, these features makes both placenta and umbilical cord a convincing alternate advantageous source of autologous and allogenic cell transplant when compared to postnatal adult stem cell sources. Unlike embryonic stem cells, neither their procurement nor their use is deemed controversial.

Why Feto-Maternal Stem Cells for Diabetic Treatment?

Diabetes is a devastating glucose metabolic disorder caused either by autoimmune destruction of pancreatic beta-cells or by decreased insulin sensitivity in peripheral tissues. To overcome the limitations of conventional treatment, replenishing the lost insulin-producing cells (IPCs) either by islet transplantation or by differentiated stem cells is the preferred approach to achieve glucose homeostasis [20–22]. Among different starting cell sources, both adult and embryonic stem cells (ESCs) have received an enormous attention as possible sources of IPCs. ESCs, owing to their intrinsic pluripotency, have been blessed with increased efficiency to differentiate into different cell types [23, 24]. However, these cells have the limitations of ethical concern due to their blastocyst origin and are also subjected to immunological surveillance, prompting the inhibition of its usage. A breakthrough was generation of iPSC which exhibited similar characteristics as that of ESCs and avoided the ethical constraints faced by human embryos, a major hurdle met with human ESCs [25].

If only at a small scale, islet transplantation has successfully addressed what ought to be the primary endpoint of any cell therapy: the functional replenishment of damaged tissue in patients. Limited organ availability is the main hurdle that stands in the way of the widespread clinical utilization of this pioneering intervention. Progress in stem cell research over the past decade, coupled with our decades-long experience with islet transplantation, is shaping the future of cell therapies for the treatment of diabetes. Despite these advantages, these cells still face few serious obstacles like teratoma formation similar to ESCs, viral-mediated iPSC derivation, becoming antigenic to an autologous host, accumulating karyotypic abnormalities with passages, and retaining epigenetic memory of the cell type of origin [24]. Though adult stem cells lack pluripotency as that of their embryonic counterparts, they are devoid of ethical debate and teratomas, the main hurdles imbibed by ESCs.

Other MSC-rich, easily bankable tissues that could potentially be used for the treatment of diabetes are the perinatal tissues like umbilical cord (blood and Wharton's jelly) and the placenta [26] which is considered as fetal waste after parturition. Fetally derived cells from these sources are thought to be more malleable and potent than most of their adult counterparts [27]. MSC-like cells adhere to plastic and form colonies not only from the stroma of the cord proper, the chorion, and the amniotic membrane but also from the blood itself. The advantages and ray of hope of using these stem cells for diabetic treatment are demonstrated below.

Placental Mesenchymal Stem Cells

An important yet neglected potentially rich source of perinatal MSCs for the generation of insulin-producing cells which may prove to be beneficial is the placenta. Besides playing a fundamental and essential role in fetal development, nutrition, and tolerance, placenta may also represent a reserve of progenitor/stem cells. Placental tissue draws great interest as a source of cells for regenerative medicine because of the phenotypic plasticity of many of the cell types isolated from this tissue. Fukuchi et al. demonstrated the existence of placenta-derived MSCs (PL-MSCs) equipped with favorable properties like ethically conducive, easily accessible, high-yielding source of stem cells, multilineage differentiation potential and having immunomodulatory properties [28]. Apart from expressing typical MSC surface antigens, due to their close ontogenic relationship to embryonic stem cells, PL-MSCs at molecular level also express ESC markers like SOX2, REX-1, and OCT4 [29]. Surprisingly, there are also transcripts of insulin, glucagon, somatostatin, Ngn3, and Isl1 in undifferentiated hPDMSCs; such transcripts were not detected in human umbilical cord and amnion-derived MSCs [16, 17].

Placenta has already been recognized as highly vascularized organ among the other perinatal sources of MSCS such as amnion which is avascular, and umbilical cord Wharton's jelly acts as a matrix to support two arteries and a vein. This anatomical variation makes placenta an endocrine organ compared to other fetal membranes. These PDMSCs has been well characterized for the closely related peptide insulin and the insulin-like growth factors (IGF)-I and II [30]. This was further supported by Giddings and Carnaghi who demonstrated the presence of insulin mRNA in extra-placental membranes prior to pancreatic differentiation [31]. Apart from this, high amount of insulin receptors relative to other tissues in the body [32] facilitates the transport of nutrients to the developing fetus across the placenta [33]. These salient features confirms that the anatomical architecture of placenta bestows it with endocrine properties, which is absent in MSCs of other origins. The above observation is further supported by the restoration of normoglycemia in STZ-induced diabetic mice on grafting of undifferentiated PL-MSCs [16]. Reversal of experimental diabetes through hPDMSC transplant was also confirmed by IPGTT performed on normoglycemic mice. Thus, placenta proved overwhelmingly effective in treating diabetes due to its superior properties and inbuilt insulin-producing capacity.

Amnion-Derived MSCs (AM-MSCs)

One of the enticing sources to join the ever-growing armamentarium of tissue-specific MSCs is the amniotic fluid. These fibroblastic-like cells, shed by the fetus to the surrounding liquid as it develops, share many properties with placental, chorionic, and cord blood/tissue MSCs. Amniotic membrane delineates the gestational sac that carries the developing fetus during gestation. Amniotic fluid and amniotic membrane together constitute the major source for stem cells of amniotic membrane. These stem cells resemble several characteristics of placental, chorionic, and cord blood-/tissues-derived cells [34]. The human amnion mesenchymal stromal cells (hAM-MSCs) have an embryonic mesodermal origin [35]. De Coppi and colleagues recently described a subpopulation of amniotic fluid cells with multigerm layer (mesoderm, ectoderm, and endoderm) specification potential [36].

Amniotic membrane is of interest as a source of cells for regenerative medicine because of its ease of availability, plasticity, and inexhaustible source that does not violate the sanctity of independent life. Although researchers have shown the stem cell-like potential of human amniotic epithelial cells, the mesenchymal part of amnion has remained less explored. Kaviani et al. [37] first described the presence of a subpopulation of amniotic fluid cells with mesenchymal features, able to proliferate in vitro more rapidly than comparable fetal and adult cells. Zhou et al. showed that AM-MSCs could be nudged down to pancreatic islet lineage by ectopically expressing key pancreatic transcription factor PDX1. The transplanted hAM-MSC-derived IPCs normalized the blood glucose levels for 210 days in experimental diabetic mice. By sequential addition of growth factors, Kadam et al. showed that a full-term human amnion could be differentiated into insulin-producing cells and reverse the hyperglycemia without graft rejection when xenotransplantated into diabetic mice using an immunoisolatory biocompatible capsule. Mice in the ILC-transplanted group showed a reduction in BG levels and reversal of experimental diabetes after 15 days [5].

However, due to its laborious and high cost factors involved in islet induction, even amniotic membrane-derived stem cells have been used as a tool to treat diabetes. A study demonstrated a successful transplantation of amniotic-derived MSCs into a type I diabetic patient with reversal of hyperglycemic condition [4, 38, 39]. Thus, it is evident that amniotic membrane stem cell transplantation may improve islet function in response to glucose in vivo; however, its long-term effects in normoglycemic recovery are yet to be explored. Although it has a higher utility in treating diabetes, a recent report has shown the epigenetic changes in these MSCs in latter passages, thus restricting its use after passage 8 [40]. Therefore, some more work is required to validate their potential for using them as an alternate source for diabetic therapy.

Umbilical Cord

Other MSC-rich, easily bankable tissues that could potentially be used for the treatment of diabetes are the umbilical cord (cord blood and Wharton's jelly). UC-MSC's immediate availability, easier isolation and expansion procedures, and their enhanced disposition toward endoderm than BM-MSCs have garnered large attention as an alternative cell source for islet therapy [41]. Romanov et al. [42] suggested that umbilical cord contains a high number of MSC-like elements forming colonies of fibroblastoid cells that may be successfully expanded in culture. UC-MSCs express consistent mesenchymal surface markers CD10, CD13, CD29, CD44, and CD90 and also several stem cell markers including c-Kit, NANOG, OCT4, and SOX2 [43]. Few reports have attempted to generate beta cell-like cells from UC-MSCs [17, 44, 45]. Recent studies showed that no rejections occur even after xenotransplantation of post-differentiated umbilical cord mesenchymal stem cells without immunosuppression therapy [46].

The cord contains proteoglycan-rich connective tissue called Wharton's jelly, a rich source of MSCs that have the potential to differentiate into all the trilineages [47]. WJ-MSCs possess immunomodulatory and immunosuppressive properties. Along with the absence of class II HLA-DR molecule, inhibition of T-cell proliferation, and dendritic cell differentiation, these cells also express the nonclassical HLA-G molecule at both mRNA and protein levels [48]. HUMSCs in Wharton's Jelly of the umbilical cord seem to be a favorable source of stem cells for conversion into insulin-producing cells, because of its large potential donor pool, rapid availability, absence of discomfort to the donor, and low risk of rejection. Therefore, HUMSCs in Wharton's Jelly of the umbilical cord have the potential to become an excellent candidate in B-cell replacement therapy of diabetes [46].

Subsequently, a comparative study carried out by Wu and colleagues compared the differentiative ability of WJ-MSC and BM-MSC in order to acquire an IPC phenotype. Both cellular types were able to form islet-like clusters. The researchers found a higher expression of Pdx-1 in differentiated WJ-MSC than in differentiated BM-MSC. Secretion of insulin and mRNA expression of insulin and C-peptide were comparably higher in the differentiated WJ-MSC. Wang and co-workers further corroborated these data with in vitro and in vivo experiments using differentiated human WJ cells to treat diabetes in NOD mice. After transplantation, IPCs were located in the liver and were able to normalize glycemia [49]. Taken collectively, these promising data suggest that WJ-MSC possess the ability, both in vitro and in vivo, to differentiate into insulin-secreting cells. Recent results highlight that differentiated beta-cells may engraft and survive in organs other than pancreas (e.g., in the liver) and survive maintaining their differentiated state, thus contributing to glycemia normalization.

Apart from islet cells, long-term effect of the implantation of WJ-MSCs studied in 29 newly onset T1DM patients showed improved C-peptide levels compared to saline-treated patients. These data clearly suggested the safe and beneficial role of WJ-MSCs in diabetes treatment [50]. However, further similar studies on the efficacy of WJ-MSCs in treating diabetes in regard with its insulin secretion capacity and reversal of normoglycemic status are warranted. It might be a breakthrough study due to the inbuilt constitutive higher expression of nestin and with the presence of pluripotent markers such as oct-4 and SSEA-4 NANOG, OCT4, DNMT3B, and GABRB3 [8, 49] in hUCMSCs. Suggest a connecting link between embryonic stem cell and adult stem cell, thus signifying them as a more potential candidate for the generation of mature β-cells.

Umbilical Cord Blood-Derived Cells

Human umbilical cord blood is a readily available source of mesenchymal stem cell as there are rising number of public and private bank for the storage of cord blood, thus allowing optimal matching of HLA for allogenic as well as autologous transplantation. Besides this HUCB stem cells contain telomerase enzyme and have long telomeres as compared to BM cells [51]. Thus, interest continues to grow regarding the therapeutic potential for umbilical cord blood therapies to modulate autoimmune disease including diabetes. Practical matters provide an additional rationale for umbilical cord blood-based therapies. First, the lack of low-risk (i.e., safe) diabetes intervention trials seeking to reverse disease, especially for young children with type 1 diabetes, renders the potential use of umbilical cord blood particularly appealing. The study reported that umbilical cord blood-derived cells expressed basal levels of genes involved in the progression of pancreatic endocrine development [13]. This lends support to the intriguing notion that these tissues may contain cells that are somehow primed to give rise to beta-cells. Pessina and colleagues had demonstrated the expression of endocrine pancreatic progenitor markers such as nestin and cytokeratin 19 and transcription factors such as neurogenin-3 (Ngn-3), Pax-4, and Isl-1 in HUCB cells [52]; later on it was confirmed by Prabakar K R et al. [53], which shows it can be used as another alternative source of MSCs for β-cell differentiation.

Pessina and co-workers were the first to highlight that naïve UBC-MSC, cultured in a medium with serum but without specific cytokines or growth factors, expressed a panel of typical markers in the pancreatic differentiation pathway (Ngn3, Nestin, Cytokeratin-18 Cytokeratin-19, Isl-1, and Pax-4). This was later confirmed by Prabakar et al. [53] who demonstrated that cord blood MSCs could effectively recapitulate all the stages of the standard protocol used to differentiate beta-cells from hES cells, including the first critical step of definitive endoderm specification. This showed that it can be used as another alternative source of MSCs for β-cell differentiation. Yoshida and co-workers investigated in vivo the capacity of human UCB mononuclear cells to give rise to IPC when transplanted in newborn NOD/SCID/β2-m null mice. These cells, 1–2 months after transplantation, were able to generate IPC, as testified by the presence of human insulin at the RNA level and human chromosome-containing insulin-positive cells in situ. In addition, this study clarified that UCB cells could generate IPC by both fusion-dependent and fusion-independent mechanisms in vivo [54]. Parekh and co-workers, in fact, confirmed that UCB-derived mononuclear cells can express some pancreatic genes and when transplanted in NOD/SCID (nonobese diabetic/severe combined immunodeficiency).

Human cord blood-derived stem cells (CB-SCs) have been shown to modulate immune activity in vitro [55–62]. Subsequent studies have demonstrated that CB-SCs can be used to alter immune function and improve markers of T1D in nonobese diabetic mice (NOD) [59] and CB-SCs have been shown to modulate the immune function of T1D patient-derived islet B-cell-specific pathogenic T-cell clones in coculture.

Summary

Although there is vast availability of both postnatal and perinatal sources of MSCs for the generation of in vitro functional islet-like clusters, it remains to be elucidated which is the best source. However, it is apparent from the aforesaid attributes that feto-maternal prenatal sources themselves might give a ray of hope for treating diabetes, thereby paving way for emerging non-laborious, cost-effective, efficient, long-term benefit, largely available, non-immunogenic source for treating diabetes. Therefore, current attention should be paid not on generating the mature β-cells but on the utility of these undifferentiated feto-maternal sources and its long-term maintenance in treating diabetes. Although there is substantial evidence that MSCs harbor immunomodulatory properties, additional studies are needed to validate the long-term stability of the MSCs and furthermore to what extent these in vitro cultured MSCs might form endogenous islets after transplantation needed to be elucidated. Although the prenatal source proves efficient as compared to postnatal stem cell sources, the ideal candidate among these feto-maternal sources is uncertain, thus, warranting further in-depth research in this area to qualify the criteria laid down for effective therapeutics with respect to safety, viability, tissue architecture, and functionality. The novel approaches enabling long-term survival of these MSCs and their banking will pave the path for transplantation of MSCs as a routine procedure to combat type I diabetes.

References

1. Evangelista M, Soncini M, Parolini O et al (2008) Placenta derived stem cells, new hope for cell therapy? Cytotechnology 58:33–42
2. Blanco O, Leno-Duran E, Morales JC, Olivares EG, Ruiz-Ruiz C et al (2009) Human decidual stromal cells protect lymphocytes from apoptosis. Placenta 30:677–685
3. Witkowska-Zimny M, Wróbel E et al (2011) Perinatal sources of mesenchymal stem cells: Wharton's jelly, amnion and chorion. Cell Mol Biol Lett 16:493–514
4. Indumathi S et al (2013) Comparison of feto-maternal organ derived stem cells in facets of immunophenotype, proliferation and differentiation. Tissue Cell 45(6):434–442
5. Kadam SS, Sudhakar M, Nair PD, Bhonde RR (2010) Reversal of experimental diabetes in mice by transplantation of neo-islets generated from human amnion-derived mesenchymal stromal cells using immuno-isolatory macrocapsules. Cytotherapy 12:982–991
6. Sabapathy V, Ravi S, Srivastava V, Srivastava A, Kumar S (2012) Long term cultured human term placenta-derived mesenchymal stem cells of maternal origin displays plasticity. Stem Cells Int 419. doi:10.1155/2012/174328
7. Ilancheran S, Michalska A, Peh G, Wallace EM, Pera M, Manuelpillai U et al (2007) Stem cells derived from human fetal membranes display multilineage differentiation potential. Biol Reprod 77: 577–588
8. Wang HS, Hung SC, Peng ST, Huang CC, Wei HM, Guo YJ et al (2004) Mesenchymal stem cells in the Wharton's jelly of the human umbilical cord. Stem Cells 22:1330–1337
9. Roura S, Bago' JR, Soler-Botija C, Pujal JM, Ga'lvez-Monto'n C et al (2012) Human umbilical cord blood-derived mesenchymal stem cells promote vascular growth in vivo. PLoS One 7(11):e49447
10. Nascimento DS, Mosqueira D, Sousa LM, Teixeira M et al (2014) Human umbilical cord tissue-derived mesenchymal stromal cells attenuate remodelling after myocardial infarction by proangiogenic, antiapoptotic, and endogenous cell-activation mechanisms. Stem Cell Res Ther 5:5
11. Lim JY, Jeong CH, Jun JA, Kim SM, Ryu CH et al (2011) Therapeutic effects of human umbilical cord blood-derived mesenchymal stem cells after intrathecal administration by lumbar puncture in a rat model of cerebral ischemia. Stem Cell Res Ther 2(5):38
12. Jiao H, Guan F, Yang B, Li J, Shan H, Song L et al (2011) Human umbilical cord blood-derived mesenchymal stem cells inhibit C6 glioma via downregulation of cyclin D1. Neurol India 59(2): 241–247
13. Parekh VS, Joglekar MV, Hardikar AA (2009) Differentiation of human umbilical cord blood-derived mononuclear cells to endocrine pancreatic lineage. Differentiation 78(4):232–240
14. Lee EJ, Choi EK, Kang SK, Kim GH, Park JY et al (2012) N-cadherin determines individual variations in the therapeutic efficacy of human umbilical cord blood-derived mesenchymal stem cells in a rat model of myocardial infarction. Mol Ther 20(1):155–167
15. Yen BL, Huang HI, Chien CC, Jui HY, Ko BS et al (2005) Isolation of multipotent cells from human term placenta. Stem Cells 23:3–9
16. Kadam S, Muthyala S, Nair P, Bhonde R (2010) Human placenta derived mesenchymal stem cells and islet-like cell clusters generated from these cells as a novel source for stem cell therapy in diabetes. Rev Diabet Stud 7:168–182
17. Kadam SS, Bhonde RR (2010) Islet neogenesis from the constitutively nestin expressing human umbilical cord matrix derived mesenchymal stem cells. Islets 2:112–120
18. Malgieri A, Kantzari E, Patrizi MP, Gambardella S (2010) Bone marrow and umbilical cord blood human mesenchymal stem cells: state of the art. Int J Clin Exp Med 3(4):248–269
19. Chen G, Yue A, Ruan Z, Yin Y, Wang R et al (2014) Monitoring the biology stability of human umbilical cord-derived mesenchymal stem cells during long-term culture in serum-free medium. Cell Tissue Bank. doi:10.1007/s10561-014-9420-6
20. Shapiro AM, Lakey JR, Ryan EA, Korbutt GS, Toth E et al (2000) Islet transplantation in seven patients with type 1 diabetes mellitus using a glucocorticoid-free immunosuppressive regimen. N Engl J Med 343(4):230–238
21. Shapiro AM, Ricordi C, Hering BJ, Auchincloss H, Lindblad R et al (2006) International trial of the Edmonton protocol for islet transplantation. N Engl J Med 355(13):1318–1330
22. Couri CE, Voltarelli JC (2011) Stem cell based therapies and immunomodulatory approaches in newly diagnosed type 1 diabetes. Curr Stem Cell Res Ther 6(1):10–15
23. Liew CG (2010) Generation of insulin producing cells from pluripotent stem cells: from the selection of cell sources to the optimization of protocols. Rev Diabet Stud 7(2):82–92
24. Hebrok M (2012) Generating Beta cells from stem cells-the story so far. Cold Spring Harb Perspect Med 2(6):a007674
25. Takahashi K, Tanabe K, Ohnuki M, Narita M, Ichisaka T, Tomoda K et al (2007) Induction of pluripotent stem cells from adult human fibroblasts by defined factors. Cell 131(5):861–872
26. Bhonde RR, Sheshadri P, Sharma S, Kumar A (2014) Making surrogate β-cells from mesenchymal stromal cells: perspectives and future endeavors. Int J Biochem Cell Biol 46:90–102
27. Zhao Y, Mazzone T (2010) Human cord blood stem cells and the journey to a cure for type 1 diabetes. Autoimmun Rev 10(2):103–107
28. Fukuchi Y, Nakajima H, Sugiyama D, Hirose I, Kitamura T, Tsuji K (2004) Human placental derived cells have mesenchymal stem/progenitor cell potential. Stem Cells 22(5):649–658

29. Murphy SV, Atala A (2013) Amniotic Fluid and placental membranes: unexpected sources of highly multipotent cells. Semin Reprod Med 31(1):62–68
30. Bhaumick B, Danilkewich AD, Bala RM (1988) Altered placental insulin and insulin-like growth factor-I receptors in diabetes. Life Sci 42(17):1603–1614
31. Giddings SJ, Carnaghi L (1989) Rat insulin II gene expression by extraplacental membranes. A non-pancreatic source for fetal insulin. J Biol Chem 264(16):9462–9469
32. Desoye G, Hauguel-de Mouzon S (2007) The human placenta in gestational diabetes mellitus. The insulin and cytokine network. Diabetes Care 30 Suppl 2:S120–S126
33. Han VK, Carter AM (2000) Spatial and temporal patterns of expression of messenger RNA for insulin-like growth factors and their binding proteins in the placenta of man and laboratory animals. Placenta 21(4):289–305
34. Priest RE, Marimuthu KM, Priest JH (1978) Origin of cells in human amniotic fluid cultures: ultrastructural features. Lab Invest 39(2):106–109
35. Pappa KI, Anagnou NP (2009) Novel sources of fetal stem cells: where do they fit on the developmental continuum? Regen Med 4(3):423–433
36. De Coppi P, Bartsch G Jr, Siddiqui MM et al (2007) Isolation of amniotic stem cell lines with potential for therapy. Nat Biotechnol 25:100–106
37. Kaviani A, Perry TE, Dzakovic A, Jennings RW, Ziegler MM, Fauza DO (2001) The amniotic fluid as a source of cells for fetal tissue engineering. J Pediatr Surg 36:1662–1665
38. Liu Y, Cao D-L, Guo L-B, Guo S-N, Xu J-K et al (2013) Amniotic stem cell transplantation therapy for type 1 diabetes: a case report. J Int Med Res 41:1370
39. Villani V, Milanesi A, Sedrakyan S, Da Sacco S, Angelow S et al (2014) Amniotic fluid stem cells prevent β-cell injury. Cytotherapy 16(1):41–55
40. Zulewski H, Abraham EJ, Gerlach MJ, Daniel PB et al (2001) Multipotential nestin-positive stem cells isolated from adult pancreatic islets differentiate ex vivo into pancreatic endocrine, exocrine, and hepatic phenotypes. Diabetes 50(3):521–533
41. Montesinos JJ, Flores-Figueroa E, Castillo-Medina S, Flores-Guzman P, Hernadez-Estevez E, Fajardo-orduna G et al (2009) Human mesenchymal stromal cells from adult and neonatal sources: comparative analysis of their morphology, immunophenotype, differentiation patterns and neural protein expression. Cytotherapy 11(2):163–176
42. Romanov YA, Svintsitskaya VA, Smirnov VN (2003) Searching for alternative sources of postnatal human mesenchymal stem cells: candidate MSC like cells from umbilical cord. Stem Cells 21:105–110
43. Weiss ML, Medicetty S, Bledsoe AR, Rachakatla RS, Choi M, Merchav S et al (2006) Human umbilical cord matrix stem cells: preliminary characterization and effect of transplantation in a rodent model of Parkinson's disease. Stem Cells 24(3):781–792
44. Gao F, Wu DQ, Hu YH et al (2008) In vitro cultivation of islet-like cell clusters from human umbilical cord blood-derived mesenchymal stem cells. Transl Res 151(6):293–302
45. Prabhakar KR, Domiguez-Bendala J, Molano RD, Pileggi A, Villate S, Ricordi C et al (2012) Generation of glucose responsive, insulin producing cells from human umbilical cord blood derived mesenchymal stem cells. Cell Transplant 21(6):1321–1339
46. Chao KC, Chao KF, Fu YS, Liu SH (2008) Islet-like clusters derived from mesenchymal stem cells in Wharton's jelly of the human umbilical cord for transplantation to control type 1 diabetes. PLoS One 3(1):e1451
47. Prasanna SJ, Jahnavi VS (2011) Wharton's jelly mesenchymal stem cells as off the shelf cellular therapeutics: a closer link into closer into their regenerative and immunomodulatory properties. Open Tissue Eng Regen Med J 4:28–38
48. Anzalone R, Farina F, Zummo G et al (2011) Recent patents and advances on isolation and cellular therapy applications of mesenchymal stem cells from human umbilical cord Wharton's jelly. Recent Patents Regen Med 1:216–227
49. Wang HS, Shyu JF, Shen WS, Hsu HC, Chi TC et al (2011) Transplantation of insulin producing cells derived from umbilical cord stromal mesenchymal stem cells to treat NOD mice. Cell Transplant 20(3): 455–466
50. Ying Hu, Jun Liang, Hongping Cui, Xinmei Wang, Hua Rong et al (2013) Wharton's jelly mesenchymal stem cells differentiate into retinal progenitor cells. Neural Regen Res 8:1783–1792
51. Vaziri H, Dragowska W, Allsopp RC, Thomas TE, Harley CB et al (1994) Evidence for a mitotic clock in human hematopoietic stem cells: loss of telomeric DNA with age. Proc Natl Acad Sci U S A 91(21):9857–9860
52. Pessina A, Eletti B, Croera C et al (2004) Pancreas developing markers expressed on human mononucleated umbilical cord blood cells. Biochem Biophys Res Commun 323(1):315–322
53. Prabakar KR, Domínguez-Bendala J, Molano RD, Pileggi A, Villate S et al (2012) Generation of glucose-responsive, insulin-producing cells from human umbilical cord blood-derived mesenchymal stem cells. Cell Transplant 21(6):1321–1339
54. Yoshida S, Ishikawa F, Kawano N, Shimoda K, Nagafuchi S, Shimoda S et al (2005) Human cord blood derived cells generate insulin producing cells in vivo. Stem Cells 23(9):1409–1416
55. Harris DT, Rogers I (2007) Umbilical cord blood: a unique source of pluripotent stem cells for regenerative medicine. Curr Stem Cell Res Ther 2(4): 301–309
56. Zhao Y, Huang Z, Qi M, Lazzarani P, Mazzone T (2007) Immune regulation of T lymphocyte by a newly characterized human umbilical cord blood stem cell. Immunol Lett 108(1):78–87

57. Abdi R, Fiorina P, Adra CN, Atkinson M, Sayegh MH (2008) Immunomodulation by mesenchymal stem cells: a potential therapeutic strategy for type 1 diabetes. Diabetes 57(7):1759–1767
58. Ucceli A, Moretta L, Pistoia V (2008) Mesenchymal stem cells in health and disease. Nat Rev Immunol 8(9):726–736
59. Zhao Y, Lin B, Darflinger R, Zhang Y, Holterman MJ, Skidgel RA (2009) Human cord blood stem cell modulated regulatory T lymphocytes reverse the autoimmune caused type 1 diabetes in non obese diabetic mice. PLoS One 4(1):e4226
60. Aguayo-Mazzucato C, Bonner-Weir S (2010) Stem cell therapy for type 1 diabetes mellitus. Nat Rev Endocrinol 6(3):139–148
61. Phadnis SM, Joglekar MV, Dalvi MP et al (2011) Human bone marrow-derived mesenchymal cells differentiate and mature into endocrine pancreatic lineage in vivo. Cytotherapy 13(3): 279–293
62. Domínguez-Bendala J, Lanzoni G, Inverardi L, Ricordi C (2012) Concise review: mesenchymal stem cells for diabetes. Stem Cells Trans Med 1: 59–63

Stem Cell Therapy for Acute and Chronic Liver Failure

Meghnad G. Joshi, Apurva Gadgil, and Ramesh R. Bhonde

Introduction

Liver disease is an exceptionally common cause of morbidity and mortality worldwide. The acute and chronic liver diseases are still treated with supportive rather than curative approaches. Orthotropic liver transplantation has so far been the only available therapy for patients with end-stage liver diseases. Unfortunately, the availability of donor organs is limited, and many patients die each year waiting for liver transplants. Prevalence rates and the demand for liver transplantation are rising annually worldwide. In Europe, nearly 5,500 orthotropic liver transplantations (OLTs) are performed [1]. In the UK, active transplantation list in March 2012 was 7,636 patients at the end of March 2012 and 508 patient died waiting for suitable transplant [2]. Similarly, in USA as of October 2013, 123,392 liver transplants had been reported and as of today 16,479 patients are on waiting list [3].

Cell-based therapies with stem cells and their progeny is a promising new approach to this largely unmet medical need. Recent advances in the liver repopulation field and the use of alternative cell sources are under investigation. The major drawbacks and the most important advantages of adult or fetal hepatocytes and hepatocytes generated from different stem cells in the wide range of experimental and clinical applications are published. Even after 40 years of extensive experimental research, none of the procedures represent a "gold standard" in the clinical practice. The achievement of positive outcomes in many experimental and clinical studies involving liver progenitor cells has been handicapped by the limited engraftment of cells that can be of therapeutic significance. If stem cells are to be used as an alternative or bridge to organ transplantation, it is very important to reduce the cell loss during the transplantation, repeated cell infusions, and large numbers of cells for transplantation.

Acute Liver Failure

Acute liver failure (ALF) or fulminant hepatic failure (FHP) is characterized by rapid deterioration of hepatocyte function that leads to hepatic encephalopathy, coagulopathy, cerebral edema, infection, and multiorgan dysfunction syndrome [4]. ALF still has high mortality rate, and

M.G. Joshi (✉)
Department of Stem Cells and Regenerative Medicine, D.Y. Patil University, D. Y. Patil Vidyanagar, Kasba Bawda, Kolhapur 416006, Maharashtra, India
e-mail: cmd@stemplusbiotech.com

A. Gadgil
Stem Plus Biotech Pvt. Ltd., Shiv Meridian, Block #11, MG Road, Opp. Collector Bunglow, Sangli 416416, MS, India
e-mail: info@stemplusbiotech.com

R.R. Bhonde
School of Regenerative Medicine, Manipal Institute of Regenerative Medicine, Manipal University, GKVK Post, Bellary Road, Allalasandra, Near Hotel Royal Orchid, Yelahanka, Bangalore 560065, India
e-mail: rr.bhonde@manipal.edu

I. Somasundaram (ed.), *Stem Cell Therapy for Organ Failure*,
DOI 10.1007/978-81-322-2110-4_16, © Springer India 2014

orthotropic liver transplantation (OLT) is the only available treatment that gives satisfactory results [4, 5]. However, clinicians face difficulties in making a decision about the transplant due to limited patient history and rapid deterioration of the patient. Thus, delay results in a patient becoming un-transplantable due to other contraindications like multiorgan failure. Urgent OLT has become a standard treatment for ALF patients in USA where survival rates have shown improvement and 1-year survival exceeding 80 % [6].

The common causes of ALF are viral hepatitis, idiosyncratic drug reactions, acetaminophen, and mushroom ingestion [7]. Viral hepatitis B is the most common cause of ALF worldwide, responsible for about 70 % of cases, and it produces significant morbidity and mortality [8, 9]. Another causative factor for ALF in Western world is acetaminophen (APAP) overdose which is more commonly encountered. APAP intake leads to excessive production of its active metabolite *N*-acetyl-p-benzoquinone imine in the liver, causing depletion of the glutathione stores followed by centrilobular necrosis [10].

However, OLTs have numerous limitations like shortage of donor organs, the high costs, and the lifelong immunosuppressive treatments [11]. Various alternatives to OLT have been evaluated, such as split liver, cross circulation, plasma exchange, hemofiltration, hemodialysis, and hemoperfusion without any significant improvement [12, 13]. The alternatives such as stem cell transplant or artificial liver support systems can be helpful, as bridge to transplant that will increase the availability of suitable donor for patients who would otherwise have died might survive until transplantation.

Safety and efficacy of hepatocyte transplantation procedure has been studied in several animal models of ALF. The galactosamine-induced liver failure is the most commonly used models that include mice, rats, rabbits, guinea pigs, and dogs [14–16]; thioacetamide-induced liver injury in rabbits and rats [17, 18]; other models include complete hepatic devascularization [9, 19] and total [20] or subtotal (95 %) hepatectomy [21]. These models showed replacement of about 1–5 % of total hepatocyte mass, which is the limiting factor for treatment of ALF [22]. In experimental animal models, improved survival rate in ALF was documented by these studies.

Chronic Liver Failure

Chronic liver failure (CLF) death toll is about 1.4 million annually, and nearly 150,000 patients die due to CLF in India [23]. The predominant reasons for CLF were cirrhosis due to hepatitis C, B, and D viral infection, followed by HCC, and alcoholic cirrhosis with or without concomitant infection with HCV. Autoimmune causes include primary biliary cirrhosis (PBC), primary sclerosing cholangitis (PSC), and biliary atresia. Non-alcoholic steatohepatitis (NASH) is associated with diabetes, protein malnutrition, obesity, coronary artery disease, and corticosteroid treatment. Other inherited causes are alpha-1 antitrypsin deficiency, hemochromatosis, Wilson's disease, galactosemia, and glycogen storage diseases [24].

Stem cell therapy for CLF patients has more limitations than acute or metabolic liver diseases. During the diseases progression, there is a major loss of hepatocytes and abnormal hepatic architecture due to scar formation in the liver.

Study of CLF in animal models is difficult due to lack of suitable animal model that can mimic the human situation. There are toxin-induced animal models such as carbon tetrachloride (CCl_4) cirrhosis, phenobarbital, retrorsine, and end-to-side portacaval shunt [25–29]. In the animal experiments, liver toxins were injected to normal liver, and 4 weeks after the discontinuation of liver toxins, animals were subjected to cell therapy. During the experimental studies, different cell types applied intrasplenic, namely, fetal hepatocytes and mesenchymal, stem cells [29], rat or porcine hepatocytes [26], syngeneic rat hepatocytes [27], or immortalized rat hepatocytes [25]. These cell therapy experimental models clearly improved liver function and prolonged survival.

It is believed that at least 20 % of the normal hepatocyte mass is required to carry out normal physiological parameters [30]. Hepatocyte recovery from an average liver is about 2.8×10^{11} hepatocytes, occupying almost 80 %

of the total liver volume. Considering the safety of the cell transplantation, only 2.4×10^6 hepatocytes per gram of liver can be transplanted, suggesting replacement of approximately 10 % of functional liver mass. The therapeutic mass of hepatocytes actually required to restore adequate liver function for CLF patient is extremely high and is not possible to transplant into the scar liver or spleen. In this situation, transplantation hepatocytes into other sites will accommodate the large number of therapeutic hepatocyte mass and carryout metabolic function for temporary support.

Types of Stem Cells Used for Liver Failures

Embryonic Stem Cells

Embryonic stem cells (ES) are capable of generating unlimited hepatocytes, which can be used for transplantation. The study of ES is of great interest to clinicians, but it is also surrounded by controversies about its use. First critical step in generating ES from the definitive endoderm (DE) tissue can be achieved by treating the cultures with TGFβ family ligand activin [31]. Many investigators used DE as starting material to generate hepatocyte-like cells [32–34]. The hepatic maturation can be achieved by the combination of HGF, OSM, FGF, and dexamethasone to expand the hepatoblast population and to promote hepatic maturation. The mouse and human ES cells can be differentiated in to "hepatocyte-like" cells. These cells showed hepatocyte morphology, glycogen storage, express hepatic enzymes, and drug metabolism and secrete albumin [34–39].

Yamamoto et al. and Heo et al. demonstrated mouse ES have potential to differentiate in to hepatocyte-like cells. These cells, when transplanted in to mouse model of liver injury, engrafted and showed improved survival without malignancy [40, 41]. Bin et al. showed the therapeutic effect of embryonic hepatocytes in Wistar rats with CCl_4-induced cirrhosis [42]. Transplantation of EC in different liver injury models showed evidence of repopulation of recipient liver; however, undifferentiated ES showed teratogenic property in immunodeficient mice [40, 43].

The major obstacle in the use of ES in the clinic is safety issue and ethical concerns. ECs are capable of unlimited proliferation and able to produce hepatocyte-like cells. They show reasonable functional capacity, although there is no consensus about which functional parameters are necessary for clinical use. ES-derived hepatocytes into animals with various liver injury models showed rescue of liver function, although engraftment of the cells was low.

Fetal Hepatocytes

Human fetal liver (FL) has alternative source of liver progenitor that can be procured from donated tissues after indicated abortions. The FL, between gestation weeks 5–18, contains a large number of actively dividing hepatic progenitor cells that are termed as hepatoblasts. FL cells are committed progenitors, and the development is directed to differentiate to liver cells. FL can be an ideal candidate for cell therapy without concern of teratoma or cancer formation after transplantation.

It has been hypothesized that FL arise from bipotential mesendoderm precursor from mesoderm and endoderm [44]. Cells from the ventral midline of the endoderm lip, probably originating from the mesendodermal prechordal plate, give rise to part of the liver bud [45]. Many studies have shown that mesodermal multipotent adult progenitor cells in both mice and humans can differentiate into hepatocytes [46, 47]. FL tissue obtained during the first trimester mainly exhibits hematopoietic and endothelial markers [48]. During the second trimester development, the hematopoietic cells (HSC) shift towards the hematopoietic cells. Using markers originally expressed on HSC, including c-Kit (CD117) and CD34, can be useful for the identification and isolation of FH. Isolated FH were found to co-express hepatocytic or biliary phenotypic markers implying lineage progression [48]. We have previously isolated and characterized FH from

Fig. 1 Primary fetal hepatocytes culture (Passage 3)

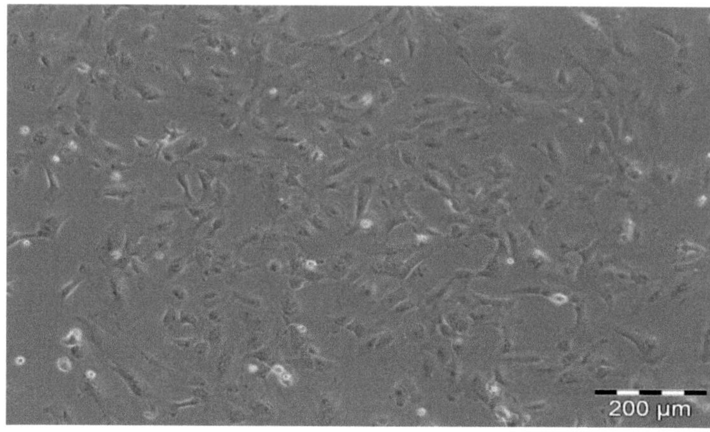

human FL expressing the markers CD34 and CD117 [29, 49, 50]. Flow cytometric and IHC analysis FH in early passages were positive for the hepatic stem cell markers EPCAM, CD133, and CD90, but not CD34 or CD45, indicating the non-hematopoietic origin of these cells. We have confirmed with IHC staining of CK19, CK18, and albumin the committed progenitors for biliary and hepatocytic lineages [29, 49, 50]. The isolation of fetal hepatocytes (FH) consists of mechanical disruptions into small fragment and then incubation with collagenase for digestion of connective tissue. Mechanical disruptions may yield about 10^4–10^6 cells per donation (unpublished data). Our own experiments have yielded an average of 5×10^4 cells (data not published) from 12 human fetal tissue isolations (gestation week 5–11) by mechanical disruption (Fig. 1).

There has been fair success in using FH in different animal models. These experiments show successful repopulation of immunodeficient mouse and rat liver with rodent fetal hepatocytes following liver injury [51–53]. In our previous study, we used retrorsine-induced liver injury mouse model, followed by 30 % partial hepatectomy. We infused 2×10^6 intrasplenic injection of hFH and/or MSC. Overall animal survival rates were 100 % during the experiments. We confirmed the presence of engraftment and repopulation in the experimental model by immunohistochemistry and PCR. Immunohistochemical analysis showed cell colonies positive for human albumin, alpha-feto protein, c-met, human hepatocyte antigen, and the human cytokeratins CK 8, 18, and 19. We also detected higher levels of human serum albumin and mouse serum. RT-PCR analysis showed intense expression of the important hepatic factors HNF 1α, β, HNF 4α which provides useful information of persistence hepatocyte function and differentiation [29].

Isolated FH under GMP (good manufacturing practices) compliance can be useful for clinical transplant. In several clinical studies, FH was assessed for infectious diseases such as HCV, HBV, HIV, EBV, HEV, HDV, toxoplasmosis, rubella, cytomegalovirus, parvovirus, herpes simplex type 1 and 2, TPHA, sterility, and endotoxin testing. Therapeutic advantage of FH is useful for a variety of genetic/metabolic liver diseases and acute and chronic liver failure. In recent studies, a significant improvement of all liver serum biochemistry has been achieved after transplantation of FH in a patient with Crigler–Najjar syndrome and biliary atresia [54, 55]. FH isolated from second trimester (EpCAM + cells) and labeled with Tcd, l-hexamethyl-propylene-amine oxime (Tc-HMPAO) were infused into the hepatic artery of 25 patients with end-stage liver cirrhosis. All patients showed marked improvement of clinical and biochemical parameters during 6 months follow-up [23]. In another study, FH was transplanted in a patient with end-stage chronic liver failure. The patient received 2 intrasplenic infusions (5 × 108) on day 0 and 80. The patient's Model for End-Stage Liver Disease (MELD) score improved from 15 to 10 within the first 18 months of observation [56].

Presently available clinical case reports on use of FH for liver disease give very limited information. This may be due to ethical concerns in getting donors. Larger clinical studies are essential to prove the clinical efficacy of FH.

iPSC

Induced pluripotent stem cells (iPSCs) are adult cells that are genetically reprogrammed to an embryonic-like stem cell. This is achieved by introducing genes and factors important for maintaining the properties of ES. The reprogramming of adult cells into ES cells enables the generation of patient-specific stem cells and thus has enormous potential for the treatment of degenerative diseases.

In 2006, Kazutoshi Takahashi and Shinya Yamanaka first introduced factors for reprogramming of mouse fibroblasts through retroviral transduction with 24 transcription factors highly expressed in ES [57]. Similarly, Park et al. [58] generated iPSCs from human skin through ectopic expression of four genes (Oct3/4, Sox2, c-Myc, and Klf4). Human iPSCs and ES show similar morphologies, proliferation rates, and expression of a number of stem cell markers. However, the specific difference between ESCs and iPSCs is the origin of cell type. iPSCs are derived from adult tissue. In addition, genomic analyses of two types of cells show that hundreds of genes are differentially expressed [59]. Importantly, considering the adult origin, iPSCs can contain an epigenetic "memory" of the donor tissue, which restrict their differentiation capacity and therefore utility [60]. Jang and colleagues generated human iPSCs from a variety of adult human cells, including the liver cells, fibroblasts, bone marrow stem cells, and skin cells [61]. They found that though the iPSCs overall were molecularly similar to each other and to embryonic stem cells, they retained a distinct molecular "signature" inherited from the cell of origin. Regardless of their origin, the different iPSCs showed the same ability to develop into liver cells. A comprehensive study by Miura et al. using various mouse iPSC, has demonstrated that the origin of the iPSC has a profound influence on the tumor-forming propensities in a cell therapy animal model. Mouse tail-tip fibroblast iPSC cells have shown the highest tumorigenic propensity, whereas gastric epithelial cells and hepatocyte iPS cells have shown lower propensities [62].

Human iPSCs can be directed to differentiate into hepatocyte-like cells using different differentiation methods [63–66]. Jozefczuk et al. [67] demonstrated the 80 % similarities in gene expression responsible for normal liver physiology between human ES and iPSC. Hepatocyte-like cells generated from iPSC have been shown to secrete human albumin, synthesize urea, and express human cytochrome P450 enzymes [65]. Espejel et al. [68] demonstrated the iPSC-derived hepatocytes have both the functional and proliferative capabilities needed for liver regeneration in mice with fumarylacetoacetate hydrolase deficiency. Asgari et al. performed a growth factor-mediated differentiation of iPSCs and evaluated their potential for recovery of CCl_4-injured mouse liver following transplantation [69]. In another study, transplantation iPSCs are engrafted, integrated, and proliferated in livers of an immune-deficient mouse model. iPSCs secreted human albumin and cells function was similar to primary human hepatocytes, including metabolic function [70].

The recent success in generating iPSC without viral vectors has brought iPSC one step closer to therapeutic application [71]. However, the suitability of individual iPSC for generating cell for therapy needs to be demonstrated. In spite of these, limitations iPSC-derived hepatocytes are a very promising population for cell therapies.

Adult Hepatocytes

Adult hepatocyte (AH) transplantation has led its scientific foundations over 40 years. Transplantation of AH in animals has shown effectiveness in defective hepatic enzymes in metabolic models, improving survival rate in acute hepatic and chronic liver failure [72–79]. Animal models have given insight of mechanism of AH engraftment. However, limitations of animal models

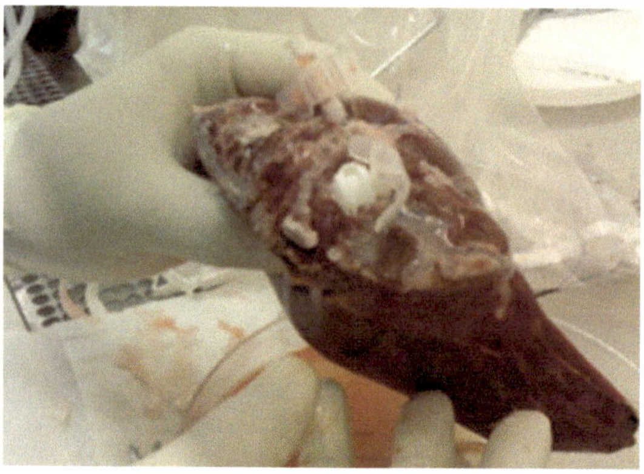

Fig. 2 Cannulation of resected liver lobe for hepatocyte isolation

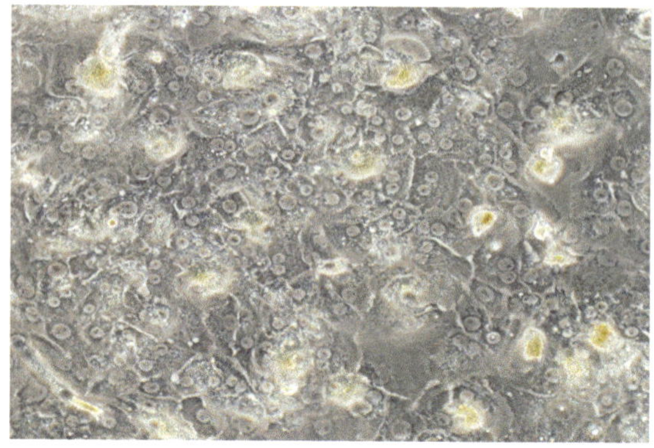

Fig. 3 Primary culture of adult hepatocytes

reflected on the clinical AH transplant. Animal studies showed that the AH get engrafted in the recipient liver and function normally, even when the engraftment accounted for only 1–5 % of the total hepatocyte mass [22]. This low percentage of engraftment is unlikely to support acute or chronic damage unless a significant liver repopulation is achieved.

The source of AH for the experimental and clinic use is livers discarded for liver transplantation, liver biopsies, or livers after tumor resection [80–82]. Isolation of AH from liver biopsies is quite difficult due to size variations and the paucity of visible vessels available for catheterization. Nationwide study by Baccarani et al.[82] found that organs rejected have steatosis more than 30 %, resulting in reduced hepatocyte yield, decreased viability, and reduced hepatocyte engraftment. Overall cell yield $9.3 \times 10^9 \pm 8 \times 10^9$ hepatocytes, achieving $7.2 \times 10^6 \pm 7 \times 10^6$ hepatocytes/g of liver tissue digested with an average viability of 73 % ± 14 % (Figs. 2 and 3).

Over the years, many researchers have attempted to optimize a hepatocyte isolation protocol. Unfortunately, there is no consensus on a single isolation protocol for use in all laboratories, so it is difficult to compare among the outcomes of different studies [83]. AH isolation process starts from surgical removal to transportation to the laboratory and the isolation process. Hepatocytes are exposed to a number of variables that affect their functional ability and viability. The surgical technique of resection, for example, necessitates restriction of blood supply to the organ, leading to warm ischemia and hypoxia of the affected portion of the liver and potentially reducing the viability of hepatocytes. Thus, a protocol that preserves the liver, minimizing the damage to hepatocytes, is essential. Berry and Friend in 1969 demonstrated a protocol involving a two-step collagenase perfusion method. This protocol has become the basis of all current hepatocyte isolation protocol [84]. To standardize the protocol of human AH isolation and functionality testing, European Centre for Validation of Alternative Methods (ECVAM) has recommended the protocol. One of the most debited recommendations by ECVAM is the source of the liver, which will be never accepted by the clinicians.

Furthermore, culture of primary AH is difficult to maintain and cannot be efficiently expanded. Primary AH during the culture looses the metabolic and biotransformation capacity [85, 86]. Cryopreservation of hepatocytes allows the banking of hepatocytes that could be used upon need. Various cryopreservation protocols successfully used in animal experiments when applied human hepatocytes results in loss of 60–65 % hepatocyte viability and 5–90 % attachment rate [30, 87, 88].

Mito et al. in 1993 attempted the first human AH transplantation, till date more than 60 clinical cases have been treated with AH transplantation either to bridge patients to OLT or to improve hepatic metabolic deficiencies [30]. Strom et al. reported the results of AH transplantation into 25 patients diagnosed with acute liver failure from a number of clinical trials in USA. Complete recovery without organ transplantation occurred in 2 patients, 6 were bridged to OLT (within 1–10 days), and the remaining 10 died between 18 h and 52 days after the first hepatocyte transplantation [89, 90]. Fisher et al. reported a 37-year-old woman with FHF who was infused with 8.8×10^8 allogeneic AH into the liver through a catheter placed into the portal vein. This patient fully recovered, with a rapid fall in serum ammonia levels and was discharged from the hospital after 2 weeks [91]. Pareja et al. reported two patients received AH transplant as a bridge to whole-organ retransplantation. Patient already received a liver transplant (LT) in the past, with an end-stage liver disease due to recurrent hepatitis C virus cirrhosis while on the waiting list for an OLT. Both patients showed low blood ammonia levels and clinically improved the degree of hepatic encephalopathy, thus serving as a bridge to liver retransplantation in 1 patient [92].

The clinical use of hepatocyte transplantation is currently limited to support for the inborn errors of metabolism mainly because none of the "hepatocyte-like cells" were able to give metabolic support as AH. Importantly, treatment of acute and chronic liver failures required larger number of hepatocyte dose, which can be useful as temporary bridge to OLT. Successful cryopreservation protocol that could maintain the metabolic capacity after thawing is absolute clinical necessity.

Bone Marrow

Petersen et al. first described the contribution of bone marrow-derived stem cells (BMSC) to liver regeneration [93]. Cell populations in BMSC that contribute to regeneration are hematopoietic stem cells, mesenchymal stem cells, multipotent adult progenitor cells, and very small embryonic-like cells.

Lagasse et al. reported that intravenous injection of adult bone marrow cells in the $FAH^{-/-}$ mouse (animal model of tyrosinemia type I) rescued the mouse and restored the biochemical function of its liver. Moreover, hematopoietic stem cells gave rise to donor-derived hematopoietic and hepatic regeneration [94]. There is increasing evidence in the literature, suggesting BMS transplantation can be useful in liver rescue after acute of chronic injury [93–100]. Jang et al. reported that transplantation of BMS via the portal vein promoted functional improvement in mice with CCl_4-induced acute liver injury. Liver function was restored 2–7 days after transplantation and fibrosis reduction was also reported [101]. Shizhu et al. transplanted BMS cells via tail veins of mice. BMSCs were found to populate the damaged liver around the portal and centrilobular regions, and they appeared to differentiate into albumin-producing hepatocyte-like cells. Animals showed toward improved liver enzymes as well as enhanced survival rates [102]. Many investigators showed successful BMSC engraftment in recipient liver, but repopulation was low.

Several clinical trials found that BMSCs were beneficial in the treatment of the patients with end-stage liver failure. Autologous BMSCs transplantation resulted in improvement of liver function tests [103–108]. Gasbarrini et al. reported the use of autologous unsorted BMSCs as rescue treatment for hepatic failure in a 67-year-old man who was ineligible for liver transplantation. Patient showed rapid improvement in hepatic synthetic function after the portal venous infusion of the cells. A liver biopsy performed 20 days after cell transplant was reported to show increased hepatocyte replication around necrotic foci [109]. Salama et al. conducted a study of 90 patients with end-stage liver disease. The patients received G-CSF for 5 days followed by autologous $CD34^+$ and $CD133^+$ stem cell infusion in the portal vein. Study reported that 54 % patient showed near-normal levels of liver enzymes. Couto et al. investigated BMSC therapy in patients with severe liver disease. BMMCs were isolated from autologous bone marrow and 10 % of the cells were labeled with $^{99m}Tc\text{-}SnCl_2$. Eight patients received $2.0–15.0 \times 10^8$ cells. A patient developed a cutaneous immunomediated disorder, and another patient developed hepatocellular carcinoma (HCC) 12 months after infusion. A reduction in bilirubin was shown at 1 week, while serum albumin increased above baseline up to 1 month after infusion. Couto et al. pointed out that the early improvement of liver function should be interpreted with caution, and controlled studies are needed to determine whether BMMCs infusion affects HCC development in cirrhosis [105].

Spahr et al. reported the feasibility of autologous bone marrow mononuclear cells (BMMNCs) for the treatment of patients with decompensated alcoholic liver disease (ALD). Administration of G-CSF and followed by the $CD34^{+}$ cells from BMSC showed significant improvement in liver function in many clinical trials [103, 110, 111]. In the study, 58 patients (mean age 54 years, mean MELD score 19, all with cirrhosis, 81 % with alcoholic steatohepatitis at baseline liver biopsy) were randomized. The procedure includes the combination of G-CSF injections and autologous BMMNCs into the hepatic artery. Adverse events were observed in BMNCs and standard medical treatment (SMT) groups. After 3 months, 2 and 4 patients died, respectively, in the BMNCs and standard medical treatment groups, respectively. The MELD score improved in parallel in both groups during follow-up with 18 patients (64 %) from the BMMCT group and 18 patients (53 %) from the SMT group [112].

Vast amount of experimental and clinical data denotes that transplantation of BMSC brings functional outcome into liver parenchyma, either by fusion or transdifferentiation of MSCs, though the amount of fusion or transdifferentiation is extremely low. One of the mechanisms behind transient hepatoprotective effect is different soluble growth factors produced by MSCs. Considering the outcome of many studies, the knowledge of biological properties and plasticity of BMSC is incomplete. In clinical settings, controlled studies are needed to determine the effectiveness of BMMNCs. Presently, 108 clinical trials are going on for the evaluation of therapeutic efficacy of BMSCs for end-stage liver diseases [113]. Overall, accessibility and availability of HSCs for transplant is an attractive tool for the liver regenerative therapies.

Umbilical Cord Blood

Umbilical cord blood (UCB) cells are rich source of hematopoietic stem cells without any ethical concerns. Recent studies on UCB-derived stem cells (UCBSc) showed that these cells are capable of differentiating into adipocytes, osteocytes, chondrocytes, cardiomyocytes, neurons, and hepatocytes in vitro [114].

Newsome et al. demonstrated that human UCBSc could differentiate into hepatocytes after transplantation into immunodeficient mice without evidence of cell fusion. The percentage of human hepatocytes reached an average of 0.011 % after 16 weeks compared with mouse [115]. Kögler et al. reported that UCBSc differentiate into hepatocytes after transplantation into a pre-immune fetal sheep model. In vitro UCBSs lack HLA class II and costimulatory molecule expression [116]. Many experimental animal models of liver injury by carbon tetrachloride (CCl_4), 2-acetylaminofluorene (AAF), and the Fas ligand showed that the repopulation of recipient liver by UCBSc is extremely low [117–123]. Two studies do report that UCBSc transplantation significantly reduces the mortality caused by induced liver injury [117, 118]. Burra et al. evaluated the therapeutic potential of UCB-derived mesenchymal stem cells (UCMSCs) in a murine model of acute liver injury using CCl_4. UCMSCs-transplanted mice showed a more rapid damage resolution, lower inflammation level, and an increased catalase activity compared to CCl_4-treated mice alone [124]. Chen Li demonstrated the therapeutic potential of human umbilical cord matrix stem cells (hUCMSCs) into nonobese diabetic-severe combined immunodeficient (NOD-SCID) mice with CCl_4-induced ALF. hUCMSCs were engrafted in to recipient liver and showed the survival benefit and prevented the release of liver injury biomarkers. These data suggest that direct transplantation of native hUCMSCs can rescue ALF and repopulate livers of mice through paracrine effects to stimulate endogenous liver regeneration rather than hepatic differentiation for compensated liver function [125].

Zhang et al. evaluated the therapeutic use of hUCMSCs in 45 chronic hepatitis B patients with decompensated LC. During the study, 30 patients who received hUCMSCs showed significant reduction in the volume of ascites and improved liver function indicated by the increase of serum albumin levels, decrease in total serum bilirubin levels, and decrease in the sodium model for end-stage liver disease scores [126]. Presently, there are 4 ongoing clinical trials (phase I and II) using UCBSc, investigators thus studying the safety and efficacy of UCBSc on patients with liver cirrhosis.

Thus, there will be much excitement about the use of the UCBSc for the treatment of acute and chronic liver diseases. Collation of UCBSc is done by noninvasive methods, and there is no problem in the availability of donor. Importantly, UCBSc can be cryopreserved without major loss during cryopreservation. Furthermore, UCBSc are nonimmunogenic, which make them more suitable for allogeneic transplant.

Route of Administration

There are different approaches used for hepatocyte transplant in animal models. The most accepted approach for hepatocyte transplantation is intrasplenic transplantation. Hepatocytes injected in splenic pulp migrate through hepatic artery in liver parenchyma. The flow of the injection helps hepatocyte to invade the sinusoidal boundaries and entrap in vascular spaces [127]. Entrapped hepatocytes can integrate and proliferate into hepatic lobule. During intrasplenic transplantation, about 40 % of hepatocytes are entrapped in splenic pulp, and this serves as extra hepatic site for hepatocytes. The hepatocytes showed synthetic, metabolic, and biliary transport function [128] (Figs. 4 and 5).

Major limiting factors in using intrasplenic or intraportal route are portal vein thrombosis, portal hypertension, and pulmonary embolism due to transplanted cells [129]. Importantly, there are limitation of cellular dose that can be transplanted single time; thus, repetitive infusion of cells is not possible. However, portal hypertension usually resolves within hours after transplantation, and other measures taken to reduce these complications are by hepatic artery ligation to decrease the sinusoidal blood flow prior to hepatocytes infusion or the slow infusion of hepatocytes over a longer period [119, 130]. In our studies, we have used 70 % partial hepatectomy to mouse model of liver injury followed by 2×10^6 fetal hepatocytes. During the experiments, we have seen portal hypertension and lung embolism as a

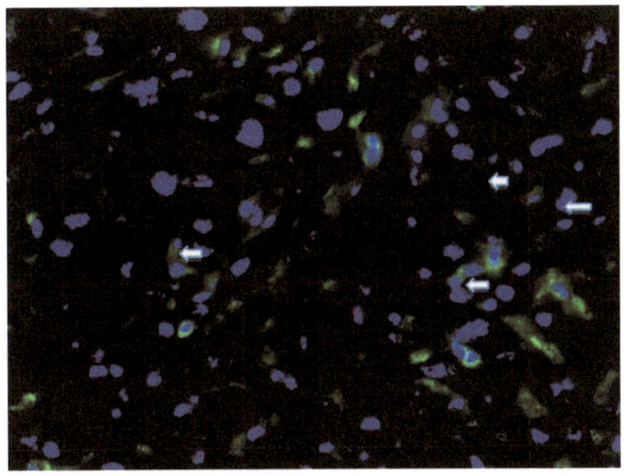

Fig. 4 D-galactose intoxicated C57BL/6 nude mice had 40 % partial hepatectomy followed by transplantation of FH (2×10^6/mice). Fresh-frozen section of mice liver showing albumin (*green* staining), nuclei were counterstained with DAPI (*blue*)

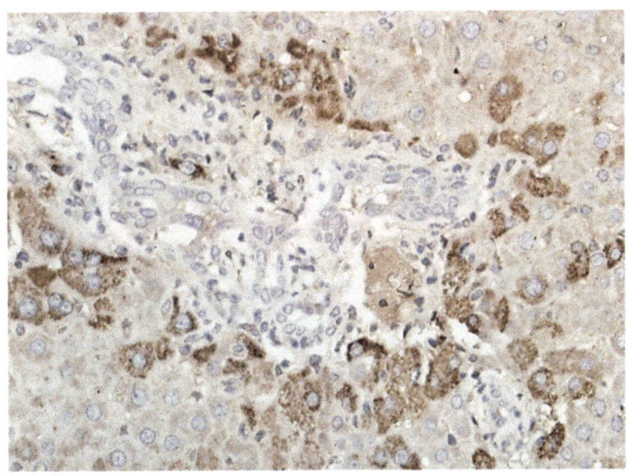

Fig. 5 D-galactose intoxicated C57BL/6 nude mice had 40 % partial hepatectomy followed by transplantation of FH (2×10^6/mice). Paraffin section of mice liver showing human hepatocyte antigen (*brown* staining)

result of cell infusion. We could achieve survival rate of 50 % (unpublished data). Therefore, to overcome the limitation, we followed 30–40 % partial hepatectomy followed by cell infusion. From this modification, we could get 100 % survival. We have observed transient hypertension due to lung embolism, but it was transient [29].

Presently, different alternative sites to encourage hepatocyte attachment, proliferation, and survival are under investigation. In several animal studies, cells were directly injected into the liver parenchyma to avoid cell loss during the transplantation. But transplanted cells were observed in central veins indicating an increased risk of lung embolism [131–133]. Therefore, in the clinical setting, direct injection of cells into the liver may not be feasible. Ectopic hepatocyte transplantation is defined as a transplantation site for hepatocytes other than the liver or spleen. Cell therapy directed towards the liver may not be feasible in cirrhotic and fibrotic liver during end-stage disease. Thus, transplantation of cells in ectopic sites may give temporary relief for the metabolic and synthetic stress on hepatocytes. Researchers have evaluated different ectopic sites for hepatocyte transplant, including the intraperitoneal cavity, pancreas, mesenteric leaves, intrapulmonary artery, lung parenchyma, kidney capsule, interscapular fat pads, and lymph nodes [134–141]. Unfortunately, transplanted cells survive for a short time. The specific reason for short-term survival of cells is unknown; it is likely due to lack of liver specific growth factors or lack of neovascularization. To overcome this limitation, the use of growth factors, like SDF-1, HGF, EGF, or VEGFR, can play a crucial role for initial support for hepatocyte survival and proliferation. Yokoyama et al. injected fibroblast growth factor (bFGF) in subcutaneous space followed by transplantation of rat and mouse hepatocytes by providing a polyethylene terephthalate matrix. Transplanted hepatocytes survived from 4–8 months and retained their albumin synthetic and drug metabolizing capacity. Furthermore, the authors were able to transplant ten times the usual number of hepatocytes into the subcutaneous cavity [136].

Future Strategies

Stem cell therapy holds promises for acute and chronic liver diseases, but extensive research for the last 2 decades could not improve the outcome. Major limitation of the therapy is the availability of good quality hepatocytes for transplantation. Typically, hepatocytes are complex cells able to do multiple metabolic and secretary functions. The term "hepatocyte-like cells" fails to show major metabolic and secretory characteristics of adult hepatocyte [30]. There are ongoing effects to make "hepatocyte-like cells" more functional like adult hepatocytes. One of the important problem clinicians face during the fibrotic or chronic liver failure is the availability of space to accommodate functional hepatocyte. To overcome this issue, study of ectopic sites where large number of functional hepatocytes can be transplanted without any other complication is ideal need to cell therapy. Considering the variables in the pathogenesis of acute and chronic liver diseases, it is highly important to tailor made the therapies to specific patient using different stem cells (ES, iPSCs, FH, BMMNCs, CBSCs).

Development in the strategies of generating hepatocytes, banking of hepatocytes, hepatocyte engraftment, therapeutic cell number, and functionality of hepatocytes will change the present scenario of liver-related cell transplant.

Conclusion

Available data on clinical hepatocyte transplantation indicated the limitations of cell therapy. Realistically, cell therapy can be advantageous to acute liver failure or metabolic support where therapeutic cell dose is less as compared to chronic liver failure. Considering the limitations on availability and storage of adult hepatocytes, alternative sources of hepatocytes generated from stem cell can be available for treatment. There are numerous advantages of stem cells generated hepatocyte transplant: (1) unlimited supply of hepatocytes; (2) cryopreservation and banking of hepatocytes; (3) complete metabolic and secretory profile; (4) noninvasive treatment, without much hospitalization, and economical; (5) no need for posttransplant medication or immunosuppressant.

Many of the stem cell sources such as iPSCs, FH, BMMSCs can be alternative sources of hepatocytes and should be evaluated in larger clinical study. The present experimental and clinical studies demonstrate the advantages and limitations of each cell types used in therapy. Considering the information, revised clinical trials can be conducted in larger study groups; and this will take hepatocyte transplant in to clinical reality.

References

1. Evolution of liver transplantations in Europe from 05/1968 to 12/2009. European Liver Transplant Registry. Available from: http://www.eltr.org
2. NHS Blood and Transplant (2012) Organ donation and transplantation: activity report 2011/12. http://www.organdonation.nhs.uk/statistics/transplant_activity_report/archive_activity_reports/pdf/ukt/activity_report_2011_12.pdf
3. Organ Procurement and Transplantation Network (2014) Waiting list candidates: liver. transplants: liver. Organ Procurement and Transplantation Network [Web site]. Accessed Jan 2014
4. Meier M, Woywodt A, Hoeper MM, Schneider A, Manns MP, Strassburg CP (2005) Acute liver failure: a message found under the skin. Postgrad Med J 81:269–270
5. Rajiv J (2005) Acute liver failure: current management and future prospects. J Hepatol 42(Suppl 1):S115–S123
6. Ostapowicz G, Fontana RJ, Schiodt FV, Larson A, Davern TJ, Han SH et al (2002) Results of a prospective study of acute liver failure at 17 tertiary care centers in the United States. Ann Intern Med 137:947–954
7. Ezzat T, Dhar DK, Malago M, Olde Damink SW (2012) Dynamic tracking of stem cells in an acute liver failure model. World J Gastroenterol 18(6):507–516
8. Lee WM (2008) Etiologies of acute liver failure. Semin Liver Dis 28:142–152
9. van den Broek MA, van Dam RM, Malagó M, Dejong CH, van Breukelen GJ, OldeDamink SW (2009) Feasibility of randomized controlled trials in liver surgery using surgery-related mortality or morbidity as endpoint. Br J Surg 96:1005–1014
10. Martin-Murphy BV, Holt MP, Ju C (2010) The role of damage associated molecular pattern molecules in acetaminophen-induced liver injury in mice. Toxicol Lett 192:387–394
11. Fiegel HC, Lange C, Kneser U, Lambrecht W, Zander AR, Rogiers X, Kluth D (2006) Fetal and adult liver stem cells for liver regeneration and tissue engineering. J Cell Mol Med 10(3):577–587
12. Chamukean RAFM (1997) Hepatocyte transplantation for acute hepatic failure. In: Mito M, Sawa M (eds) Hepatocyte transplantation. Karger Landes Systems, New York, pp 157–167
13. Di Campli C, Nestola M, Piscaglia AC et al (2003) Cell-based therapy for liver diseases. Eur Rev Med Pharmacol Sci 7(2):41–44
14. Patil PB, Begum S, Joshi M et al (2014) Phenotypic and in vivo functional characterization of immortalized human fetal liver cells. Scand J Gastroenterol 49(6):705–714
15. Zenoroli MI (1985) Hepatic encephalopathy. Experimental studies in a rat model of fulminant hepatic failure. J Hepatol 2:301–312
16. Dixit V, Chang TMS (1990) Brain edema and the blood brain barrier in galactosamine-induced fulminant hepatic failure rats. An animal model for evaluation of liver support systems. ASAIO Trans 36:21–27
17. Peeling J, Schoemaker L, Gauthier T (1993) Cerebral metabolic and histologic effects of thiocetamide-induced liver failure. Am J Physiol 265:G572–G578
18. Yurdaydin C, Hortnagl H, Steindl P (1990) Increased serotoninergic and noradrnergic activity in hepatic encephalopathy in rats with thioacetamid-induced acute liver failure. Hepatology 12:695–700
19. Maziotti A, Bernardi M, Antonini L et al (1981) Plasma amino acid patterns in experimental acute hepatic failure: comparison between hepatectomy and liver devascularization in pigs. Surgery 90: 527–534
20. Vogels BAPM, Maas MAW, Bosma A et al (1996) Significant improvement of survival by intrasplenic hepatocyte transplantation in totally hepatactomised rats. Cell Transplant 5:369–378
21. Roger V, Balladur P, Honiger J et al (1995) A good model of experimental acute hepatic failure: 95 % hepatectomy; treatment by transplantation of hepatocytes. Transplant Proc 27:2504–2505
22. Gupta S, Bhargava KB, Novikoff PM (1973) Hepatocyte transplantation: emerging insights into mechanisms of liver repopulation and their relevance to potential therapies. J Hepatol 30:162–170
23. Khan AA, Shaik MV, Parveen N et al (2010) Human fetal liver-derived stem cell transplantation as supportive modality in the management of end-stage decompensated liver cirrhosis. Cell Transplant 19:409–418
24. 2009 Annual Report of the U.S. Organ Procurement and Transplantation Network and the Scientific Registry of Transplant Recipients: Transplant Data 1999–2008 (2010) U.S. Department of Health

and Human Services, Health Resources and Services Administration, Healthcare Systems Bureau, Division of Transplantation
25. Cai J, Ito M, Nagata H, Westerman KA, LaFleur D, Chowdhury JR et al (2002) Treatment of liver failure in rats with end-stage cirrhosis by transplantation of immortalized hepatocytes. Hepatology 36:386–394
26. Nagata H, Ito M, Cai J, Edge AS, Platt JL, Fox IJ (2003) Treatment of cirrhosis and liver failure in rats by hepatocyte xenotransplantation. Gastroenterology 124:422–431
27. Kobayashi N, Ito M, Nakamura J, Cai J, Gao C, Hammel JM et al (2000) Hepatocyte transplantation in rats with decompensated cirrhosis. Hepatology 31:851–857
28. Lauterbourg BH, Sautter V, Preising R et al (1976) Hepatic functional deterioration after portacaval shunt rat. Gastroenterology 71:221–227
29. Joshi M, Patil PB, He Z et al (2012) Fetal liver derived mesenchymal stromal cells augment engraftment of transplanted hepatocytes. Cytotherapy 14(6):657–669
30. Akhter J, Johnson LA, Morris DL (2009) Hepatocyte transplantation: a new approach to treat liver disorders. In: Dionigi R (ed) Recent advances in liver surgery. Landes Bioscience, Austin, pp 331–351
31. Yasunaga MT, Nishikawa ST, Nakano S et al (2005) Induction and monitoring of definitive and visceral endoderm differentiation of mouse ES cells. Nat Biotechnol 23:1542–1550
32. Lavon N, Yanuka O, Benvenisty N (2004) Differentiation and isolation of hepatic-like cells from human embryonic stem cells. Differentiation 72: 230–238
33. Cai J, Zhao Y, Liu Y et al (2007) Directed differentiation of human embryonic stem cells into functional hepatic cells. Hepatology 45:1229–1239
34. Oertel M, Shafritz DA (2008) Stem cells, cell transplantation and liver repopulation. Biochim Biophys Acta 1782:61–74
35. Dan YY, Yeoh GC (2008) Liver stem cells: a scientific and clinical perspective. J Gastroenterol Hepatol 23:687–698
36. Agarwal S, Holton KL, Lanza R (2008) Efficient differentiation of functional hepatocytes from human embryonic stem cells. Stem Cells 26:1117–1127
37. Hay DC, Zhao D, Ross A et al (2007) Direct differentiation of human embryonic stem cells to hepatocyte-like cells exhibiting functional activities. Cloning Stem Cells 9:51–62
38. Zhao CJ, Liu Y, Ye Y et al (2007) Directed differentiation of human embryonic stem cells into functional hepatic cells. Hepatology 45:1229–1239
39. Baharvand H, Hashemi SM, Shahsavani M (2007) Differentiation of human embryonic stem cells into functional hepatocyte-like cells in a serum-free adherent culture condition. Differentiation 76(5): 465–477
40. Yamamoto H, Quinn G, Asari A et al (2003) Differentiation of embryonic stem cells into hepatocytes: biological functions and therapeutic application. Hepatology 37:983–993
41. Heo J, Factor VM, Uren T et al (2006) Hepatic precursors derived from murine embryonic stem cells contribute to regeneration of injured liver. Hepatology 44:1478–1486
42. Bin WT, Ma LM, Xu Q (2012) Embryonic hepatocyte transplantation for hepatic cirrhosis: efficacy and mechanism of action. World J Gastroenterol 28(4):309–322
43. Choi D, Oh HJ, Chang UJ et al (2002) In vivo differentiation of mouse embryonic stem cells into hepatocytes. Cell Transplant 11:359–368
44. Rodaway A, Patient R (2001) Mesendoderm: an ancient germ layer? Cell 105:169–172
45. Tremblay KD, Zaret KS (2005) Distinct populations of endoderm cells converge to generate the embryonic liver bud and ventral foregut tissues. Dev Biol 280:87–99
46. Schwartz RE, Reyes M, Koodie L et al (2002) Multipotent adult progenitor cells from bone marrow differentiate into functional hepatocyte-like cells. J Clin Invest 109:1291–1302
47. Lee KD, Kuo TK, Whang-Peng J et al (2004) In vitro hepatic differentiation of human mesenchymal stem cells. Hepatology 40:1275–1284
48. Nava S, Westgren M, Jaksch M (2005) Characterization of cells in the developing human liver. Differentiation 73:249–260
49. Nowak G, Ericzon BG, Nava S et al (2005) Identification of expandable human hepatic progenitors which differentiate into mature hepatic cells in vivo. Gut 54:972–979
50. Begum S, Joshi M, Ek M et al (2009) Characterization and engraftment of long-term serum-free human fetal liver cell cultures. Cytotherapy 24: 201–211
51. Mahieu-Caputo D, Allain JE, Branger J et al (2004) Repopulation of athymic mouse liver by cryopreserved early human fetal hepatoblasts. Hum Gene Ther 15:1219–1228
52. Sandhu JS, Petkov PM, Dabeva MD, Shafritz DA (2001) Stem cell properties and repopulation of the rat liver by fetal liver epithelial progenitor cells. Am J Pathol 159:1323–1334
53. Dabeva MD, Petkov PM, Sandhu J et al (2000) Proliferation and differentiation of fetal liver epithelial progenitor cells after transplantation into adult rat liver. Am J Pathol 156: 2017–2031
54. Khan AA, Parveen N, Mahaboob VS et al (2008) Treatment of Crigler-Najjar Syndrome type 1 by hepatic progenitor cell transplantation: a simple procedure for management of hyperbilirubinemia. Transplant Proc 40:1148–1150
55. Khan AA, Parveen N, Mahaboob VS et al (2008) Management of hyperbilirubinemia in biliary atresia

by hepatic progenitor cell transplantation through hepatic artery: a case report. Transplant Proc 40:1153–1155
56. Gridelli B, Vizzini G, Pietrosi G et al (2012) Efficient human fetal liver cell isolation protocol based on vascular perfusion for liver cell-based therapy and case report on cell transplantation. Liver Transpl 18:226–237
57. Takahashi K, Yamanaka S (2006) Induction of pluripotent stem cells from mouse embryonic and adult fibroblast cultures by defined factors. Cell 126(4):663–676
58. Park IH, Zhao R, West JA et al (2008) Reprogramming of human somatic cells to pluripotency with defined factors. Nature 451:141–146
59. Chin MH, Mason MJ, Xie W et al (2009) Induced pluripotent stem cells and embryonic stem cells are distinguished by gene expression signatures. Cell Stem Cell 5:111–123
60. Kim K, Doi A, Wen B et al (2010) Epigenetic memory in induced pluripotent stem cells. Nature 467:285–290
61. Liu H, Ye Z, Kim Y et al (2010) Generation of endoderm-derived human induced pluripotent stem cells from primary hepatocytes. Hepatology 51(5):1810–1819
62. Miura K, Okada Y, Aoi T et al (2009) Variation in the safety of induced pluripotent stem cell lines. Nat Biotechnol 27:743–745
63. Sullivan GJ, Hay DC, Park IH et al (2010) Generation of functional human hepatic endoderm from human induced pluripotent stem cells. Hepatology 51:329–335
64. Rashid ST, Corbineau S, Hannan N et al (2010) Modeling inherited metabolic disorders of the liver using human induced pluripotent stem cells. J Clin Invest 120:3127–3136
65. Si-Tayeb K, Noto FK, Nagaoka M et al (2010) Highly efficient generation of human hepatocyte-like cells from induced pluripotent stem cells. Hepatology 51:297–305
66. Zhou WL, Medine CN, Zhu L et al (2012) Stem cell differentiation and human liver disease. World J Gastroenterol 18(17):2018–2025
67. Jozefczuk J, Prigione A, Chavez L et al (2011) Comparative analysis of human embryonic stem cell and induced pluripotent stem cell-derived hepatocyte-like cells reveals current drawbacks and possible strategies for improved differentiation. Stem Cells Dev 20:1259–1275
68. Espejel S, Roll GR, McLaughlin KJ et al (2010) Induced pluripotent stem cell-derived hepatocytes have the functional and proliferative capabilities needed for liver regeneration in mice. J Clin Invest 120(9):3120–3126
69. Asgari S, Moslem M, Bagheri-Lankarani K et al (2013) Differentiation and transplantation of human induced pluripotent stem cell-derived hepatocyte-like cells. Stem Cell Rev Rep 9(4): 493–504
70. Ma X, Duan Y, Tschudy-Seney B, Roll G et al (2013) Highly efficient differentiation of functional hepatocytes from human induced pluripotent stem cells. Stem Cells Transl Med 2(6):409–419
71. Huang P, He Z, Ji S et al (2011) Induction of functional hepatocyte-like cells from mouse fibroblasts by defined factors. Nature 475:386–389
72. Rajvanshi P, Kerr A, Bhargava KK et al (1996) Studies of liver repopulation using the dipeptidyl peptidase IV – deficient rat and other rodent recipients: cell size and structure relationships regulate capacity for increased transplanted hepatocyte mass in the liver lobule. Hepatology 23:482
73. Groth CG, Arborgh B, Bjorken C et al (1977) Correction of hyperbilirubinemia in the glucuronyltransferase-deficient rats by intraportal hepatocyte transplantation. Transplant Proc 9:313–316
74. Yoshida Y, Tokusashi Y, Lee GH et al (1996) Intrahepatic transplantation of normal hepatocytes prevents Wilson's disease in Long—Evans cinnamon rats. Gastroenterology 11:1654–1660
75. De Vree JM, Ottenhoff R, Bosma PJ et al (2000) Correction of liver disease by hepatocyte transplantation in a mouse model of progressive familial intrahepatic cholestasis. Gastroenterology 119:1720–1730
76. Michel JL, Rabier D, Rambaud C et al (1993) Intrasplenic transplantation of hepatocytes in spf-ash mice with congenital ornithine transcarbamylase deficiency. Chirurgie 119:666–671
77. Wiederkehr JC, Pollak R (1990) Hepatocyte transplantation for the low-density lipoprotein receptor-deficient state. A study in the Watanabe rabbit. Transplantation 50:466–471
78. Sugiyama K, Emori T, Nagase S (1982) Synthesis and secretion of plasma proteins by isolated hepatocytes of analbuminemic rats. J Biochem 92:775–779
79. Gagandeep S, Sokhi R, Slehria S et al (2000) Hepatocyte transplantation improves survival in mice with liver toxicity induced by hepatic over expression of Mad 1 transcription factor. Mol Ther 1:358–365
80. David P, Viollon C, Alexandre E et al (1998) Metabolic capacities in cultured human hepatocytes obtained from non-wedge small liver biopsies. Hum Exp Toxicol 17:544–553
81. Ballet F, Bouma M, Wang S et al (1984) Isolation, culture and characterization of adult human hepatocytes from surgical liver biopsies. Hepatology 4:849–854
82. Baccarani U, Sanna A, Cariani A et al (2003) Isolation of human hepatocytes from livers rejected for liver transplantation on a national basis: results of a 2 year experience. Liver Transpl 9:506–512
83. Richert L, Alexandre E, Lloyd T et al (2004) Tissue collection, transport and isolation procedures required to optimize human hepatocyte isolation from waste liver surgical resections. A multi-laboratory study. Liver Int 24:371–378

84. Berry MN, Friend DS (1969) High-yield preparation of isolated rat liver parenchymal cells: a biochemical and fine structural study. J Cell Biol 43: 506–520
85. de Sousa G, Dou M, Barbe D et al (1991) Freshly isolated or cryopreserved human hepatocytes in primary culture: influence of drug metabolism on hepatotoxicity. Toxicol In Vitro 5:483–486
86. Ruegg CE, Silber PM, Mughal RA (1997) Cytochrome P-450 induction and conjugated metabolism in primary human hepatocytes after cryopreservation. In Vitro Toxicol 10:217–222
87. Chesne C, Guillouzo A (1988) Cryopreservation of isolated rat hepatocytes: a critical evaluation of freezing and thawing conditions. Cryobiology 25:323–330
88. Ostrowska A, Bode CD, Pruss J et al (2000) Investigation of functional and morphological integrity of freshly isolated and cryopreserved human hepatocytes. Cell Tissue Bank 1:55–68
89. Strom SC, Chowdhury JR, Fox IJ (1999) Hepatocyte transplantation for the treatment of human disease. Semin Liver Dis 19:39–48
90. Strom SC, Fisher RA, Rubinstein WS et al (1997) Transplantation of human hepatocytes. Transplant Proc 29:2103–2106
91. Fisher RA, Bu D, Thompson M et al (2000) Defining hepatocellular chimerism in a liver failure patient bridged with hepatocyte infusion. Transplantation 69:303–307
92. Pareja E, Gomez-Lechon MJ, Cortes M et al (2013) Human hepatocyte transplantation in patients with hepatic failure awaiting a graft. Eur Surg Res 50:273–281
93. Petersen BE, Bowen WC, Patrene KD et al (1999) Bone marrow as a potential source of hepatic oval cells. Science 284(5417):1168–1170
94. Lagasse E, Connors H, Al-Dhalimy M et al (2000) Purified hematopoietic stem cells can differentiate into hepatocytes in vivo. Nat Med 6(11):1229–1234
95. Alison MR, Poulsom R, Jeffery R et al (2000) Hepatocytes from non-hepatic adult stem cells. Nature 406(6793):257
96. Theise ND, Badve S, Saxena R et al (2000) Derivation of hepatocytes from bone marrow cells in mice after radiation-induced myeloablation. Hepatology 31(1):235–240
97. Quintana-Bustamante O, Alvarez-Barrientos A, Kofman AV et al (2006) Hematopoietic mobilization in mice increases the presence of bone marrow-derived hepatocytes via in vivo cell fusion. Hepatology 43:108–116
98. Theise ND, Nimmakayalu M, Gardner R (2000) Liver from bone marrow in humans. Hepatology 32(1):11–16
99. Cantz T, Sharma AD, Jochheim-Richter A et al (2004) Reevaluation of bone marrow-derived cells as a source for hepatocyte regeneration. Cell Transplant 13(6):659–666
100. Kanazawa Y, Verma IM (2003) Little evidence of bone marrow-derived hepatocytes in the replacement of injured liver. Proc Natl Acad Sci U S A 100(1):11850–11853
101. Jang YY, Collector MI, Baylin SB et al (2004) Hematopoietic stem cells convert into liver cells within days without fusion. Nat Cell Biol 6(6):532–539
102. Shizhu J, Xiangwei M, Xun S et al (2012) Bone marrow mononuclear cell transplant therapy in mice with CCl_4-induced acute liver failure. Turk J Gastroenterol 23:344–352
103. Gordon MY, Levicar N, Pai M et al (2006) Characterization and clinical application of human $CD34^+$ stem/progenitor cell populations mobilized into the blood by granulocyte colony-stimulating factor. Stem Cells 24:1822–1830
104. Salama H, Zekri AR, Zern M et al (2010) Autologous hematopoietic stem cell transplantation in 48 patients with end-stage chronic liver diseases. Cell Transplant 19:1475–1486
105. Couto BG, Goldenberg RC, da Fonseca LM et al (2011) Bone marrow mononuclear cell therapy for patients with cirrhosis: a Phase 1 study. Liver Int 31(3):391–400
106. Lyra AC, Soares MB, da Silva LF, Fortes MF, Silva AG, Mota AC, Oliveira SA, Braga EL, de Carvalho WA, Genser B et al (2007) Feasibility and safety of autologous bone marrow mononuclear cell transplantation in patients with advanced chronic liver disease. World J Gastroenterol 13:1067–1073
107. Pai M, Zacharoulis D, Milicevic MN, Helmy S, Jiao LR, Levicar N, Tait P, Scott M, Marley SB, Jestice K et al (2008) Autologous infusion of expanded mobilized adult bone marrow-derived $CD34^+$ cells into patients with alcoholic liver cirrhosis. Am J Gastroenterol 103:1952–1958
108. Lyra AC, Soares MB, da Silva LF, Braga EL, Oliveira SA, Fortes MF, Silva AG, Brustolim D, Genser B, Dos Santos RR et al (2010) Infusion of autologous bone marrow mononuclear cells through hepatic artery results in a short-term improvement of liver function in patients with chronic liver disease: a pilot randomized controlled study. Eur J Gastroenterol Hepatol 22:33–42
109. Gasbarrini A, Rapaccini GL, Rutella S, Zocco MA, Tittoto P, Leone G, Pola P, Gasbarrini G, Di Campli C (2007) Rescue therapy by portal infusion of autologous stem cells in a case of drug-induced hepatitis. Dig Liver Dis 39:878–882
110. Terai S, Ishikawa T, Omori K et al (2006) Improved liver function in patients with liver cirrhosis after autologous bone marrow cell infusion therapy. Stem Cells 24:2292–2298
111. Evicar N, Pai M, Habib NA et al (2008) Long-term clinical results of autologous infusion of mobilized adult bone marrow derived CD34+ cells in patients with chronic liver disease. Cell Prolif 41 Suppl 1:115–125

112. Spahr L, Chalandon Y, Terraz S et al (2013) Autologous bone marrow mononuclear cell transplantation in patients with decompensated alcoholic liver disease: a randomized controlled trial. PLoS One 8(1):e53719
113. Bone marrow stem cells and liver http://www.clinicaltrials.gov/ct2/results?term=bone+marrow+stem+cells+AND+liver. Accessed on 17 Jan 2014
114. van de Ven C, Collins D, Bradley MB et al (2007) The potential of umbilical cord blood multipotent stem cells for nonhematopoietic tissue and cell regeneration. Exp Hematol 35:1753–1765
115. Newsome PN, Johannessen I, Boyle S et al (2003) Human cord blood-derived cells can differentiate into hepatocytes in the mouse liver with no evidence of cellular fusion. Gastroenterology 124:1891–1900
116. Kögler G, Sensken S, Airey JA et al (2004) A new human somatic stem cell from placental cord blood with intrinsic pluripotent differentiation potential. J Exp Med 200:123–135
117. Di Campli C, Piscaglia AC, Rutella S et al (2005) Improvement of mortality rate and decrease in histologic hepatic injury after human cord blood stem cell infusion in a murine model of hepatotoxicity. Transplant Proc 37:2707–2710
118. Nonome K, Li XK, Takahara T et al (2005) Human umbilical cord blood-derived cells differentiate into hepatocyte-like cells in the Fas-mediated liver injury model. Am J Physiol Gastrointest Liver Physiol 289:G1091–G1099
119. Beerheide W, von Mach MA, Ringel M et al (2002) Downregulation of beta2-microglobulin in human cord blood somatic stem cells after transplantation into livers of SCID-mice: an escape mechanism of stem cells? Biochem Biophys Res Commun 294:1052–1063
120. Brulport M, Schormann W, Bauer A et al (2007) Fate of extrahepatic human stem and precursor cells after transplantation into mouse livers. Hepatology 46:861–870
121. Shyu MK, Yuan RH, Shih JC et al (2007) Kinetics and functional assay of liver repopulation after human cord blood transplantation. Dig Liver Dis 39:455–465
122. Wulf-Goldenberg A, Eckert K, Fichtner I (2008) Cytokine-pretreatment of CD34(+) cord blood stem cells in vitro reduces long-term cell engraftment in NOD/SCID mice. Eur J Cell Biol 87:69–80
123. Burra P, Arcidiacono D, Bizzaro D et al (2012) Systemic administration of a novel human umbilical cord mesenchymal stem cells population accelerates the resolution of acute liver injury. BMC Gastroenterol 12:88
124. Chen Li, Liu T, Zhang B et al (2012) Human umbilical cord matrix stem cells efficiently rescue acute liver failure through paracrine effects rather than hepatic differentiation. Tissue Eng Part A 18(13–14):1352–1364
125. Zhang Z, Lin H, Shi M et al (2012) Human umbilical cord mesenchymal stem cells improve liver function and ascites in decompensated liver cirrhosis patients. J Gastroenterol Hepatol 27(Suppl 2):112–120
126. Umbilical Cord Blood Therapy in Liver Cirrhosis http://www.clinicaltrials.gov/ct2/results?term=Umbilical+Cord+Blood+Therapy+in+Liver+Cirrhosis&Search=Search. Accessed 17 Jan 2014
127. Mito M, Ebata H, Kusano M et al (1979) Morphology and function of isolated hepatocytes transplanted into rat spleen. Transplantation 28:499–505
128. Strom SC, Jirtle RL, Jones RS et al (1982) Isolation, culture and transplantation of human hepatocytes. J Natl Cancer Inst 68:771–778
129. Rajvanshi P, Kerr A, Bhargava KK et al (1996) Efficacy and safety of repeated hepatocyte transplantation for significant liver repopulation in rodents. Gastroenterology 111:1092–1102
130. Soriano HE, Wood RP, Kang DC et al (1997) Hepatocellular transplantation in children with fulminant liver failure. J Hepatol 26:239A
131. von Mach MA, Hengstler JG, Brulport M et al (2004) In vitro cultured islet-derived progenitor cells of human origin express human albumin in severe combined immunodeficiency mouse liver in vivo. Stem Cells 22:1134–1141
132. Ruhnke M, Ungefroren H, Nussler A et al (2005) Reprogramming of human peripheral blood monocytes into functional hepatocyte and pancreatic islet-like cells. Gastroenterology 128:1774–1786
133. Yokoyama T, Ohashi K, Kuge H et al (2005) In vivo engineering of metabolically active hepatic tissue in a neovascularised subcutaneous cavity. Am J Transplant 6:50–59
134. Demetriou AA, Levenson SM, Novikoff PM et al (1986) Survival, organization and function of microcarrier-attached hepatocytes transplanted in rats. Proc Natl Acad Sci U S A 83:7475–7479
135. Jaffe V, Darby H, Selden C et al (1988) The growth of transplanted liver cells within the pancreas. Transplantation 45:497–498
136. Sano K, Cusick RA, Lee H et al (1996) Regenerative signals for heterotopic hepatocyte transplantation. Transplant Proc 28:1857–1858
137. Selden C, Gypta S, Johnstone R et al (1984) The pulmonary vascular bed as a site for implantation of isolated liver cells in inbred rats. Transplantation 38:81–83
138. Then P, Sandbichler P, Erhart R et al (1991) Hepatocyte transplantation into lung for treatment of acute hepatic failure in the rat. Transplant Proc 23: 892–893
139. Ricordi C, Lacy PE, Callery MP et al (1989) Trophic factors from pancreatic islets in combined hepatocyte-islet allografts enhance hepatocellular survival. Surgery 105:218–223
140. Jirtle RL, Biles C, Michalopoulos G (1980) Morphologic and histochemical analysis of hepatocytes transplanted into syngeneic hosts. Am J Pathol 101:115–126
141. Komori J, Boone L, DeWard A et al (2012) The mouse lymph node as an ectopic transplantation site for multiple tissues. Nat Biotechnol 30(10):976–983

Use of Stem Cells to Block the Activation of Hepatic Stellate Cells in Diseased Liver

Syamantak Majumder, Palanivel Gajalakshmi, and Suvro Chatterjee

Introduction

The liver is a distinct organ which shows a high degree of regenerative capacity. Major functions of liver include metabolism, digestion, storage, synthesis, and release of vitamins, carbohydrates, proteins, and lipids. Detoxification of toxic substances and metals are the other important functions of liver [1]. The anatomy of the liver shows that it is divided into lobules, which are hexagonal in shape and at each corner, there is a portal vein, hepatic artery and bile duct. The liver is a richly perfused organ receiving approximately 1/4th of the cardiac output. The blood flows from the portal vein to the central vein through the sinusoidal capillaries and finally leaves the liver via the hepatic vein (Fig. 1a). Most of the drugs consumed and absorbed by gastrointestinal (GI) tract by humans undergo first-pass effect, i.e., it is metabolized by the liver and then passes on to the general circulation [2]. Circulation of blood in the liver is unique because it receives both oxygenated and deoxygenated blood. The liver harbors several types of cells such as hepatic cells or hepatocytes, hepatic stellate cells, Kupffer cells, endothelial cells, and liver stem cells. Hepatocytes are highly polarized epithelial cells that are responsible for most of the metabolic functions. Hepatic stellate cells (HSCs) and Kupffer cells are found in between the hepatocyte cord and sinusoids (Fig. 1b). Hepatocytes produce bile salts which are excreted into the space (called as canaliculus) outside the hepatocytes that lead to bile duct. The region that connects the terminal bile canaliculus and hepatocytes is known as the canals of Hering, from where the liver stem cells are formed [3]. The liver possesses a strong self-healing system by nature. The property of the liver to regenerate by itself dates back to Greek Titan Prometheus, whose liver was believed to be eaten by an eagle every day, and it was grown back each night. Indeed, the regenerative potential is so high that the hepactectomized liver can grow back to normal within 45 days. Because of high pressure and heavy drainage organ with multiple tasks, a strong self-healing system is in place for the liver. Strong regenerative potential of the liver makes it a very special model system to study regenerative medicine.

S. Majumder
Aab Cardiovascular Research Institute, School of Medicine and Dentistry, University of Rochester, Rochester, NY 14642, USA
e-mail: sya.sam2005@gmail.com

P. Gajalakshmi
Life Sciences Division, Vascular Biology Lab, AU-KBC Research Centre, Anna University, Chennai, India
e-mail: gajvig@yahoo.com

S. Chatterjee (✉)
Department of Biotechnology, Vascular Biology Lab, AU-KBC Research Centre, Anna University, Chennai, India

Centre for Biotechnology, Anna University, Chennai, India
e-mail: soovro@yahoo.ca

I. Somasundaram (ed.), *Stem Cell Therapy for Organ Failure*,
DOI 10.1007/978-81-322-2110-4_17, © Springer India 2014

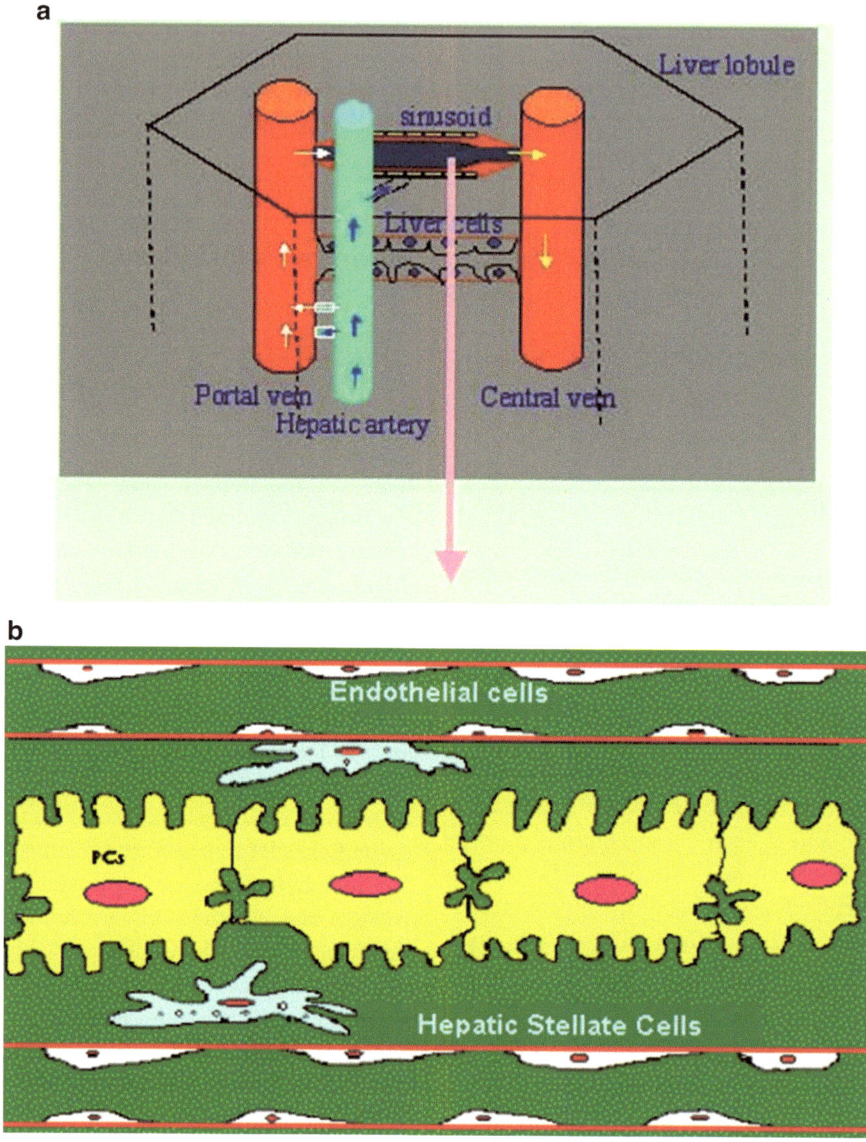

Fig. 1 (**a**, **b**) Shows the schematic representation of liver architecture. (**a**) Shows an individual liver lobule with portal system, sinusoid, and hepatocytes. (**b**) Shows a magnified image of sinusoid where different types of cells such as hepatic stellate cells, endothelial cells, and hepatic cells are shown (Adapted from http://bio.au-kbc.org/faculty/suvro/liver)

Liver Diseases: The Fibrotic Template

The liver is bombarded with mechanical, chemical, and pathological insults throughout life. There are more than 100 types of liver diseases in the world which are caused due to viral infection (hepatitis A, B, and C), alcohol consumption, obesity, genetics, autoimmune diseases, drugs, toxins, and cancer. Chronic liver diseases are life-threatening to humans. Hepatitis is also reported in about 25 % of people with HIV infection, thus they are coinfected with two viruses at the same time. They are subjected to the risk for serious, life-threatening complications through the development of liver disease, liver failure, and liver-related death. The age-standardized death rate due to liver disease in India is 23.6 per 100,000 individuals according to WHO, 2011 (http://www.worldlifeexpectancy.com/cause-of-death/liver-disease/by-country/). Liver fibrosis is nothing but a histological change in the liver due to inflammation that causes HSCs to undergo activation with synthesis of excess ECM proteins. In chronic viral hepatitis, and in nonalcoholic fatty liver diseases, hepatic cells lose the ability to undergo division and instead lead to fibrosis and ultimate development of liver cirrhosis. Cirrhosis is a condition of severe damage to the liver that impairs its ability to function normally. HSCs play an important role that facilitates the process of fibrosis and thereby liver cirrhosis. HSCs are liver-specific pericytes that usually store fat and vitamin A in normal liver. Under inflammation, HSCs undergoes a trans-differentiated stage leading to liver cirrhosis [4]. Key cytokines which promotes liver fibrosis include TGF-β1, PDGF, norepinephrine, angiotensin II, endothelin-1, NADPH oxidase enzyme, and leptin. In contrast, IL-10, IFN-γ, nitric oxide, relaxin, and adiponectin are known to attenuate fibrotic liver. Nevertheless, there are reports about reversibility of liver fibrosis in a hepatitis B viral-infected patient treated with lamivudine. Lamivudine is an analogue of cytidine capable of inhibiting reverse transcriptase enzyme of hepatitis B virus and acts by terminating DNA elongation [5].

Treatment of Liver Fibrosis

The best treatment available for liver fibrosis is to make lifestyle changes such as modification of diet, prohibition of alcohol, and taking regular exercise, antiviral medications, and steroids. However, when the liver does not respond to the treatments due to severity of disease, liver transplantation is the best treatment available today. Transplantation is unfortunately made as a rare curative treatment due to lack of organ donors. Thus, alternative therapies based on stem and progenitor cells for treatment of chronic liver diseases are emerging today [6, 7]. Recently, cell type-specific deletion or overexpression of pro- and antifibrotic genes is a major goal of basic fibrosis research. Troeger et al. generated a transgenic mice expressing tamoxifen-inducible Cre estrogen under control of the endogenous vimentin promoter (VimCreER) to study deactivation of HSCs during liver fibrosis [8]. Such system warrants the presence of tamoxifen (an estrogen receptor antagonist) and also of vimentin, which is predominantly expressed in myofibroblast-like cells such as activated HSCs. Thus, administration of tamoxifen at a desired time point allows Cre-mediated deletion of target genes or activation of reporter genes specifically in activated stellate cells. However, the practical application of this strategy has to be evaluated in future studies. Although the removal of causative agent for liver fibrosis is the best treatment, corticosteroid drugs and antioxidants have been widely used. Substances such as pentoxifylline (phosphodiesterase inhibitor), amiloride (Na+/H+ pump inhibitor), and renin-angiotensin blocking agents and administration of vasodilators were known to exert antifibrotic activity [9, 10]. Many traditional, Asian origin herbal compounds are known to possess antifibrotic properties. There are about 300 preparations in the Indian system of medicine for the treatment of chronic liver diseases. Sylibum marianum, Glycyrrhiza glabra, Picrorhiza kurroa, and Phyllanthus amarus are well known for their hepatoprotective properties [11, 12]. Traditional Chinese medicine is still being widely used for treatment of liver disease in China. A formulation known as "Compound 861," made of 10 herbs with Salvia miltiorrhiza as its chief component, has been shown experimentally to be effective in reducing fibrogenesis and enhancing collagen degradation [13].

Hepatic Stellate Cells and Hepatopathophysiology

Under normal conditions, HSCs remain in a dormant state and have protrusions extending from the cell body that wraps around the sinusoids. In pathological conditions such as liver fibrosis or liver cirrhosis, activated HSCs undergo cell proliferation, cell migration, loss of retinoids, secretion of proinflammatory cytokines, and synthesize of a large amount of extra cellular matrix (ECM) components including collagen, proteoglycan, and adhesive glycoproteins. The morphology of these cells also changes from the star-shaped stellate cells to that of the fibroblasts or myofibroblasts and acquires contractile properties. HSCs show remarkable change in secretory pattern after activation. Leptin, PPAR, and SREBP-1c are the markers expressed in quiescent HSCs. Upon activation, HSCs express mostly alpha smooth muscle actin, c-myb, and myocyte enhancer factor-2 [5]. About 5–8 % of the total number of liver cells is represented by quiescent HSCs, while it increases to 15–20 % during cirrhotic condition [14]. Activation of HSC consists of 2 major phases: (1) initiation (also called a preinflammatory stage) and (2) perpetuation. Initiation refers to early paracrine-mediated changes in gene expression and phenotype that render the cells responsive to other cytokines and stimuli. Perpetuation results from the effects of these stimuli on maintaining thc activated phenotype and generating fibrosis [15]. Paracrine stimulation by neighboring cell types such as sinusoidal endothelium, Kupffer cells, hepatocytes, platelets, and leukocytes contributes to the activation of HSCs. Kupffer cells can stimulate matrix synthesis, cell proliferation, and release of retinoids by stellate cells through the actions of cytokines (especially TGF-β1) and reactive oxygen intermediates/lipid peroxides [16, 17]. Reactive oxygen species whether produced internally within stellate cells [18] or released into the extracellular environment [19] are capable of activating the stellate cells and collagen synthesis. Kupffer cells also produce the predominant mitogen (platelet-derived growth factor) for activated HSCs [20]. Apart from HSCs, myofibroblasts and cells of bone marrow origin have been shown to exhibit fibrogenic potential [21].

Therapeutic Strategies Targeting HSCs

Different strategies such as inhibition of HSC proliferation or induction of HSC apoptosis, downregulation of collagen production or promotion of its degradation, administration of cytokines, and infusion of mesenchymal stem cells are being used as antifibrotic therapy. Several studies were carried out to inhibit the activation of HSCs using pharmacological compounds. Angiotensin inhibitors, endothelin inhibitors, interferon-α, pentoxifylline, PPAR antagonists, and TGF-β1 inhibitors inhibit HSC activation [5]. Stimulation of death receptors in activated HSCs and a decrease in survival factors, including TIMP-1, can precipitate HSC apoptosis. Antioxidants such as vitamin E, silymarin, phosphatidylcholine, and S-adenosyl-L-methionine are also known to inhibit HSC activation. As HSC activation is the central event in hepatic fibrosis, it becomes necessary for developing and identifying targeted therapy to curb "HSC activation." Therefore, stem cell-based therapies targeting HSC inactivation could be ideal for antifibrotic approach [22]. Stem cell transplantation is unique and seems to be more efficient than other kinds of treatments bccausc of its bipotent differentiation capacity and continuous transplantation efficiency in restoring the functional liver tissue [23].

Liver Stem Cells

Most of the information on liver stem cells today is derived from animal studies due to the fact that human study involves serial liver biopsies and long-term follow-up studies to assess changes in liver fibrosis which is not always feasible. However, it is evident that functional liver stem cells do exist in humans [24]. Liver precursor cells (LPCs) are recognized as bipotential precursor cells in the damaged liver. They can rapidly proliferate, change their cellular composition, and differentiate into hepatocytes and cholangiocytes to compensate for the cellular loss and maintain liver homeostasis in animal models of liver injury [25]. Bipotential murine oval liver (BMOL) cells are the rodent counterpart of LPCs. The oval cell is a blast-like, liver-specific stem cell, capable of self-renewal and multipotent differentiation. It was initially identified that a small population of cells with a low cytoplasmic/high nuclear ratio that arose from portal cells, not from hepatocytes, in rodents came to be known as "oval cells" [26]. BMOL cells are termed so due to its oval-shaped nucleus, 7–10 μm in size (smaller than hepatocytes), and are observed in rodent models of liver injury. They arise from the juncture between hepatocyte plates and terminal bile duct cells, known as the canals of Hering in rodents [27]. It was noted that there are similarities in the requirement of growth factors such as TNF-α and IL-6 between oval cells and hepatocytes for their proliferation [28]. However, expression of markers and gene expression pattern among them differs [29]. Typical biomarkers that express on bipotential oval stem cells include AFP, Thy-1, CD34, c-kit, SOX9, CD133, CD90, CD44, EpCAM, CD13, GGT, OV-6, CK-19, OC.2, and HEA-125 [30–32].

Harnessing HSC Activation Using Stem Cells

ECM-producing HSC could be the ideal target to reduce the burden of liver diseases. Transplantation of bone marrow cells into fibrotic mouse led to significant decrease in the expression of galectin-3 (+) cells and proved that fibrosis regression is mediated by bone marrow cell transplantation [33]. It was observed that human amnion epithelial cells (hAEC) administered into liver fibrotic mice could establish themselves in fibrotic liver areas within 2 weeks' time, which was demonstrated through human leukocyte antigen-G expression and detection of human albumin protein in mouse sera of hAEC-treated animals. In addition, hAEC-treated mice showed fewer activated collagen-producing hepatic stellate cells and less fibrosis area and collagen content compared to controls [34]. Lam et al. studied the mechanism involved in hepatic differentiation of mesenchymal stem cells (MSCs), which are transplanted into D-galactosamine (GalN)-induced acute liver injury in rats [35]. Their results revealed that activation of interleukin-6 (IL-6)/glycoprotein 130 (gp130)-mediated signal transducer and activator of transcription 3 (STAT3) pathway through soluble IL-6 receptor is important in hepatic differentiation of MSCs. The work of Gur et al. elaborated that NKp46, one of the major nuclear killing-activating receptors expressed by nuclear killer cells in the liver, was able to kill both murine and human HSCs and showed antifibrotic activity [36].

In the present study, we attempted to understand about how bipotential murine oval cells interplay with activated HSCs and thereby support liver regeneration using in vitro cell culture system. Our results demonstrate the differentiation of BMOL cells into hepatocytes and showed that BMOL cells are capable of compromising HSC viability and induced apoptosis. In addition, we have also proved that differentiated BMOL cells are capable of functionally blocking the activated HSCs by inhibiting collagen synthesis and secretion.

Materials and Methods

Cell Culture and Differentiation of Stem Cells

Bipotential murine oval liver (BMOL) stem cells were a kind gift from Prof. George Yeoh, University of Western Australia, Perth, Australia. These cells were maintained in two different kinds of medium such as normal medium and differentiation medium. The cells grow and propagate as BMOL stem cells in normal medium, and the stem cells were induced in vitro to grow as hepatocytes by maintaining them in a differentiation medium. The composition of normal medium contains Williams' E medium supplemented with glutamine, 10 % fetal bovine serum, 1 % penicillin, and streptomycin. The ingredients such as ITS + Premix (6.25 g/ml each), 10 mM nicotinamide, and 10^{-7} M dexamethasone were added extra in the normal medium to be used as differentiation medium. The cells were maintained initially in the normal media as BMOL stem cells for 10 days and later changed to differentiation media. Cells grown to 90 % confluency under normal media served as control [27]. Activated and immortalized hepatic stellate LX2 cells were cultured in DMEM supplemented with 10 % fetal bovine serum and 1 % penicillin (*w/v*) and streptomycin (*w/v*).

Preparation of Conditioned Media (CM)

BMOL stem cells were trypsinized and seeded on collagen-plated 24-well plate with 80 % cell density and incubated for 12 h in normal media. Fresh media was added to the dish and incubated for 12 h. Next, the media was taken out and served as undifferentiated conditioned media (UDCM). Similarly, the media that was obtained from the flask in which BMOL cells were grown in differentiation media served as differentiated conditioned media (DCM).

Immunofluorescence

BMOL cells were seeded in 12-well plates and were grown under normal media and differentiation media. The cells were processed using the cold paraformaldehyde-Triton X-100 procedures as described elsewhere [37]. Paraformaldehyde fixed cells were incubated at 40 °C overnight with mouse monoclonal antibodies (dilution 1:1,000) against HNF4-α. Next, corresponding goat anti-mouse secondary antibodies (dilution 1:2,000) tagged with FITC was used. Images were taken using an Olympus IX71 epifluorescence microscopy system equipped with a DP71 camera.

Western Blot

BMOL cells were fed with normal and differentiation media and were harvested for Western blot analysis. Total protein was normalized using Biuret Assay. Proteins were detected using 1:500 dilution of HNF4-α antibody (H79, Santa Cruz) and 1:1000 dilutions of goat anti-rabbit IgG conjugated to horseradish peroxidase (Bangalore GENEI). The blots were developed by using TMB/H_2O_2 as substrate. Loading control was the expression level of β-actin gene (ab8229, Abcam, USA).

Measurement of Apoptosis

PI Incorporation Assay

PI has been widely used to measure apoptosis in different experimental systems [38]. Activated hepatic stellate cells, LX2, were treated with UDCM and DCM for 4 h, followed by incubation of cells with 1 μM of PI for 10 min. Next, cells were washed with PBS for two to three times to remove unbound PI, and the fluorescence images were acquired with DP71 camera adapted to an Olympus IX71 microscope. The protocol is based on the basic principle that late apoptotic or necrotic cells take up PI, while normal cells do not take up the dye due to the difference in their membrane permeability and rigidity. Number of PI positive cells were counted from 5 different fields of 5 individual experiments and plotted.

Annexin V-FITC Apoptosis Detection Kit

LX2 cells were treated with UDCM and DCM for 4 h and were processed as per the protocol supplied by the manufacturer (Merck, Calbiochem, EMD Chemicals Inc., Darmstadt, Germany). Fluorescence images of the cells were taken and the number of annexin V-FITC-positive cells were counted per field.

Cell Viability Assay

Cell viability was measured following the trypan blue protocol. In brief, LX2 cells were seeded in 12-well plates and incubated for 4 h in UDCM and in DCM. Next, the respective media containing trypan blue (0.04 mg/ml) was added to the cells and incubated for another 15 min. The cells were then washed with PBS before imaging. Bright-field images were taken using Olympus inverted microscope. The number of cells having a blue nucleus was counted and the percentage of dead cells was plotted as a graph. Cell viability was also measured using 5-chloromethylfluorescein diacetate probe (CMFDA). LX2 cells were subjected to UDCM and DCM treatments for 4 h, followed by incubation with 5 μM CMFDA for 2 h. Next, the cells were washed with PBS and images were taken using DP71 camera adapted to an Olympus IX71 microscope.

Collagen Formation Assay

Collagen formation assay was performed using the collagen-specific dye, Sirius red. A solution of 5 μg/ml concentration Sirius red was prepared by dissolving 0.1 g of Sirius red in 100 ml of saturated picric acid solution. LX2 cells were treated with UDCM and DCM for 4 h and probed with acidic solution of Sirius red. After 1 h of incubation, cells were washed with PBS and images were acquired with Nikon cool pix camera adapted to an Olympus inverted microscope. For colorimetric measurement of collagen-bound Sirius red, cells were lysed using lysis buffer followed by taking the absorbance of the supernatant at 540 nm.

Statistical Analysis

All experiments were performed in triplicate unless otherwise specified. Data are presented as mean ± SE. Data was analyzed using *t*-test, one-way and two-way ANOVA as appropriate. *P*-values $p \geq 0.05$ were selected as the criterion for a statistically significant difference.

Results

BMOL Cells Differentiated into Hepatocytes

BMOL cells incubated with differentiation media (dexamethasone, ITS, and selenium) for 10 days were able to differentiate to hepatic cells, and it was confirmed by identifying the expression of HNF4-α, a hepatocyte-specific marker. Morphology and biochemical characterization of the cells suggested that BMOL cells were differentiated to hepatocytes. The results of Western blot analysis confirmed the expression of HNF4-α protein in differentiated hepatocytes when compared to undifferentiated cells (Fig. 2a). Results of immunofluorescence analysis also showed that HNF4-α was expressed in cells which were grown in differentiation media, while it was absent in cells that were grown in normal medium (Fig. 2b).

BMOL Cells Compromised HSC Viability and Induced Apoptosis

Viability of the activated HSCs was estimated following trypan blue assay protocol. Analysis of the images revealed that number of dead cells were almost equal in UDCM- and DCM-treated HSCs which were comparatively higher than that of HSC incubated under normal media (Fig. 3a). Additionally, the number of live cells under CM treatment was estimated using CMFDA,

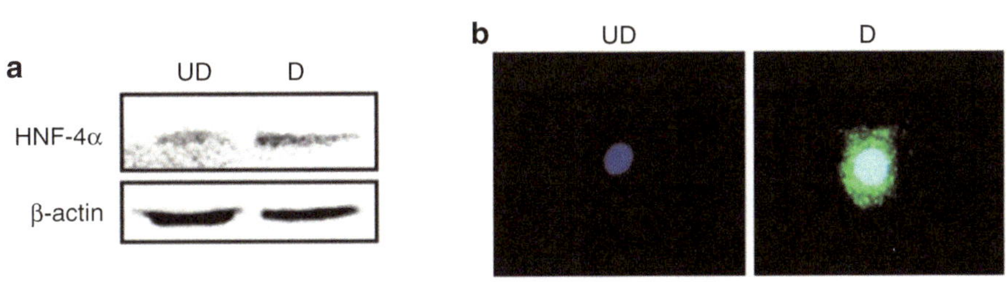

Fig. 2 (**a**, **b**) Confirming the differentiation of BMOL cells. BMOL cells were processed following the differentiation protocol. (**a**) Differentiation was confirmed by probing with HNF4-α antibody using Western blot. (**b**) Immunofluorescence image shows the differentiated BMOL cells as hepatocytes using HNF4-α antibody [37]

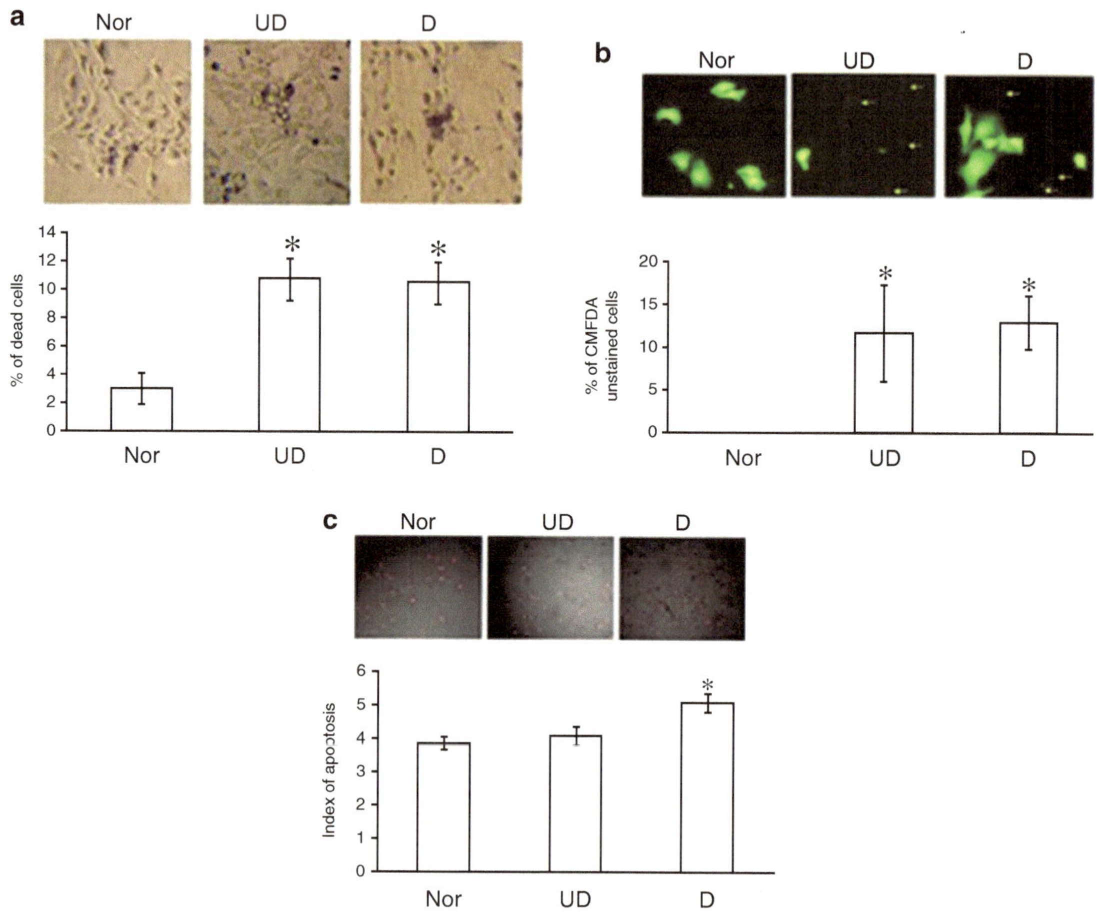

Fig. 3 BMOL cells affected cellular health of HSC. (**a**) HSC was treated with CM for 4 h followed by incubating with 0.04 mg/ml concentration of trypan blue-containing media. Cells were then imaged and the number of blue nucleus positive cells was counted. * versus Nor (*$p < 0.05$) ($n = 5$). (**b**) HSC were treated with CM for 4 h followed by incubation with CMFDA for 30 min. *Yellow arrows* in the images showed the CMFDA unstained dead cells. * indicates $p < 0.05$. (**c**) HSC were treated with CM for 4 h followed by incubation with PI (10 μM) for 15 min. Cells were then washed with PBS to remove unbound PI and then imaged with DP71 camera adapted to a Olympus IX71 camera

a cell-permeable live cell detecting fluorescent probe. A number of CMFDA unstained cells were observed in UDCM- and DCM-treated HSCs which were not seen in normal media-treated HSC suggesting that undifferentiated and differentiated CM treatments induced hepatic stellate cell death (Fig. 3b). We measured the number of late apoptotic and necrotic cells using PI incorporation assay. Membrane of late apoptotic and necrotic cells become leaky due to membrane disintegration and thus become permeable to PI, while live cells do not take up the dye. Apoptotic studies with PI showed that CM from differentiated BMOL cells induced HSC apoptosis by 1.42-fold, while CM from undifferentiated BMOL cells were unable to induce late apoptosis or necrosis (Fig. 3c).

BMOL Cells Blocked Collagen Synthesis and Secretion by HSC

To evaluate the collagen production by LX2 cells, Sirius red staining of the cells treated with UDCM and DCM were performed. Analysis of the images showed that LX2 cells lose its

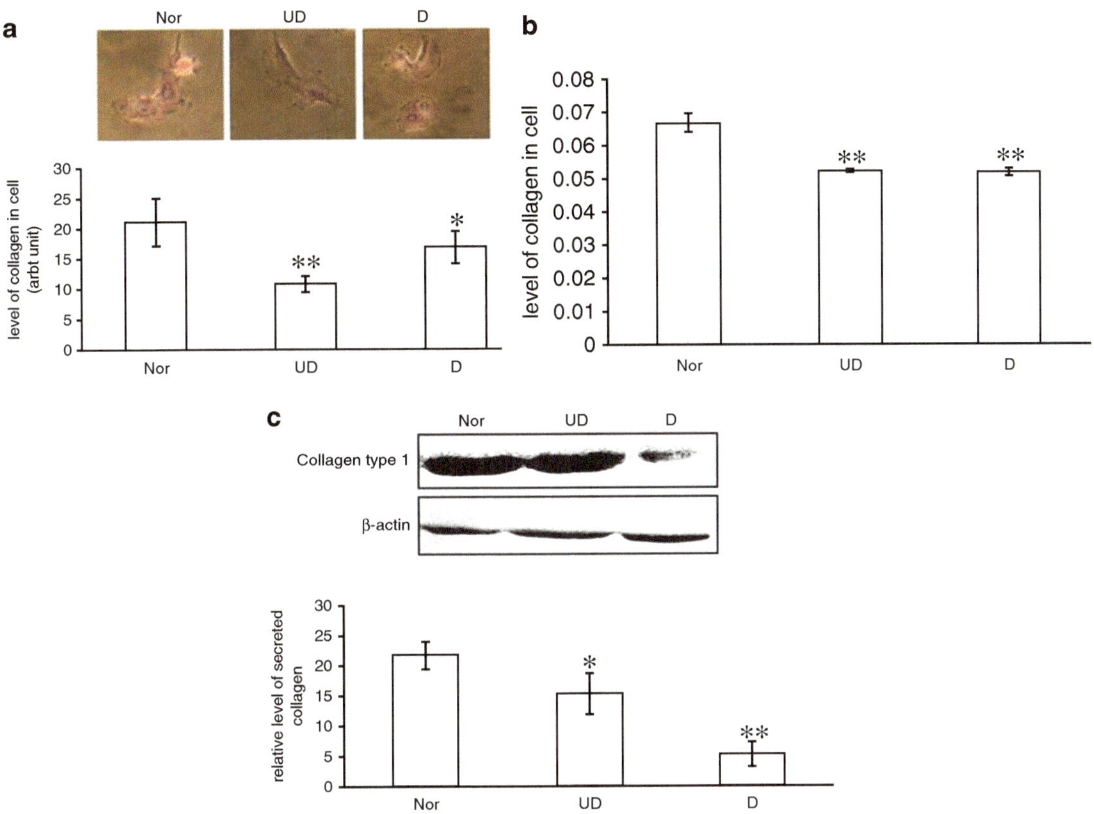

Fig. 4 BMOL cells attenuated HSC-mediated collagen production. (**a**) HSC was treated with CM for 4 h followed by incubating with Sirius red for 1 h. Next cells were washed with PBS and images were acquired with Nikon coolpix camera adapted to motic inverted microscope. Images are representative of 5 individual set of experiments. * and ** versus Nor ($*p < 0.05$;$**p < 0.001$) ($n = 5$). (**b**) HSC was treated with CM for 4 h followed by incubating with Sirius red for 1 h. Next cells were washed with PBS and scrapped and further homogenized. Homogenate were centrifuged to remove the cell debris followed by taking the absorbance at 540 nm. ** versus Nor ($**p < 0.001$) ($n = 5$). (**c**) HSC was treated with CM for 4 h followed by adding fresh FBS-free DMEM media to cells and further incubating for 2 h. Next, the media was taken and the protein was extracted from the CM using TCA protocol. Cells were also processed to prepare Western blot sample which was used to probe β-actin. Densitometry values of the bands were analyzed using ImageJ module. * and ** versus Nor ($*p < 0.05$;$**p < 0.001$) ($n = 3$)

collagen production with UDCM and DCM treatment compared to normal media-treated cells (Fig. 4a). A similar set of experiments were carried out where after UDCM and DCM treatment, cells were incubated with Sirius red followed by scraping and lysing the cells and taking absorbance at a wave length of 540 nm. CM from undifferentiated and differentiated stem cells blunted collagen production by 22 % and 23 %, respectively (Fig. 4b). Western blotting analysis of CM-treated activated HSC revealed that CM from differentiated BMOL cells maximally blocked collagen synthesis by 73 %, while undifferentiated CM also blocked collagen synthesis by 31 % compared to normal media-treated controls (Fig. 4c).

Discussion

The present study established a novel conditioned media-based model which studies the implications of liver stem cells in controlling activated HSC. The results of the present study demonstrated that BMOL stem cells are capable of inhibiting activated HSCs. Although several studies have been published using bone marrow-derived stem cells to combat liver fibrosis using animal models, ours is the first study which used simple cell culture-based model to induce cell death in activated HSCs. Cell therapy based on transplantation of bone marrow-derived cells including mesenchymal stem and embryonic stem cells was reported to reduce carbon tetrachloride (CCl4)-induced liver fibrosis [33, 39–42], while fetal liver epithelial progenitor cells have also been shown to ameliorate diethylnitrosamine-induced liver fibrosis [43]. The basic mechanism involves the entry of these stem cells to the portal circulation followed by differentiation to hepatocyte and further helping in the regeneration of liver. Earlier studies report that oval cells function as a facultative liver stem cell compartment and are induced to proliferate when the regenerative capacity of mature hepatocytes is compromised [44–46]. A strong association was observed between oval cell numbers and the severity of liver injury [32]. Inhibition of activated HSCs is an important step in preventing liver fibrosis. Transdifferentiation of HSCs was blocked by DNA methylation inhibitor 5-aza-2′-deoxycytidine (5-azadC), demonstrating the epigenetic regulation of hepatic wound healing and fibrogenesis [47]. Histone deacetylase (HDAC) inhibitors, trichostatin A, and valproic acid have been identified as potent inhibitors of HSC activation [22]. Treatment with MC1568, another HDAC inhibitor, resulted in a significant decrease in stellate cell activation in vitro through the induction of micro RNA-29 [22].

Although many antifibrotic approaches such as renin-angiotensin system blockers and antioxidants are being used [48, 49], the efficiency with which these compounds are taken up by activated HSCs remains a problem. Targeted delivery to activated HSCs could provide a solution to such problem. Accumulation of activated HSCs should be inhibited by interfering with either their activation or proliferation. Although several studies have demonstrated the use of stem cell therapy in liver diseases [50, 51], targeting activated HSCs using stem cell therapy islimited. The present work has demonstrated that activated HSCs could be killed by using conditioned media derived from differentiated oval stem cells. Further, the function of activated HSCs could be inhibited by suppressing collagen formation using differentiated conditioned medium compared to normal medium. However, the limitation of the current study is that it is an in vitro-based cell culture model. The future directions should focus on identification of factors in the UDCM and DCM which are responsible for inducing the programmed death of activated HSCs. Future studies addressing this aspect would lead to a better understanding of the mechanisms through which stem cells can attenuate activated HSC functions and thus can be used as a therapeutic strategy in liver diseases like liver fibrosis. Bisgaard et al. reported that γ-interferon (IFN-γ), a cytokine, plays an important role in stem cell-mediated liver regeneration [29]. Parekkadan et al. showed that mesenchymal stem cells (MSCs) modulate the activated HSCs through paracrine factors such as IL-6 and IL-10 which led to inhibition of collagen synthesis. They also observed that MSCs released hepatocyte growth factor which induced apoptosis of activated HSCs [52]. We propose that the present study can be extended in two different ways: (1) identification of factors in the UDCM and DCM and characterizing it further and (2) application of BMOL cells to fibrotic liver in an animal model and identifying the stage(s) at which the differentiated BMOL cells have optimal effect on inducing the apoptosis of activated HSC in vivo.

Conclusion

The present study offers an important clue that conditioned medium from the hepatocyte-differentiated BMOL is able to induce apoptosis of activated HSCs and attenuate collagen production by the cells. Further research in identifying the factors in the conditioned medium will help in developing better treatment strategies for liver fibrosis.

Acknowledgement This work was financially supported by grants from the Department of Biotechnology (DBT), Government of India (BT/SBIRI/779/34-B15/2011), and University Grants Commission (UGC), Government of India, under Faculty Recharge Program to SC. This work was partially supported by the Department of Science and Technology (DST), Government of India (SR/WOS-A/LS-14/2011(G) to PG).

References

1. Maton A, Hopkins J, McLaughlin CW, Johnson S, Warner MQ, LaHart D, Wright JD (1993) Human biology and health. Prentice Hall, Englewood Cliffs
2. Pond SM, Tozer TN (1984) First-pass elimination. Basic concepts and clinical consequences. Clin Pharmacokinet 9(1):1–25
3. Saxena R, Theise ND, Crawford JM (1999) Microanatomy of the human liver-exploring the hidden interfaces. Hepatology 30:1339–1346
4. Friedman SL (2004) Mechanisms of disease: mechanisms of hepatic fibrosis and therapeutic implications. Nat Clin Pract Gastroenterol Hepatol 1(2):98–105
5. Bataller R, Brenner DA (2005) Liver fibrosis. J Clin Invest 115(2):209–218
6. Forbes JS (2008) Stem cell therapy for chronic liver disease choosing the right tools for the job Commentaries. Gut 57:153–155
7. Espanol-Suner R et al (2012) Liver progenitor cells yield functional hepatocytes in response to chronic liver injury in mice. Gastroenterology 143(6):1564–1575
8. Troeger JS, Mederacke I, Gwak GY, Dapito DH, Mu X, Hsu CC, Pradere JP, Friedman RA, Schwabe RF (2012) Deactivation of hepatic stellate cells during liver fibrosis resolution in mice. Gastroenterology 143(4):1073–1083
9. Tomita K et al (2004) Piogliazone prevents alcohol-induced fatty liver in rats through upregulation of c-Met. Gastroenterology 126:873–885
10. Yokohama S, Yoneda M, Haneda M, Okamoto S, Okada M, Aso K, Hasegawa T, Tokusashi Y, Miyokawa N, Nakamura K (2004) Therapeutic efficacy of an angiotensin II receptor antagonist in patients with nonalcoholic steatohepatitis. Hepatology 40:1222–1225
11. Thyagarajan SP, Jayaram S, Gopalakrishnan V, Hari R, Jeyakumar P, Sripathi MS (2002) Herbal medicines for liver diseases in India. J Gastroenterol Hepatol 17(Suppl 3):S370–S376
12. Girish C, Pradhan SC (2012) Indian herbal medicines in the treatment of liver diseases: problems and promises. Fundam Clin Pharmacol 26(2):180–189
13. Wang BE (2000) Treatment of chronic liver diseases with traditional Chinese medicine. J Gastroenterol Hepatol 15(Suppl):E67–E70
14. Geerts A (2001) History, heterogeneity, developmental biology and functions of quiescent HSC. Semin Liver Dis 21:311–335
15. Friedman SL (2000) Molecular regulation of hepatic fibrosis, an integrated cellular response to tissue injury. J Biol Chem 275:2247–2250
16. Gressner AM, Lotfi S, Gressner G, Haltner E, Kropf J (1993) Synergism between hepatocytes and Kupffer cells in the activation of fat storing cells (perisinusoidal lipocytes). J Hepatol 19:117–132
17. Gressner AM (1995) Cytokines and cellular crosstalk involved in the activation of fat-storing cells. J Hepatol 22:28–36
18. Svegliati-Baroni G, Saccomanno S, van Goor H, Jansen P, Benedetti A, Moshage H (2001) Involvement of reactive oxygen species and NO radicals in activation and proliferation of rat HSC. Liver 21:1–12
19. Lee KS, Buck M, Houglum K, Chojkier M (1995) Activation of HSC by TGF-α and collagen type I is mediated by oxidative stress through c-myb expression. J Clin Invest 96:2461–2468
20. Pinzani M, Gesualdo L, Sabbah GM, Abboud HE (1989) Effects of platelet-derived growth factor and other polypeptide mitogens on DNA synthesis and growth of cultured rat liver fat storing cells. J Clin Invest 84:1786–1793
21. Forbes SJ et al (2004) A significant proportion of myofibroblasts are of bone marrow origin in human liver fibrosis. Gastroenterology 126:955–963
22. Mannaerts I, Eysackers N, Onyema OO, Van Beneden K, Valente S, Mai A, Odenthal M, van Grunsven LA (2013) Class II HDAC inhibition hampers hepatic stellate cell activation by induction of microRNA-29. PLoS One 8(1):e55786
23. Christ B, Pelz S (2013) Implication of hepatic stem cells in functional liver repopulation. Cytometry A 83(1):90–102
24. Roskams T, De Vos R, Van Eyken P, Myazaki H, Van Damme B, Desmet V (1998) Hepatic OV-6 expression in human liver disease and rat experiments: evidence for hepatic progenitor cells in man. J Hepatol 29(3):455–463
25. Kisseleva T, Gigante E, Brenner DA (2010) Recent advances in liver stem cell therapy. Curr Opin Gastroenterol 26(4):395–402
26. Grisham JW, Hartroft WS (1961) Morphologic identification by electron microscopy of "oval" cells in experimental hepatic degeneration. Lab Invest 10:317–332
27. Tirnitz-Parker JE, Tonkin JN, Knight B, Olynyk JK, Yeoh GC (2007) Isolation, culture and immortalisation of hepatic oval cells from adult mice fed a choline-deficient, ethionine-supplemented diet. Int J Biochem Cell Biol 39:2226–2239
28. Nagy P, Kiss A, Schnur J, Thorgeirsson SS (1998) Dexamethasone inhibits the proliferation of hepatocytes and oval cells but not bile duct cells in rat liver. Hepatology 28:423–429

29. Bisgaard HC, Muller S, Nagy P, Rasmussen LJ, Thorgeirsson SS (1999) Modulation of the gene network connected to interferon-gamma in liver regeneration from oval cells. Am J Pathol 155:1075–1085
30. Crosby HA, Hubscher SG, Joplin RE, Kelly DA, Strain AJ (1998) Immunolocalization of OV-6, a putative progenitor cell marker in human fetal and diseased pediatric liver. Hepatology 28(4):980–985
31. Suzuki A, Sekiya S, Onishi M, Oshima N, Kiyonari H, Nakauchi H, Taniguchi H (2008) Flow cytometric isolation and clonal identification of self-renewing bipotent hepatic progenitor cells in adult mouse liver. Hepatology 48(6):1964–1978
32. Lowes KN, Croager EJ, Olynyk JK, Abraham LJ, Yeoh GC (2003) Oval cell-mediated liver regeneration: role of cytokines and growth factors. J Gastroenterol Hepatol 18:4–12
33. de Oliveira SA, de Freitas Souza BS, Sá Barreto EP, Kaneto CM, Neto HA, Azevedo CM, Guimarães ET, de Freitas LA, Ribeiro-Dos-Santos R, Soares MB (2012) Reduction of galectin-3 expression and liver fibrosis after cell therapy in a mouse model of cirrhosis. Cytotherapy 14(3):339–349
34. Manuelpillai U, Tchongue J, Lourensz D, Vaghjiani V, Samuel CS, Liu A, Williams ED, Sievert W (2010) Transplantation of human amnion epithelial cells reduces hepatic fibrosis in immunocompetent CCl_4-treated mice. Cell Transplant 19(9):1157–1168
35. Lam SP, Luk JM, Man K, Ng KT, Cheung CK, Rose-John S, Lo CM (2010) Activation of interleukin-6-induced glycoprotein 130/signal transducer and activator of transcription 3 pathway in mesenchymal stem cells enhances hepatic differentiation, proliferation, and liver regeneration. Liver Transpl 16(10):1195–1206
36. Gur C, Doron S, Kfir-Erenfeld S, Horwitz E, Abu-Tair L, Safadi R, Mandelboim O (2012) NKp46-mediated killing of human and mouse hepatic stellate cells attenuates liver fibrosis. Gut 61(6):885–893
37. Majumder S, Siamwala JH, Srinivasan S, Sinha S, Sridhara SR, Soundararajan G, Seerapu HR, Chatterjee S (2011) Simulated microgravity promoted differentiation of bipotential murine oval liver stem cells by modulating BMP4/Notch1 signaling. J Cell Biochem 112(7):1898–1908
38. Srinivas G, Anto RJ, Srinivas P, Vidhyalakshmi S, Senan VP, Karunagaran D (2003) Emodin induces apoptosis of human cervical cancer cells through poly (ADP-ribose) polymerase cleavage and activation of caspase-9. Eur J Pharmacol 473:117–125
39. Moriya K, Yoshikawa M, Ouji Y, Saito K, Nishiofuku M, Matsuda R, Ishizaka S, Fukui H (2008) Embryonic stem cells reduce liver fibrosis in CCl4-treated mice. Int J Exp Pathol 89:401–409
40. Sakaida I, Terai S, Yamamoto N, Aoyama K, Ishikawa T, Nishina H, Okita K (2004) Transplantation of bone marrow cells reduces CCl4-induced liver fibrosis in mice. Hepatology 40:1304–1311
41. Zhao DC, Lei JX, Chen R, Yu WH, Zhang XM, Li SN, Xiang P (2005) Bone marrow derived mesenchymal stem cells protect against experimental liver fibrosis in rats. World J Gastroenterol 11: 3431–3440
42. Oyagi S, Hirose M, Kojima M, Okuyama M, Kawase M, Nakamura T, Ohgushi H, Yagi K (2006) Therapeutic effect of transplanting HGF-treated bone marrow mesenchymal cells into CCl4-injured rats. J Hepatol 44:742–748
43. Zheng JF, Liang LJ, Wu CX, Chen JS, Zhang ZS (2006) Transplantation of fetal liver epithelial progenitor cells ameliorates experimental liver fibrosis in mice. World J Gastroenterol 12:7292–7298
44. Vessey CJ, Hall PM (2001) Hepatic stem cells: a review. Pathology 33:130–141
45. Sell S (2001) Heterogeneity and plasticity of hepatocyte lineage cells. Hepatology 33:738–750
46. Lowes KN, Brennan BA, Yeoh GC, Olynyk JK (1999) Ovel cell numbers in human chronic liver disease are directly related to disease severity. Am J Pathol 154:537–541
47. Mann J, Oakely F, Akiboye F, Elsharkawy A, Thorne AW, Mann DA (2007) Regulation of myofibroblast transdifferentiation by DNA methylation and MeCP2: implication for wound healing and fibrogenesis. Cell Death Differ 14:275–285
48. Paizis G et al (2002) Up-regulation of components of the renin-angiotensin system in the bile duct ligated rat liver. Gastroenterology 123:1667–1676
49. Bataller R et al (2003) Activated human hepatic stellate cells express the renin-angiotensin system and synthesize angiotensin II. Gastroenterology 125: 117–125
50. Spee B, Carpino G, Schotanus BA, Katoonizadeh A, Vander Borght S, Gaudio E, Roskams T (2010) Characterisation of the liver progenitor cell niche in liver diseases: potential involvement of Wnt and Notch signalling. Gut 59(2):247–257
51. Español-Suñer R, Carpentier R, Van Hul N, Legry V, Achouri Y, Cordi S, Jacquemin P, Lemaigre F, Leclercq IA (2012) Liver progenitor cells yield functional hepatocytes in response to chronic liver injury in mice. Gastroenterology 143(6): 1564–1575
52. Parekkadan B, van Poll D, Megeed Z, Kobayashi N, Tilles AW, Berthiaume F, Yarmush ML (2007) Immunomodulation of activated hepatic stellate cells by mesenchymal stem cells. Biochem Biophys Res Commun 363(2):247–252

Part VII

Stem Cells and Organ Systems: Reproductive System

Stem Cells of the Reproductive System: At a Glance

Phillip G. Stubblefield

Introduction

Presently, there is great interest in stem cells from adult tissues because of the increased use of stem cells in therapy and the many practical and ethical concerns about stem cells of embryonic origin. Use of bone marrow stem cells harvested from living donors and from adipose tissue is well established, but there is great need for noninvasive sources. The human endometrium has attracted attention because it is shed monthly during menstruation and then completely regenerates. It was proposed years ago that this rapid regeneration must be based on stem cells, but it is only recently that endometrial mesenchymal stem cells were identified and characterized. Adult stem cells are identified by their functional properties: high proliferative potential, substantial self-renewal capacity, and ability to differentiate into at least one type of mature functional progeny. The menstrual blood of women volunteers has been collected, prepared in the laboratory, and studied as a nearly unlimited source for use in regenerative medicine. Adult stem cells have been looked for in other tissues in an effort to better understand the function of these tissues in health and disease. Adult stem cells have now been identified in the human female reproductive tract in the ovary, fallopian tube, decidua of pregnancy, as well as the endometrium. Adult stem cells have also been identified in the testis and the prostate of the male system [1].

Stem Cells of the Endometrium

Chan and colleagues reported the identification of clonogenic cells in the human endometrium in 2004 [2]. Endometrial samples attached to 5 mm of myometrium were obtained from 17 ovulating women who had hysterectomy for fibroids or adenomyosis. The endometrium was scraped off and prepared as a suspension of single cells after enzymatic digestion. Selection for endometrial cells was performed using Dynabeads (Dynal Biotech, Oslo, Norway) coated with BerEP4, an antibody specific for both luminal and glandular epithelium, and then cultured at clonal density in a complex medium on mouse fibroblast feeder layers as further described in their report. Two types of colonies resulted: small colonies containing large loosely associated cells and large colonies containing small, densely packed cells. Both endometrial epithelial and stromal cells were cultured. Both formed small and large colonies. Growth factors that supported clonogenic growth were identified. The clonogenic cells were a small population of the cells studied, 0.22–0.52 % of the endometrial epithelial cells and 1.25 % of stromal cells. Three growth factors strongly supported clonogenicity of both epithelial and stromal cells: TGF-alpha, EGF, and PDGF-BB. bFGF supported stromal

P.G. Stubblefield (✉)
Department of Obstetrics and Gynecology, School of Medicine, Boston University, Boston, MA, USA
e-mail: phillip.stubblefield@gmail.com

I. Somasundaram (ed.), *Stem Cell Therapy for Organ Failure*,
DOI 10.1007/978-81-322-2110-4_18, © Springer India 2014

clonogenicity but not epithelial. The epithelial phenotypes of large and small colonies differed. Cells from the small colonies expressed the three epithelial markers, BerEP4, CK, and CD49f. Cells from the large colonies weekly stained for CD49f and CK and were negative for BerEP4. The stromal cells from both small and large colonies stained for CD90, collagen type I, and the fibroblast marker 5B5 confirming that they were largely composed of fibroblasts. The authors propose that the small colonies of loosely arranged large-sized cells are putative epithelial or stromal transit activating (TA) cells already committed to differentiate after a finite number of divisions, while the large colonies of small cells which were fewer in number and had high proliferative capacity may have been initiated by a putative endometrial epithelial or stromal stem cells or committed progenitor cells.

An early method of identifying stem cells was by the ability of cells to exclude the fluorescent dyes Hoechst 33342 and rhodamine used in staining of tissues for microscopic examination. These were labeled "side population cells"(SP) as compared with main population cells which did fluoresce. It is now accepted that SP cells are stem cells and that endometrial stem cells come largely from the side population [3].

Further proof of adult stem cell activity came with the demonstration of reconstituted human endometrial tissue in xenografts. Fully dissociated unfractionated human endometrial epithelial and stromal cell suspensions were transplanted directly beneath the kidney capsule of immunocompromised mice. Well-organized endometrial and myometrial layers of functional endometrium were produced [4]. More recently it was demonstrated that endometrial SP cells were highly capable of reconstituting the entire endometrium in the mouse model, while the non-SP endometrial cells did not [3].

Biomarkers of Endometrial Cells

Many investigators have studied the presence or absence of specific biomarkers of endometrial stem [2, 3]. Their findings are not identical and it is likely that endometrial stem cells are heterologous [5]. Perhaps the most extensive study to date of markers is that of Indumathi and colleagues [6]. These investigators collected human endometrial and fallopian tube biopsies from surgical specimens from 10 women who underwent hysterectomy/tubectomy/oophorectomy for benign conditions. For comparison, human bone marrow was obtained from 10 patients undergoing experimental stem cell therapy for spinal cord injury. Single-cell suspensions were prepared in standard fashion [6]. The cells were stained with cell-surface antibodies and specific intracellular biomarkers and then characterized by fluorescence-activated cell sorting (FACS). The biomarkers studied included eight cell adhesion molecules, four hematopoietic markers, four mesenchymal stem cells, two perivascular markers, four pluripotency markers and five progenitor/endothelial cell markers. Table 1 presents the expression of the biomarkers in bone marrow, endometrium, and fallopian tube, presented as ranges of variation from sparse to robust. In gen eral endometrium and fallopian tube were similar to each other but differed from bone marrow cells in expression of several markers.

Endometrial Mesenchymal Stem Cells from Menstrual Blood

Several independent groups of investigators reported isolation and characterization of stem cells from menstrual blood. Meng and colleagues collected 5 ml of menstrual blood from healthy women after menstrual flow had begun [7]. The subject collected the specimens herself by using a urine cup-tubing method into an antibiotic-containing solution of buffered saline. The cells were then separated, washed and cultured, subcultured, and passaged twice a week for 2 weeks, and then single cells were plated for cloning. Cells were prepared for study under fluorescent microscope or flow cytometry. Four CD surface biomarkers were studied: SSEA-4, Stro-1, HLA-ABC, and HLA-DR. Three intracellular antibodies were used: Oct-4, Nanog, and telomerase. Karyotyping

Table 1 Ranges of variations in expression of biomarkers

Markers	Bone marrow	Endometrium	Fallopian tube
Cell adhesion molecules			
CD29	R	R	R
CD44	R	H	R
CD166	M	M	M
CD106	S	L	L
CD49d	H	L	L
CD31	H	L	M
CD54	H	M	M
CD13	M	L	M
Hematopoietic markers			
CD34	M	M	H
CD45	R	M	H
CD133	R	L	L
CD117	M	L	M
Mesenchymal markers			
CD90	M	M	L
CD105	M	M	M
CD73	H	H	L
Nestin	S	R	R
Perivascular markers			
CD140b	M	H	H
CD146	L	M	H
Pluripotency markers			
SSEA4	S	M	M
ABCG2	S	L	L
OCT3/4	L	M	L
SOX2	L	S	S
Endothelial progenitor/endothelial markers			
CD31	H	L	M
CD54	H	M	M
CD34	M	M	H
CD133	R	L	L
CD105	M	M	M

Source: Indumathi et al. [6] (With permission)
Variation in percentage expression of biomarkers in freshly isolated bone marrow, endometrium, and fallopian tube categorized as sparse (*S*) <10 %, low (*L*) 10–24 %, moderate (*M*) 25–49 %, high (*H*) 50–74 %, and robust (*R*) >75 %

was performed with G banding. Cytokine array analysis was performed on conditioned media from different cell lines to determine expression of proteins. Studies were done for adipogenic, osteogenic, endothelial, neurogenic, pulmonary, epithelial, hepatic/pancreatic, cardiogenic, and myogenic differentiation. The authors apply the term "endometrial-derived regenerative cells" (ERCs) to the cells obtained. The flow cytometry analysis revealed expression of CD9, CD44, CD29, CD41a, CD59, CD73, CD90, and CD105. The cells did not express CD14, CD34, CD38, CD45, CD133, STRO-1, or SSEA-4. Protein expression of the ERC was compared to commercially available cord blood-derived mesenchymal stem cells (BioE) and cord blood mesenchymal stem cells (MSCs) of their own production. Matrix metalloproteinases MMP-3 and MMP-10, cytokine growth factors GM-CSF and PDGF-BB, and angiogenic factor ANG-2 were present in the culture media of the ERC but were lacking or only minimally expressed by the control cord blood MSC. The angiogenic factors VEGF, HGF, and EGF were found in the media of all cells.

The investigators observed that the ERC shared some properties of other mesenchymal stem cells but produced some unique functional factors. Differentiation studies demonstrated differentiation not just to mesenchymal cell lines but to all three germlines: adipocytic differentiation, osteocytic differentiation, myocytic differentiation, skeletal muscle differentiation, endothelial cell differentiation, hepatocytic differentiation, pancreatic differentiation with insulin secretion, neuronal differentiation, respiratory epithelial differentiation and cardiogenic differentiation (Fig. 1). Of note, cardiomyocytes showed spontaneous contractility after 5 days of culture. No differentiation was seen in the absence of induction stimuli. Karyotype analysis was done after 38 and 68 doublings. No aneuploidy or other visible abnormality was observed. ERCs differ in several respects from tissue-resident endometrial stem cells and also proliferate more rapidly. The discovery that the ERCs are pluripotent was not expected. They speculate that the ERC may have some relationship to circulating oocyte progenitors described by Tilly's group [8].

They list the many advantages of ERC in comparison of other stem cell sources:

1. Ease of collection.

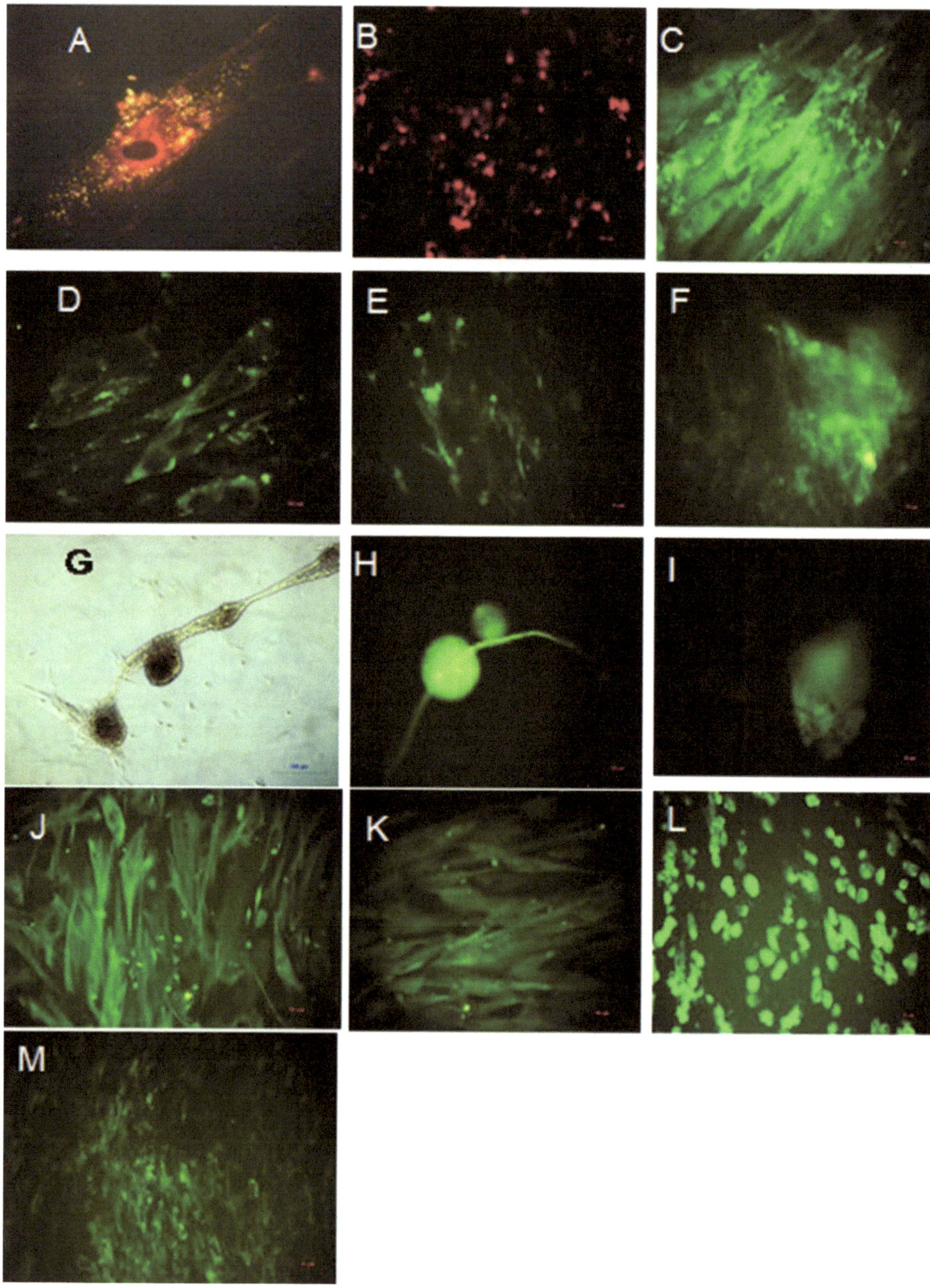
A
B
C
D
E
F
G
H
I
J
K
L
M

2. The cells are expandable and have the ability to differentiate into various tissues. They could be expanded and pre-differentiated to provide patient-specific tissues ready for use when needed.
3. The cells can be expanded for at least 68 doublings without evidence of karyotypic or functional abnormalities.
4. The cells replicate faster than bone marrow MSC.
5. They can be differentiated into cells of all three germ layers.
6. They have a unique phenotype: lack of STRO-1 expression and high expression of the embryonic stem cell marker Oct-4.
7. Proteomic characterization studies demonstrated secretion of the cytokines MMP-3, MMP-10, GM-CSF, PDGF-BB, and ANG-2.

Rossignoli and colleagues also studied the stem cells of menstrual blood [9]. The authors name the cells they discovered endometrial decidual tissue (EDT).[1] Menstrual blood was collected from three healthy female volunteers during the first few days of menstruation using a menstrual cup (DivaCup, Diva International, San Francisco, CA, USA). The blood was transferred in phosphate buffered saline containing penicillin, streptomycin, fluconazole, and heparin. The cell suspension was homogenized and then seeded into culture flasks and cultured. Cloning efficiency was measured, a proliferation assay performed, and senescence-associated beta-galactosidase staining done. FACS was performed for surface antigens using 11 different monoclonal antibodies. Multilineage differentiation assays were performed for adipogenic, osteogenic, and chondrogenic cell lines and described in detail [9]. Gene modification was studied by infecting cultured cells with a bicistronic murine stem cell virus-derived retroviral vector (pMIGR1) encoding for green fluorescent protein (GFP). The cultured plastic-adherent cells had a fibroblast shape. High clonal efficiency was observed, averaging 14 % which exceeds that of bone marrow cells and is similar or superior to adipose tissue stem cells. At least two kinds of clonal morphology were demonstrated: dense clones of small size cells and smaller clusters of large cells resembling those described by Chan and colleagues [2]. Doubling time was low, an average of 27.6 h, and the cells reached a high number of passages, more than 25 before reaching senescence.

The EDT cells expressed MSC markers. More than 90 % displayed CD90 and CD73 and more than 80 % expressed CD146. Cells expressing CD45, HLA-DR, CD31, and CD14 were below 2 % in most cases. The markers CD56, CD105, and CD146 were highly variable between cells of different donors. A small fraction of the cells expressed the pluripotency marker SSEA-4. Differentiation was induced successfully to form adipocytes with confirmed lipid content, osteocytes that formed a calcified matrix, and chondrocytes with a typical cartilaginous matrix. Studies with the retrovirus demonstrated that the EDT cells could be transduced efficiently confirming their potential for gene delivery.

Patel and colleagues are a third independent group working with stem cells from menstrual blood [7]. The cells they identified were easily expandable and expressed the multipotent markers Oct-4, SSEA-4, and c-kit. The cells were

[1] In this usage "decidual" does not refer to pregnancy endometrium but rather "falling off or shed seasonally."

Fig. 1 *Pluripotent Differentiation*. ERCs were cultured under appropriate differentiation media as described in Materials and Methods and assessed for differentiation using the indicated staining methods. (**a**) Adipocytic differentiation, *yellow* indicates lipid vacuoles stained by AdipoRed. (**b**) Osteocytic differentiation, *red* indicates calcium stained by Alizarin Red. (**c**) Myocytic differentiation, *green* indicates alpha-actinin stain. (**d**) Skeletal muscle differentiation, *green* indicates skeletal myosin stain. (**e**) Endothelial differentiation, *green* indicates CD34 stain. (**f**) Endothelial differentiation, *green* indicates CD62 stain. (**g**) Hepatocytic differentiation, morphology resembles hepatic body. (**h**) Hepatocytic differentiation, *green* indicates albumin stain. (**i**) Pancreatic differentiation, *green* indicates insulin stain. (**j**) Neuronal differentiation, *green* indicates nestin stain. (**k**) Neuronal differentiation, *green* indicates GFAP stain. (**l**) Respiratory epithelial differentiation, *green* indicates prosurfactant protein C stain. (**m**) Cardiogenic differentiation, *green* indicates troponin I stain (From: Meng et al. [7]. BioMed Central Open Access)

differentiated in culture into chondrogenic, adipogenic, osteogenic, neurogenic, and cardiogenic lineages. At about the same time, Hida and colleagues isolated and cultured endometrial stem cells from menstrual blood for differentiation into cardiomyocytes [10].

Rossignoli and colleagues studied only differentiation of EDT cells into mesenchymal cell lines. They did not seek to confirm the findings of Meng and colleagues or Patel and colleagues of the pluripotent abilities of ERG cells from menstrual blood, but other investigators have done this. Differentiation into hepatic cells was demonstrated by Yang and colleagues [11]; they studied human endometrial stromal stem cells (hESSCs) harvested from hysterectomy specimens. After the initial culture, they selected cells from large endometrial stromal colony-forming units. They demonstrated expression of the MSC markers CD73, CD146, and CD90 but not the hematopoietic markers CD34 and CD45. Osteogenic and adipogenic differentiation were demonstrated. None of the control cultured cells differentiated. Next, cultured hESSCs were exposed to hepatogenic factors following a four-step protocol to separate induction and maturation steps. The investigators used fibroblast growth factor-4 (FGF-4), hepatocyte growth factor (HGF) plus insulin-transferrin-selenium (ITS), HGF plus dexamethasone and oncostatin M (OSM), and HGF plus OSF and trichostatin A (TSA) in sequence. By the 24th day of incubation, they obtained hepatic-like cells. Functionality of the cultured cells was demonstrated by detection of urea synthesis and ammonia removal and of glycogen storage.

Two groups have induced differentiation of endometrial stem cells into insulin-producing insulin producing cells [12, 13]. Li and colleagues [12] obtained endometrial cells from hysterectomy specimen from women with uterine fibroids. Attached cells obtained after isolation and plating were expanded and subjected to adipogenic, chondrogenic, osteogenic, and also neurogenic induction. Induction to pancreatic cells was accomplished by culturing in a serum-free modified pancreatic selection medium detailed in their publication. The induced cells were shown to produce C-peptide and insulin. A microarray analysis compared gene expression of the endometrial stem cells on days 0, 7, and 14 of exposure to the pancreatic selection medium. Either upregulation or downregulation of 716 genes was demonstrated by day 14 in a pattern resembling pancreatic progenitor cells.

Santamaria and colleagues used endometrium obtained by curettage from women undergoing surgery for benign conditions [13]. Endometrial stem cells were isolated by fluorescence-activated cell sorting to obtain cells expressing markers PDGFbeta-R, CD146, and CD90 but not CD45 or CD31. They compared three different protocols for inducing insulin-producing cells and found that only the three-step protocol described by Tamagawa et al. resulted in mature insulin-producing cells [14]. By the end of step three, the induced cells assumed islet-like clusters. Reverse transcriptase PCR (RT-PCR) showed increased expression of pancreatic lineage markers compared with control undifferentiated cells. The markers PDX1, PAX4, Glut2, and insulin messenger RNA all increased markedly in the treated cells. Immunocytochemistry with antibodies specific for human insulin showed strong signals, while insulin was not detected in controls. The differentiated cells were shown to increase their insulin production when exposed to glucose. The increase was proportional to the glucose concentration, increasing tenfold when glucose concentration was increased from 5 to 25 mmol/l.

Mobarakeh and colleagues offer still more proof that endometrial stem cells can be differentiated into neural cells [15] as did Li and colleagues cited above [12]. Mobarakeh's group obtained tissue by endometrial biopsies from the uterine fundus of 10 normal ovulating women on cycle days 19–24. After dispersion and plating, the adherent cells were characterized by flow cytometry. The cells were positive for CD90, CD105, and Oct4 and negative for CD31 and CD34. Differentiation into osteocytes and adipocytes was performed with standard commercially available culture reagents. To induce neural differentiation, they cultured the cells in a cell-free medium and induced with the signaling molecules basic fibroblast growth factor (bFGF), platelet-derived growth factor (PDGF), and epidermal growth factor (EGF) in sequence. After 7 days the cells changed shape from spindle shaped to a typically neuronal appearance. After induction, the cells expressed markers of mature neural cells: neurofilament-light (NF-L), class III beta-tubulin (beta3-tub), microtubule-associated protein (MAP2), and Olig1. Studies with reverse transcriptase confirmed expression of these neural markers at the messenger RNA level and also gamma-aminobutyric acid (GABA) and nestin. Glial fibrillary acidic protein (GFAP) and Olig1 which are markers of astrocytes and oligodendrocytes, respectively, were both clearly expressed at 12 days.

Wolff and colleagues also reported induced differentiation of human endometrial stem cells into neural cells [16]. Endometrial tissue was obtained by curettage from nine reproductive aged women undergoing gynecological surgery for benign conditions. The endometrial tissue was prepared in what has become a standard fashion with mincing, digested to produce a dispersed cell solution which was then passed through a 70 μm sieve to remove glandular elements, and resuspended then plated and passaged. The stromal stem cells thus obtained were characterized by flow cytometry. Neurogenic differentiation was begun after the second passage following a two-step protocol. The resulting differentiated cells were subjected to in vitro immunofluorescent studies that confirmed the expression of the neural stem cell marker nestin and production of tyrosine hydroxylase (TH), chosen because it is the rate-limiting enzyme in dopamine synthesis. Electrophysiology studies were performed in vitro on differentiated and undifferentiated endometrial cells with whole cell patch clamp recording to look for evidence of G-protein-coupled inwardly rectifying potassium channels. These are described as characteristic of central neurons and dopaminergic cells.

The induced neurogenic cells developed neurogenic morphology with pyramidal cell bodies, long axon projections, and dendritic projections that appeared to form synapses in culture (Fig. 2). Nestin was highly expressed as was TH. None of these indicators of neuronal identity was expressed by controlled cells. The electrophysiology studies showed the differentiated cells exhibit inward potassium currents induced by voltage steps from −60 to −120 mV.

ERC in Therapy

Several groups have explored treatment of disease with transplanted differentiated human endometrial cells in animals.

Studies in Murine Models of Diabetes

Li and colleagues demonstrated successful treatment of hyperglycemic severe combined immunodeficient mice (SCID) [12]. The mice were treated with streptozotocin (STZ) to induce a diabetic state. Induced insulin-producing endometrial stem cells were injected under the renal capsules of the mice. Control mice were injected with endometrial fibroblast cells or endometrial stem cells that had not been subjected to induction. After 4 weeks the transplanted cells had produced pancreatic islet-like clusters in the renal subcapsular space of the mice. Their glucose levels were reduced by comparison to the control mice. Their levels of human insulin were significantly higher and nearly normal. This was maintained through 8 weeks. No teratoma formation was found in any mice studied at 12 weeks after transplantation.

Santamaria's group also successfully treated STZ-treated SCID mice with insulin-producing induced endometrial cells transplanted beneath the kidney capsule [13]. Mice with blood glucose levels above 220 mg/dl after STZ treatment were chosen for study. Mice receiving the insulin-producing cells were compared to three control groups: STZ-treated SCID mice transplanted with undifferentiated endometrial cells, STZ-treated SCID mice that did not undergo transplant surgery, and a group of SCID mice not treated with STZ that did not undergo surgery. The group that received the differentiated cells had glucose levels that were stable and equivalent

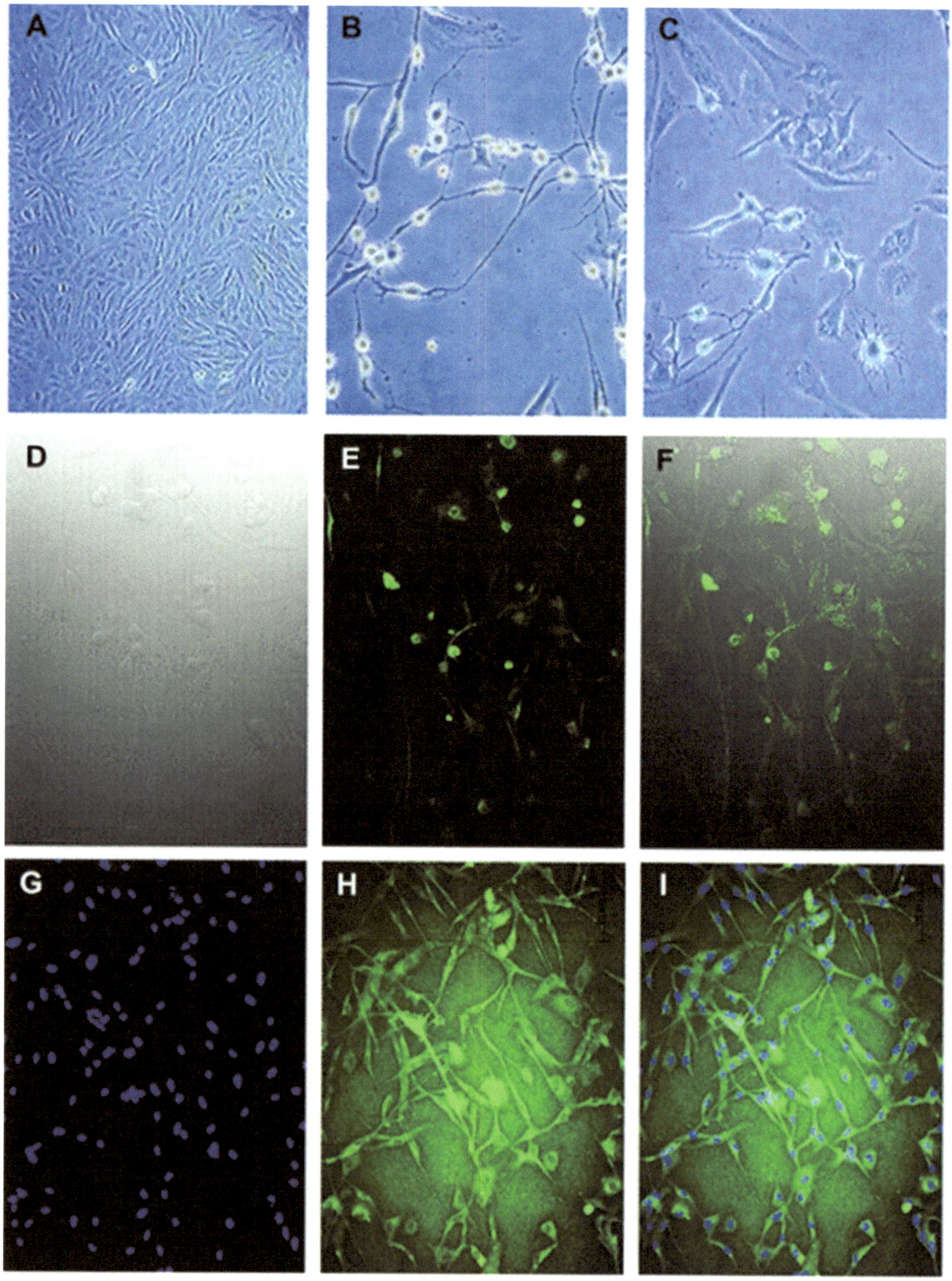
A
B
C
D
E
F
G
H
I

to the group that did not receive STZ treatment. The STZ-treated group that was transplanted with the control undifferentiated cells had rising glucose levels.

Immunofluorescence staining studies of the renal capsules of the mice treated with differentiated cells were positive for human insulin, while the controls injected with undifferentiated cells showed no human insulin signal. The STZ-treated diabetic mice that did not receive the insulin-producing cells developed many complications of diabetes during the 4-week study: dehydration, cataracts, loss of skin resilience, loss of fur sheen, passive sedate behavior, and weight loss. Those mice treated with the differentiated cells showed none of these changes.

ERC in Murine Models of Neurologic Disease

Wolff's group restored dopamine production in a Parkinson's disease model with human endometrium-derived stem cells (HEDSC) treated with a dopaminergic differentiation protocol [16]. Eight-week-old male mice were injected with intraperitoneal 1-methyl 4-phenyl 1,2,3,6-tetrahydropyridine (MPTP) on two consecutive days to induce simulated Parkinson's Disease. Both immunocompetent (57-Black 6) and immunodeficient mice (nonobese diabetic-severe combined immunodeficiency (NOD-SCID) knockout on a Black 6 background) were treated. On day 5 after MTP injection, the mice underwent intracerebral injection into the striatum of either differentiated or undifferentiated HEDSC through bilateral burr holes under stereotactic guidance. The four injection sites are described as 2 mm lateral to the bregma on both sides and both 0.5 mm rostral and 0.5 mm caudal of the bregma. The mice were sacrificed at 5 weeks for DNA analysis of one striatum to identify human cells and measurement of dopamine (DA) and its principal metabolite 3,4 dihydroxyphenylacetic acid (DOPAC) concentrations in the other. The mice brains were studied to find human DNA and to find expression of human TH. Some of the mice had transcardial perfusion of fixative after ether overdose and their brains sectioned for histologic visualization of whole human cells after staining using an antibody to human mitochondria.

Engraftment and migration were confirmed. Engraftment was successful in both immunocompetent and immunosuppressed mice and so subsequent studies were in wild-type mice only. Significant improvement in striatal DA and DOPAC concentrations was demonstrated in the MPTP-lesioned mice after HEDSC transplant as compared with MPSP-lesioned mice sham treated with only phosphate buffered saline and no cells. Human cells were found at the transplantation sites in the striatum. Undifferentiated HEDSC were found to have migrated to the lesioned areas of the substantia nigra and to have differentiated into cells with a neurogenic phenotype along the way (Fig. 3). It is interesting that the already differentiated cells did not migrate to the substantia nigra.

Peron and colleagues studied central nervous system inflammation treated with HEDSC in a murine model of experimental autoimmune encephalomyelitis (EAE) which has many of the features of human multiple sclerosis [17]. Part of the mechanism of this disease is the production of pathogenic T $CD4^+$ cells which differentiate into Th1 and Th17 cells.

An endometrial biopsy was obtained from one fertile 39-year-old woman. The tissue was

Fig. 2 In vitro neurogenic differentiation of human endometrial stem cells (HEDSC). HEDSC cultured in control media demonstrate typical stromal cell morphology (**a**), whereas cells cultured in neurogenic media demonstrated both pyramidal and dendritic cell morphology as is pictured using light microscopy (**b**, **c**). Differentiated cells visualized using differential interference contrast (**d**), IF for neural stem cell marker nestin expression (**e**), and a merge of both (**f**). Differentiated cell cultures also express TH (**h**), DAPI nuclei staining (**g**), and merge of both (**i**) (Source: Wolff et al. [16]. Figure 1 © 2011 The Authors Journal of Cellular and Molecular Medicine © 2011 Foundation for Cellular and Molecular Medicine/Blackwell Publishing Ltd. and John Wiley & Sons, Inc. with permission)

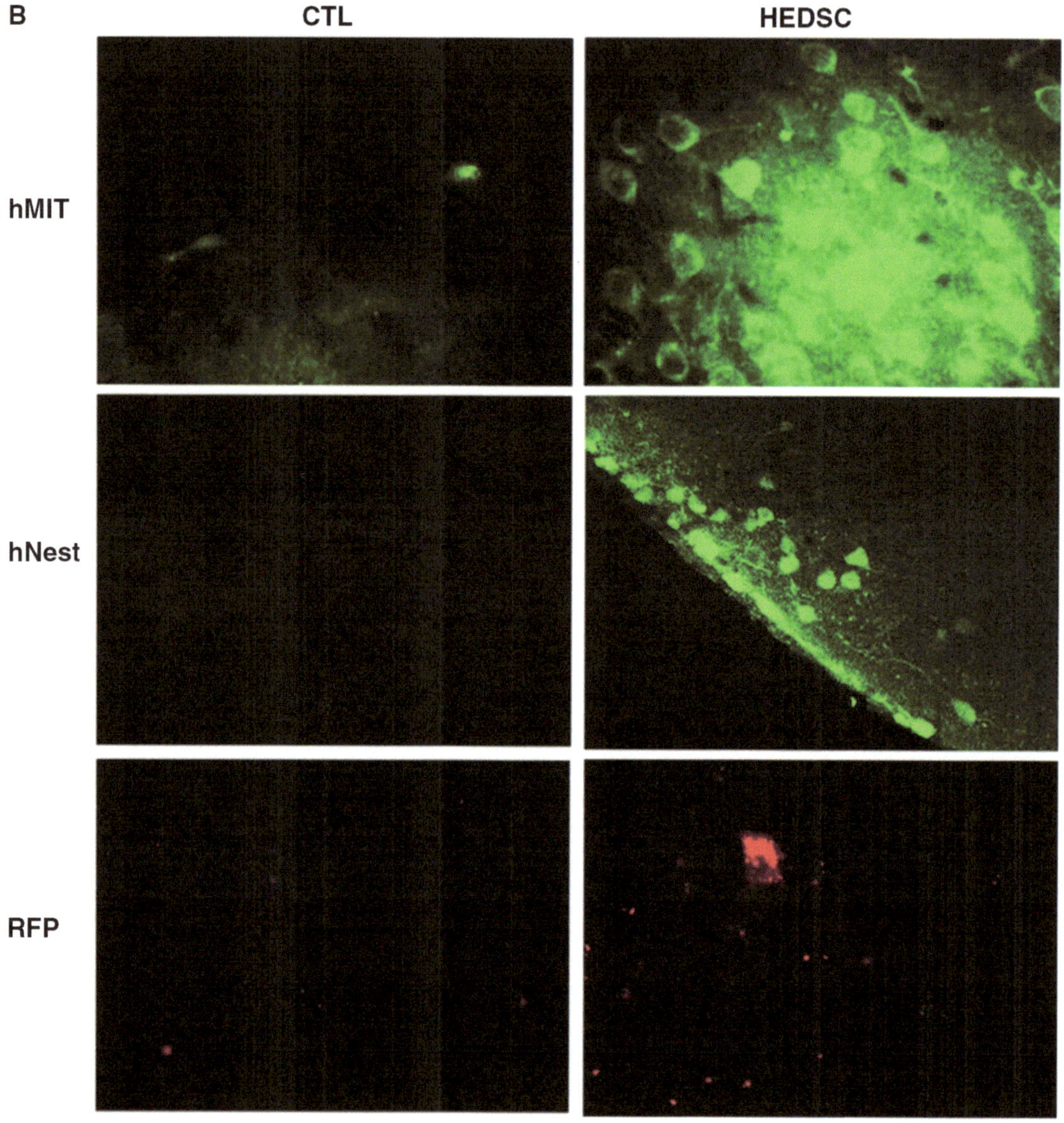

Fig. 3 Human endometrial stem cells engraft, differentiate in vivo, and migrate to the lesioned site in mice brains. (**a**) PCR detecting human DNA in mouse brain transplanted with HEDSC ($n = 14$), but not with sham transplants ($n = 8$) using PBS. (**b**) Low-power view of murine brain section from sham-treated [control (CTL)] and HEDSC-treated animals. An area that includes the substantia nigra (SN) is outlined. IHC using a human nestin antibody identifies cells localized to the SN in the transplanted animals. Human cells are visualized in mouse brains in the *right column*, and controls are shown on the *left*. In the *top panel*, all human cells are detected using a human mitochondrial antibody (hMit), which are seen here at the site of transplantation in the striatum. Spontaneous in vivo differentiation of transplanted HEDSC was observed, where they expressed nestin (hNestin). Transplanted cells adapted a neurogenic phenotype morphologically, as is visualized using red fluorescent surface labeling. Human cells were observed remote from the initial transplantation site (striatum), where they migrated to the lesioned brain area (substantia nigra) which is the area pictured in the *bottom right panel* (red fluorescent surface labeling) (Source: Wolff et al. [16]. © 2011 The Authors Journal of Cellular and Molecular Medicine © 2011 Foundation for Cellular and Molecular Medicine/Blackwell Publishing Ltd. and John Wiley & Sons, Inc. with permission)

washed, minced, and then expanded in tissue culture. The adherent cells were evaluated by flow cytometry and subjected to adipogenic, chondrogenic, and osteogenic differentiations. The cells displayed the characteristic phenotype of MSCs with high expression of CD29, CD73, CD90, HLA-ABC, and SH4 and were negative for CD14, CD31, CD45, and HLA-DR. To determine whether the human endometrial-derived MSCs (hMSCs) could modulate experimental autoimmune encephalomyelitis, they were given by intraperitoneal injection to 6–8-week D57Bl/6 female mice 1 day before induction of EAE. For EAE induction the mice were injected subcutaneously with myelin-derived antigen myelin oligodendrocyte glycoprotein (MOG35-55) with 400 μg of BCG. The mice also received two intraperitoneal doses of 200 ng of *Bordetella pertussis* toxin. The mice were examined daily for 22 days and their condition scored on a scale from 0 (no disease) to 1 (limp tail), 2 (weak/partially paralyzed hind limbs), 3 (completely paralyzed hind limbs), 4 (complete hind and partial front leg paralysis), and 5 (complete paralysis/death). The EAE scores of the treated mice were considerably reduced by pretreatment with the MSCs. After 22 days all mice were sacrificed, brains and spinal cords were excised and macerated, and cellular suspensions were prepared for study. They found a significant reduction in the absolute number of CNS-infiltrating mononuclear cells in the CNS of hMSC-treated mice compared to a control group, and the spinal cord inflammatory infiltrate was reduced as well. The HEDSC-treated mice also had lower percentage and absolute numbers of pathogenic T $CD4^+$ cells expressing TH1 and TH17 when compared to controls. The authors concluded that the HEDSC played a regulatory role and suppressed neuroinflammation in their model.

ERC in Treatment of Peripheral Vascular Disease

The angiogenic potency of bone marrow-derived stem cells has been demonstrated in a number of animal models of limb ischemia and is thought to be based on enhanced production of angiogenic cytokines [18]. Bone marrow-derived cells have been used to treat human chronic limb ischemia since 2002. However, bone marrow-derived cells have limitations for human treatment as noted earlier. A number of investigators are now replicating this work using the more recently discovered ERCs [18]. Some examples follow.

Treating Critical Limb Ischemia (CLI)

Murphy and colleagues ligated the femoral artery to induce limb ischemia and also excised the peroneal nerve to produce neurotrophic ulcer-like injury in two groups of eight immune-competent BALB/c mice [18]. One group of mice received one million ERCs by intramuscular injection into the limb distal to the ligation immediately afterwards and on days 2 and 4. The animals were followed for 14 days. All eight of the control mice developed limb ischemia followed by limb necrosis. The eight mice treated with ERC had intact limbs, though two had impeded walking. The investigator also demonstrated immunomodulatory properties of the ERC in vitro, showing that addition of ERC to ongoing mixed lymphocyte reactions suppressed proliferation, inhibited IFN-gamma, stimulated IL-4, and inhibited TNF-alpha after lipopolysaccharide (LPS) stimulation. They propose that ERC could be used as an "off-the-shelf" source of mesenchymal stem cells for treatment of CLI based on their properties: "a. High levels of growth factors and matrix metalloprotease production; b. Ability to inhibit inflammatory responses and lack of immunogenicity; and c. Expandability to great quantities without loss of differentiation ability or karyotypic abnormalities" [18].

ERC in Treatment of Stroke

There is an extensive literature on stem cells especially from bone marrow and umbilical cord sources in the treatment of stroke [19]. Borlongan and colleagues argue that ERCs might also be a potential source for autologous transplant in female stroke victims that would avoid risk of immune rejection while being safely isolated, would have high proliferative capacity, and could be differentiated into neural cells [20]. They report in vitro and in vivo small animal studies of ERC in the treatment of stroke. Menstrual blood cells were collected as described by Patel et al. [21], processed, and cryopreserved. The cells were then thawed, exposed to an antibiotic cocktail, and then expanded in culture and selected for CD117, which has been previously identified in endometrial cells and shown to be closely associated with a highly proliferative cell type with high survival and migration. For in vitro study rat neurons were cultured and then studied in an oxygen glucose deprivation model which if untreated leads to a high rate of cell death. They cultured the endometrial cells for 3 or 6 passages plus 3 additional passages and then tested the conditioned media with ELISA for possible neurotrophic factors secreted by the menstrual blood cells. They tested for vascular endothelial growth factor (VGEF), brain-derived neurotrophic factor (BDNF), neurotrophin-3 (NT-3), and glial cell line-derived neurotrophic factor (GDNF). The neural cell cultures were co-cultured with menstrual cells, or exposed to conditioned media harvested from cultured menstrual cells, or exposed to control media. The proportion of surviving cells was calculated for each group. Their findings were dramatic. Cell survival was much improved in both the group cultured with menstrual cells and the group cultured with conditioned media by comparison with the control media. Survival rates in the two treatment groups were equivalent. The ELISA results showed strong expression of growth factors VEGF, DNF and NT-3, but no detectable expression of GDNF. The second part of their study was in vivo, using an established ischemic stroke model. An intraluminal suture technique was used to produce transient middle cerebral artery occlusion for 60 min in anesthetized Sprague-Dawley rats. Transplantation surgery was performed within 2 h post stroke. The animals were given menstrual blood-derived cells by either intracerebral (IC) injection with a 28 mm catheter directly into the striatum or intravenously (IV) into the jugular vein. The cell volume was 400,000 cells by the IC route and 4,000,000 cells IV. Control rats received IC or IV vehicle. Fourteen days after transplantation, the rats were subjected to standardized behavioral assessment. The rates were sacrificed at 14 days for immunohistochemistry studies and morphological analysis. Again the results were impressive. Rats receiving the stem cells by either route, IC or IV, had significant amelioration of impairment by comparison to the controls. Improvement was equivalent for two of the three tests. Those receiving cells by the IC route did better on a test of motor coordination than those treated by the IV route. Immunohistochemical studies showed more surviving host cells in the penumbra of stroke animals of either treatment group than controls. There was no difference between the IC and IV groups. Grafted menstrual blood-derived stem cells were detected by immunofluorescent microscopic evaluation in the brains of IC and IV transplanted rats. Approximate graft survival rates were 15 % in the IC-delivered group and <1 % for IV-delivered cells. The investigators observe that behavioral recovery at this early time after transplantation suggests a bystander effect rather than cell replacement from neuronal differentiation of the grafted cells. They conclude that neurotrophic factors secreted by the graft, the host, or a combination may account for the neuroprotective effect they found.

Treating ischemic heart disease

Bone marrow-derived MSCs have been injected into humans with ischemic hearts with some success, but benefits have been limited [10]. This may be because bone marrow-derived MSCs have an extremely low rate of cardiomyogenesis in culture. Endometrial-derived stem cells, in contrast, have high transdifferentiation potential for cardiomyocytes and can be obtained in large quantities. ERCs appear to have other advantages over bone marrow MSC in addition to ease of acquisition. Stroncek and colleagues compared MSC and ERC generated from six donors. The angiogenic cytokines PDGF-BB and angiopoietin were expressed at 27- and 14-fold higher levels in the ERC, respectively. Matrix metalloproteinase-3 was expressed at 29-fold higher in ERC than MSC [22].

Hida and colleagues studied endometrial-derived cells in a rat myocardial infarction model [10]. They collected 10 ml of menstrual blood on the first day of menstruation from 6 women aged 20–30 and from a hysterectomy specimen from a 52-year-old woman. Another human donor supplied bone marrow for culture of bone marrow-derived mesenchymal cells (BMMSC) for comparison. The human menstrual blood cells were co-cultured with fetal cardiomyocytes from the hearts of day 17 mouse fetuses. The human cells from menstrual blood and from bone marrow were infected with enhanced green fluorescent protein (EGFP) adenovirus. Cells from the human and murine sources were separated by an atelocollagen membrane permeable only to molecules less than 5,000 MW. On day 5 of cocultivation, about half of the human menstrual blood-derived mesenchymal cells (MMC) were beating strongly in a synchronized manner. Immunohistochemical studies demonstrated troponin I in the differentiated MMC. The investigators recorded the action potentials from the contracting cells and noted a pattern typical of pacemaker potentials (Fig. 4). Twenty-seven to thirty-two percent of the MMCs became positive for cardiac troponin I antibody as a result of cocultivation. The investigators also established a line of monoclonal endometrial gland-derived mesenchymal cells (EMCs) of which 76–97 % transdifferentiated into cardiac cells in vitro.

In vivo studies were done using a nude rat model of myocardial infarction (MI) produced by ligation of the left coronary artery. Two weeks after induced myocardial infarction, MMC or BMMSC cells in suspension were transplanted into the center and margin of the rat's infarcted myocardium via a 31-guage needle. Two weeks after transplantation, echocardiography showed significantly greater left ventricular fraction shortening in the MI plus MMC (MI + MMC) rats than in the rats treated with BMMSC. Histologic staining showed the MI area to be significantly lower in the MI + MMC group than in the MI + BMMSC rats. The EGFP-positive mass of MMCs observed in the MI area expressed a clear striation staining pattern of cardiac troponin I and sarcomeric alpha-actinin suggesting extremely high in situ cardiomyogenic transdifferentiation of MMCs, which contributed to the improved cardiac function. The investigators propose that MMCs are the most suitable source for cardiac stem cell therapy and that it should be possible to obtain MMCs of all the HLA types to establish a banking system for therapy.

Bockeria and colleagues describe a human study currently in progress using ERC for human heart failure patients. Patients with severe congestive heart failure (NYHA classification, stage III/IV) are receiving ERCs or placebo administered by retrograde administration into the coronary sinus. As of 2013, 17 patients had been enrolled with no serious adverse events reported [23].

Stem Cells of the Fallopian Tubes

Another potential source of mesenchymal stem cells for therapy comes from human fallopian tubes. Segments of the tubes are commonly excised in operations for voluntary sterilization, and entire tubes are excised as part of hysterectomy/oophorectomy for a variety of gynecological complaints. Jazedje and colleagues have isolated stem cells from human tubes, expanded the cells in vitro, and characterized them. The cells exhibited a mesenchymal phenotype [24]. They were differentiated in vitro into adipogenic, chondrogenic, osteogenic, and myogenic cell lines. Indumathi and colleagues obtained human fallopian tube stem cells and compared them to human endometrial stem cells and human bone marrow cells for expression of a wide range of mesenchymal biomarkers as previously noted [6].

Fallopian Tube Stem Cells in Therapy

Jazedje and colleagues studied bone regeneration in immunosuppressed rats with bilateral cranial defects produced by full thickness excision of the

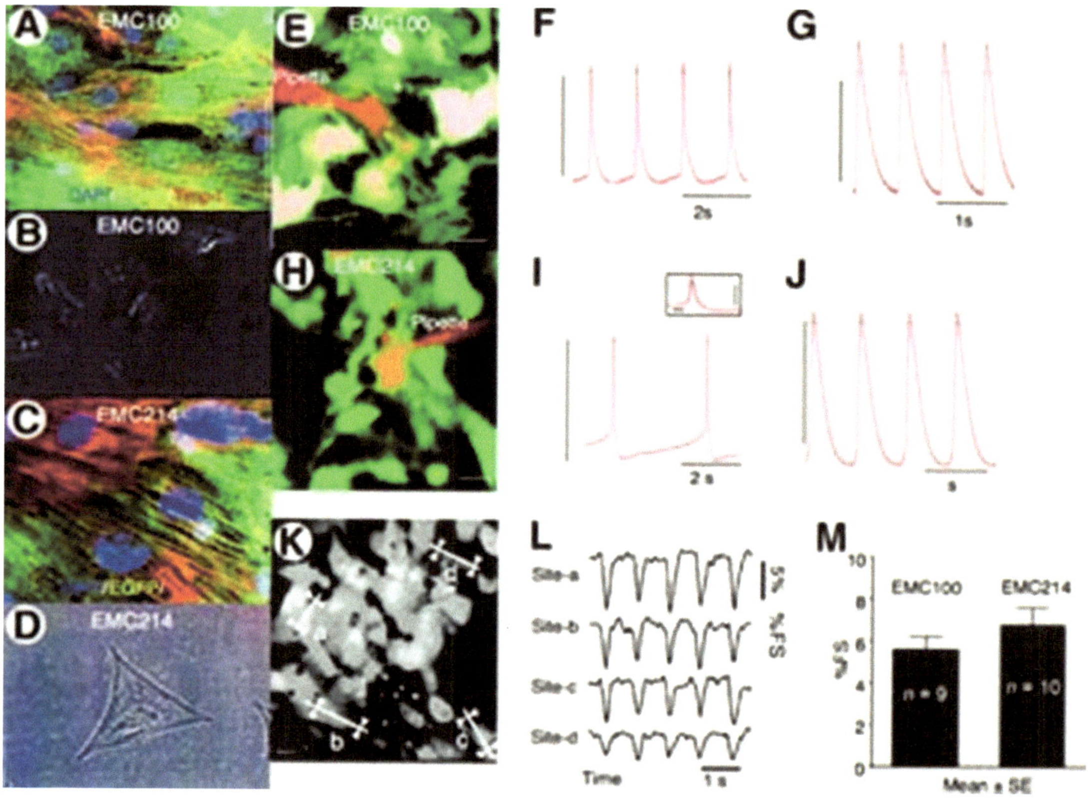

Fig. 4 Cardiomyogenic differentiation of endometrial gland-derived mesenchymal cells (EMCs) in vitro. (**a**, **c**): Immunocytochemistry of differentiated EMC100s (**a**) and EMC214s (**c**) with anticardiac troponin I (Trop-I) antibody. The cells were stained with 4′-6-diamidino-2-phenylindole (DAPI; *blue*) and anticardiac troponin I antibody (*red*). Enhanced green fluorescent protein (EGFP)-positive (*green*) human EMCs expressed Trop-I (*red*). Please note clear striation staining pattern of Trop-I (A, C) in EMCs. *Scale bar* denotes 20 μm. (**b**, **d**): Phase-contrast images of EMC100s (**b**) and EMC214s (**d**) before the cardiomyogenic induction. (**e**, **h**): EGFP-labeled EMC100s and EMC214s (*green*) were injected with Alexa 568 solution (*red*) through a microelectrode (**e**, **h**), and a recorded signal was obtained from the cells. Representative action potential traces are shown ((**f**, **g**), EMC100; (**i**, **j**), EMC214). Action potential of E is expanded in the inset (the *vertical line* denotes 100 ms). The vertical line denotes 50 mV and dotted *horizontal line* denotes 0 mV levels. (**k**–**m**): A representative still image (**k**) and detected fractional shortening (% FS) along the *white line* obtained from sites *a*, *b*, *c*, and *d* in (**l**) are shown in (**m**). (**m**): The measured % FS was averaged and is shown (Hida et al. [10]. Figure 3, Copyright © 2008 AlphaMed Press and John Wiley and Sons, Inc. with permission)

bone [25]. Human tubal stem cells were obtained from healthy 30–50-year-old women. The bony defects in the rats' crania were covered with a bioabsorbable ceramic composite. One of the two bony defects in each animal was covered with the composite containing human fallopian tube stem cells. The other defect was covered with the composite alone as a control. When the rats were euthanized at 30–60 days later, mature bone was demonstrated on the side that received the stem cell mixture, and not on the control sides.

Stem Cells of the Ovary

Stem cells have been described in most compartments of the adult ovary [26]. These include the ovarian surface epithelium (OSE) [27], the ovarian cortex [8], and the follicles [28]. The ovarian cortex appears to contain two different stem cells, somatic stem cells [26] and stem cells of the germ cell line [29]. The latter have attracted the most attention. For half a century or more, scientific opinion was that the mammalian ovary at birth contained all the oocytes it would ever have. This dogma was disproven in 2004 with the publication by Tilly and colleagues in Boston of the discovery of oogonial stem cells in the ovaries of adult mice [8]. These investigators were studying follicular atresia in the mouse ovary and noted that follicles were dying at a rate that would result in complete depletion much sooner than actually occurs in these mice. Their studies revealed large ovoid cells resembling germ cells in the ovarian surface epithelium. The cells stained positive for the marker mouse vasa homologue (MVH), a gene expressed only in germ cells. The group proved these were proliferative cells because they incorporated 5-bromodeoxyuridine (BrdU). The cells also expressed the specific protein synaptonemal complex protein 3 (SCP3) that is required to initiate meiosis to produce oocytes. In a further set of experiments, they transplanted ovarian tissue from wild-type mice into the ovaries of mice expressing GFP. After 3–4 weeks the grafted wild-type ovarian tissue contained GFP-positive oocytes surrounded by wild-type follicles [8]. Subsequently, Zou's group identified germline cells of the mouse ovary using a cell-sorting approach with the antibody Ddx4 (DEAD box polypeptide 4, another name for MVH). In the definitive study the group transplanted GFP-positive germline stem cells into the ovaries of infertile mice and produced live GFP-positive offspring [30].

Very Small Embryonic-Like Stem Cells

In the last several years, very small embryonic-like stem cells (VSELs) have been identified in many adult tissues with the highest numbers being found in the brain, kidneys, muscles, pancreas, and bone marrow [31, 32]. VSELs have now been identified in the ovarian surface epithelium and the testis [33]. As nicely summarized by Bhartiya and colleagues, VSELs are small, 3–5 μm in diameter, and are diploid (Fig. 5). They have high telomerase activity and express pluripotent markers (Rex-1, Nanog, SSEA, and Klf-4) and germ cell markers (Mvh, Stella, Fragilis, Nobox, and Hdac-6). They decrease in number as the animal ages. They differentiate into all three germ layers in vitro. Embryonic cells develop into teratomas when injected into immunodeficient mice, but VSELs do not. They are quiescent, DNA label-retaining stem cells with a low metabolic rate. Current thinking is that VSELs are descendants of epiblast stage pluripotent stem cells deposited in early development in many organs including the gonads where they produce tissue-committed stem cells. In the adult, VSELs respond to injury, moving into the circulation to the site of injury producing tissue-committed stem cells to begin regeneration. VSELs have been reported to resist cancer chemotherapy and restore germ cell depleted ovaries.

Potential Clinical Applications of Oogonial Stem Cells

Potential clinical applications have been reviewed by Dunlop and colleagues [34]. These include a new approach to treating age-related ovarian insufficiency by isolating the patient's own germline OSCs rather than relying on donor oocyte as at present. Another potential use would be for young women needing gonadotoxic treatment for cancer. Their own OSCs could be harvested and preserved prior to the cancer treatment without delaying treatment. Conception and live birth have occurred with use of cryopreserved ovarian tissue, but there is concern that use of blocks of ovarian tissue as a graft might reintroduce malignant cells into the patient.

Somatic Ovarian Stem Cells

Stimpfel and colleagues isolated and studied ovarian cortical stem cells that are not of the germ cell line [26]. These were isolated from human ovaries surgically removed to prevent breast cancer or at the time of surgery for uterine myomata. Cells from a suspension prepared from macerated ovarian biopsies were plated

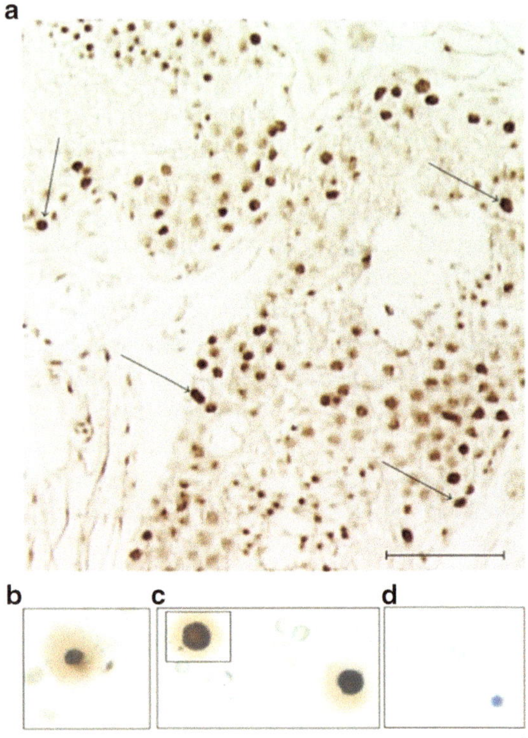

Fig. 5 Immunolocalization of 5-methylcytosine on adult human testicular section and perimenopausal ovary surface epithelium smear (using standard protocol published earlier, [54]). Note dense staining in the spermatogonial stem cells (SSCs, *arrow*, (*a*)), while the spermatocytes showed minimal staining. In few tubules spermatids showed positive staining. Similarly, the ovarian germ stem cells (OGSCs) stain positive (*b* and *c*). Negative control (*d*). The results indicate that A_{dark} SSCs in testis and OGSCs in ovaries, derived by asymmetric cell division of VSELs, undergo nuclear reprogramming associated with extensive methylation—suggesting that similar basic stem cell biology exists in both the sexes. *Scale bar* = 20 μm (Bhartiya et al. [33]. Figure 1 Open Access through Creative Commons Attribution License (creativecommons.org/license/by/3.0/))

and cultured and passaged at 12–18 day intervals as long as 8 months. Ability to produce alkaline phosphatase (AP) was measured, and expression of pluripotent and mesenchymal stem cells markers were used to demonstrate stemness. Flow cytometry was performed using anti-SSE-4 antibodies. Gene expression analysis was done on single-cell colonies. Telomerase assay, TaqMan protein expression assay, and Western blot assay for protein expression were also performed. Cells were also cultured in appropriate media to induce differentiation into adipogenic, osteogenic, neural, and pancreatic cell lines. Ability to form teratomas in vivo was tested by injection into NOD-SCID mice. The cells expressed a number of multipotent/mesenchymal markers and were successfully differentiated into cells of all three germ layers. They did not form teratomas when injected into immunodeficient mice.

Potential for Clinical Use

Stimpfel concluded that the somatic stem cells his group isolated show a high degree of plasticity and can be stored and could be safely used in regenerative medicine because of their low potential for teratoma formation.

Stem Cells of the Testis

Histologists have been describing the progression of changes in the germ cells of the seminiferous tubules of the testis for more than a century. The identification of the spermatogonia in histologic specimens as the earliest cell in the progression goes back to 1901 [35]. The details of the process were described in a series of papers by Clermont [36]. He described three phases: (1) Spermatogonia proliferate to form spermatocytes while at the same time replenishing themselves; (2) primary and secondary spermatocytes undergo reduction division to form haploid cells, the spermatids; and (3) spermatids undergo a complex series of cytological transformations leading to production of mature spermatozoa. More recently, a fourth phase is recognized, spermiation, the release of spermatozoans into the lumen of the seminiferous tubule [37]. Clermont was the first to suggest that A_{dark} spermatogonia were reserve stem cells. Isolation and characterization of the spermatogonia as a stem cell has been accomplished in animals by several groups of investigators [38–42]. More recently, human spermatagonial stem cells (SSCs) have been isolated, cultured, and characterized [41, 42]. He et al. obtained testes from organ donors within 1–2 h of their removal [41]. The seminiferous tubules were isolated. Germ cells were separated by enzymatic digestion followed by differential plating to separate germ cells from Sertoli and myoid cells. Cells were sorted by magnetic-activated cell sorting (MACS) for the presence of the G-protein-coupled receptor125 (GPR125) a marker known from animal studies to be associated with SSC. The GPR125-positive spermatogonia were seeded in dishes coated with gelatin and incubated in media consisting of StemPro SFM medium (Invitrogen) plus human GDNF, CFRA-Fc, NUDT6, EGF, and Nodal. Proliferation assays over 14 days, immunocytochemistry using markers for SSCs and progenitors in other species, and Western blot studies were performed. The principal findings were that GPR125 was expressed in a subset of spermatogonia, only one or two per human seminiferous tubule cross section, and not at all in Sertoli cells or differentiated germ cells. Similarly UCHL1 was localized in spermatogonia along the basement membrane of the seminiferous tubules but not in Sertoli cells or differentiated germ cells. The marker MAGEA4 was expressed by SSC and preleptotene spermatocytes but not in Sertoli or Leydig cells. He and colleagues concluded that the GPR125-positive spermatogonia express markers for mouse SSCs and progenitors (GPR125, ITGA6, ZBTB16, UCHL1, GFRA1, and Thy1) and could proliferate in vitro over a 2-week period.

Piravar and colleagues [42] studied testicular biopsies from 10 men with nonobstructive azoospermia. The presumptive spermatogonial stem cells they obtained were tested for expression of the marker ubiquitin carboxyl-terminal esterase L1 (UCHL1) chosen because it is expressed in spermatogonia of many species. Minced testicular tissue was subjected to a two-step enzymatic digestion process and cultured for 16 h in uncoated dishes. The floating cells were isolated and transferred to new dishes containing media based on StemPro SFM with GDNF, EGF, and LIF as described by He et al. [41], but with additional additives. Microscopic examination at 1 week revealed cells that were round or oval with a large nucleus and little cytoplasm. After 2 weeks beginning cell clusters were seen. The clusters were transferred to laminin-coated plates where they continued to proliferate during 6 weeks of observation. RT-PCR confirmed the expression of spermatogonial specific gene UCHL1 throughout the multiple passages and demonstrating self-renewal in vitro of human SSC. Laminin is produced by Sertoli cells.

A further development in study of human SSGs is their ability to revert back to pluripotent embryonic stemlike pluripotent embryonic stem-like cells without the addition of exogenous genes. This has been confirmed by three independent groups of investigators [43–45]. Two of the groups first isolated SSGs that were then further cultured and dedifferentiated to form pluripotent ES-like cells by culturing with media containing leukemia inhibiting factor (LIF) [43], or in a human ESC medium on mouse embryonic fibroblasts [44]. Golestaneh and colleagues proceeded directly from isolation of cells from testicular biopsies by dissection and enzymatic dispersion to culture in gelatin-coated plates in media previously developed to support human embryonic cells [45]. They used DMEM high glucose medium containing 15 % knockout serum replacement for ES cells (Gibco), glutamine, beta-mercaptoethanol, nonessential amino acids, penicillin-streptomycin, and TGF-beta. They manually removed the resulting cell colonies, trypsinized them, and cultured them in a new dish weekly. This differential plating removed most of the somatic cells.

The resulting cells, described by the authors as ES-like cells, were successfully expanded and passaged for more than 20 passages. The ES-like cells were differentiated in culture into cells of endodermal islet-like cells, mesodermal cardiac cells, and ectodermal neural lineages with commercially available media. The differentiation was confirmed by demonstration of the appropriate cell markers of the different lineages. The phenotype of the ES-like cells after 7–10 days and 4 weeks in human ES media resembled human ES cell colonies. Alkaline phosphatase was strongly expressed as were ES transcription factors such as POU domain containing class 5, transcription factor 1 (POU5F1), and NANOG, and they had high telomerase activity. They also expressed the known surface markers for undifferentiated human ES cells: stage-specific embryonic antigen-4 (SSEA-4) and tumor rejection antigens TRA-1-81 and TRA-1-60, while human germ cells freshly isolated from the testes were negative. A change in gene expression during the dedifferentiation in culture was loss of the germ cell-specific marker VASA by the ES-like cells. Embryonic cells are expected to form teratomas in the nude mouse model after subcutaneous injection. When the ES-like cells were injected subcutaneously into nude mice, small teratomas formed, after one month at the highest cell dose tested, 2×10^6 cells. A similar number of mouse ES cells resulted in large teratomas in 3 weeks.

Potential Clinical Applications of Testicular Stem Cells

An important potential application is in fertility preservation or restoration for males after cancer chemotherapy. These topics are thoroughly discussed in two reviews [46, 47].

Possible options include:

1. Cryopreservation of spermatozoa, an already well-established method for males after puberty.
2. Testicular biopsy for cryopreservation of testicular tissue and later transplantation. This has been accomplished in mice but not yet in humans. There is a risk of reintroducing cancer cells if any are present in the grafted tissue. Transplanted testicular tissue may have an advantage over transplanted spermatogonia since the whole SSC niche of Sertoli cells, myoid cells, and Leydig cells is also transplanted [48].
3. Obtain testicular tissue by biopsy, isolate spermatogonia for propagation in vitro, and then cryopreserve them for later transplantation, presumably preventing transmission of cancer cells [47].

The availability of large numbers of spermatogonia or mature spermatocytes by in vitro culture methods opens up possibilities for treatment of male infertility of a variety of causes. Piravar's group [42] and also Koruji's group [49] successfully cultured and propagated SSC from infertile men with nonobstructive azoospermia. Koruji and colleagues transplanted BrdU-labeled SSC clusters into the seminiferous tubules of the left testis of mice that had been treated with busulfan to produce azoospermia. Two months later transplanted cells were localized in the basal seminiferous tubules of the recipient testes as single cells. The cells had survived, but had not propagated.

Another important potential use of EC-like cells obtained by culture of SSCs is as a source of histocompatible pluripotent cells for autotransplantation, providing individual cell-based therapy in the treatment of a wide variety of illnesses [42, 43]. This would avoid the immunological problems and ethical issues seen with other sources of stem cells. One example of these possibilities is the work of Bojnordi and colleagues [50]. The ES-like cells were generated from neonatal mouse testis and then were dedifferentiated to oligoprogenitor (OP)-like cells using neural differentiation methods. They studied these cells as a treatment in a mouse model for demyelinative disease. Treated mice had higher amounts of re-myelinization when compared to controls treated with the vehicle.

Yet another example is the work of Iwasa and colleagues [51]. They established a cell line of multipotent germline stem cells from mouse neonatal testis and differentiated them into cardiomyocytes and endothelial cells by selection with the mesodermal cell-surface marker fetal liver kinase 1 (Flk1). They injected Flk1 cells into acute ischemic hearts in a mouse model. They found significant improvement in cardiac function and increased cardiomyocyte survival producing a much thicker cardiac wall in the infarcted region than in control mice. These studies suggest that the prediction by Golestaneh and colleagues that "... in the future, male patients will be cured of disease with a biopsy of their own testis" has merit [45].

Stem Cells of the Prostate

The prostate surrounds the male urethra as it exits the bladder. It has a zonal architecture with central, transition, and peripheral zones and anterior fibromuscular stroma [52]. It has a pseudostratified epithelium with three types of terminally differentiated cells: luminal, basal, and neuroendocrine. Luminal cells are the major component of the normal adult gland and are also the major component of benign prostatic hypertrophy (BPH) and most prostate cancers. They are polarized columnar cells. They secrete proteins which liquefy the semen and also secrete the proteins prostate-specific antigen (PSA) and prostatic acid phosphatase. The luminal cells express characteristic markers including cytokeratin 8 (CK8), CK18, NKX3.1, and high levels of androgen receptor (AR). The basal cells are located beneath the luminal epithelium, adherent to the basement membrane. They express the markers p63, CK5, and CK14 and do not express AR [50]. Neuroendocrine cells are rare and their function is unknown. They express neuroendocrine markers such as chromogranin A and synaptophysin.

The prostate normally has a slow turnover of cells. Castration in animals leads to rapid involution of the luminal cells, with only basal epithelial cells surviving. Rapid and complete regeneration of the gland occurs when androgen is replaced. Reasoning that regeneration of the gland must result from proliferation and differentiation of the surviving basal cells, Collins and colleagues set about isolating prostatic stem cells from surgical specimens obtained at suprapubic prostatectomy for benign prostatic hypertrophy or bladder cancer [53]. The investigators prepared single-cell suspensions and then used MACS microbeads linked to anti-CD44 followed by incubation with MACS goat anti-mouse IgG microbeads to isolate putative stem cells from the basal epithelium. The CD44 positive basal cells were plated onto dishes coated with type I collagen, type IV collagen, and laminin 1. After 5 min the dishes were washed with PBS and adherent cells were harvested. The putative stem cells were integrin bright and were selected directly from tissue on the basis of rapid adhesion to type 1 collagen. The selected population had a basal cell phenotype, expressing $\alpha 2\beta 1$integrin ($\alpha 2\beta 1^{hi}$) CK5 and CK14, and had high ability to form colonies. Some of the selected cells were trypsinized and combined with cultured prostate stromal cells and injected subcutaneously into 6–8-week-old male nude mice. The mice were sacrificed after 6 weeks and sections were taken. Fully differentiated prostate epithelium was formed at the injection sites. Adherence to type I collagen distinguished prostate stem cells from transit-amplifying daughter cells by the presence of 2–3 higher surface levels of integrin $\alpha 2\beta 1$.

In 2004, the same investigators reported an improved method of identifying prostate stem cells using antibody to CD133 in addition to adherence to type I collagen to select putative prostatic basal cells [54]. Prostatic tissue was also obtained from surgical specimens. Tissue was prepared by collagenase digestion and trypsin/EDTA incubation as described by Collins et al. [53] and was then fractionated based on cell adhesion to type I collagen after 5 min exposure. MACS macrobeads linked to antihuman CD133 were used to further purify the nonadherent population. Direct immunofluorescent staining with mouse anti-CD monoclonal antibody demonstrated the $CD133^{+}$ cells to be within the basal layer of the prostatic acini. Cell kinetic studies found the $CD133^{+}$ cells to be mostly quiescent, while the $CD133^{-}$ cells were actively cycling, and likely to be transit-amplifying cells. The proliferative potential of the $\alpha 2\beta 1^{hi}$ basal cells was studied after dividing them into two groups CD^{133+} and CD^{133-}. Both groups were passaged in culture until proliferative potential was exhausted. The total cell output of the $\alpha 2\beta 1^{hi}$ /CD^{133+} cells was more than double that of the $\alpha 2\beta 1^{hi}$/$CD133^{-}$ cells. In a final experiment, the group injected $\alpha 2\beta 1^{hi}$/$CD133^{+}$ cells into the flanks of nude mice. The cells produced prostatic epithelium at the injection sites, demonstrating their stem cell capability.

Prostate Stem Cells and Human Disease

Prostate epithelial cells are involved in both benign prostatic hypertrophy (BPH) and prostate cancer. Prostate cancer is the most frequently diagnosed cancer among men in the United States and one of the leading causes of death worldwide [52]. BPH occurs in most men as they age.

Cancer Stem Cells of the Reproductive Tract

Because cancers multiply rapidly, it was presumed that populations of cancer stem cells must exist within them. This was the first modern evidence of cancer stem cells that was provided by Lapidate and colleagues in their 1994 report of causing leukemia by transplanting myeloid leukemia stem cells into severe combined immunodeficient mice [55]. Since then, cancer stem cells (CSCs) have been described in many tumors in many organs and from most tumors of the reproductive tract. They are important to understanding the origin of cancers and to better means for treatment. CSCs have been isolated in vivo by serial transplantation into immunodeficient mice and in vitro by colony-forming cell assay, microsphere assay, SP assay, staining for surface antigens, ALDH activity, PKH label-retaining cell assay, and expression of specific markers when cells are sorted in flow cytometry [56]. CSCs have been repeatedly found to have increased resistance to chemotherapeutic agents and also to radiotherapy [56–58]. Therapies that could target CSCs while not harming normal stem cells would offer new hope for cure of cancers. There is intense interest in CSCs of ovarian epithelial tumors, which are the most lethal of female reproductive track cancers, 5th most common cause of cancer deaths in women, and kill about 14,000 women a year in the United States. Eighty percent of women with ovarian epithelial cancer are not diagnosed until the disease is advanced, and only 30 % of these women survive 5 years with the best presently available therapy, extensive surgery plus multi-agent chemotherapy [58].

A 2013 review by Lopez et al. summarizes much of this work in the female reproductive system [57]. CSCs have been identified in ovarian epithelial cancers (OEC) and uterine and cervical cancers. Many cell-surface markers have been identified with cancers of the different female reproductive organs. However, some of these markers are also expressed by normal stem cells of these tissues. Several of the endometrial cancers studied contained small quantities of SP cells. The SP cells were highly clonogenic and reproduced the cancers in the nude mouse model. In a study of serous adenocarcinomas of the ovary, cultures of cells from the tumors yielded a population of spheroid-forming nonadherent cell clusters. When xenografted into nude mice, as few as 100 cells selected from spheroid-forming dissociated cells reproduced the original tumor fully, while $>1 \times 10^5$ unselected cells did not produce a tumor [57]. The Tomao review provides additional information about ovarian CSCs [58].

Fallopian tube stem Fallopian tubal stem cells are concentrated near the distal ends of the fallopian tubes, a site of great interest because of the recent appreciation that many of the epithelial cancers attributed to ovarian origin may in reality have originated in the distal tubal epithelium which is closely associated with the ovaries [59, 60]. Karst and colleagues have provided evidence for this theory with a multistage in vitro process that transformed fallopian tube stem cells into high grade Mullerian carcinomas [60].

Putative prostatic CSCs were isolated by Collins et al. from human cancer biopsies [61]. They used the cell-surface markers $CD44^{+}$ $\alpha2\beta1$ $integrin^{high}$ $CD133^{+}$. Other markers associated with prostate cancer stem/progenitor cells include $ALDH^{high}$ and $CD44^{+}$ and more. Cells with many of these markers have reproduced the original tumors when injected in the immunodeficient mouse model [62]. Hopefully better understanding of the CSCs and their interaction with genes will allow better means for identifying indolent from aggressive prostate cancers and guide therapy for castration-resistant prostate castration resistant prostate cancer.

Tumors of the testis are comparatively rare, accounting for only 1 % of all cancers, but they are the most common malignancies in males aged 15–34 years. These are germ cell tumors. They are unique in that the normal fetal germ cell has stem cell characteristics that are shared with pluripotent human embryonic stem cells (hESSCs), but are not pluripotent [63]. A related area of great interest is the cancer/testis antigens (CTAs). These are proteins expressed in many human tumors that are not present or only weakly present in normal human tissues with the exception of the testis and the placenta. CTAs are viewed as possible markers for cancer stem cells that might provide new targets for anticancer therapy [64].

References

1. Kyurkchiev S, Galdolfi F, Hayrabedyan S, Brevini TAL, Dimmitrov R, Fitzgerald JS et al (2012) Stem cells in the reproductive system. Am J Reprod Immunol 67:445–462
2. Chan RWS, Schwab KE, Gargett CE (2004) Clonogenicity of human endometrial epithelial and stromal cells. Biol Reprod 70:1738–1750
3. Gargett CE, Masuda H (2010) Adult stem/progenitor cells of the endometrium. Mol Hum Reprod 16: 813–834
4. Masuda H, Maruyama T, Hiratsu E, Yamane J, Iwani A, Nagashima T et al (2007) Noninvasive and real-time assessment of reconstructed functional human endometrium in NOD/SCID/Gc^{null} immunodeficient mice. Proc Natl Acad Sci USA 104:1925–1930
5. Patel AN, Silva F (2008) Menstrual blood stromal cells: the potential for regenerative medicine. Regen Med 3:443–444
6. Indumathi S, Harikrishnan R, Rajkumar JS, Sudarsanam D, Dhanasekaran M (2013) Prospective biomarkers of stem cells of human endometrium and fallopian tube compared with bone marrow. Cell Tissue Res 352:537–549
7. Meng X, Ichim TE, Zhong J, Rogers A, Yin Z, Jackson J et al (2007) Endometrial regenerative cells. A novel stem cell population. J Transl Med 5:57–67
8. Johnson J, Bagley J, Skaznik-Winkiel M, Lee HJ, Adams GB, Nikura Y et al (2005) Oocyte generation in adult mammalian ovaries by putative germ cells in bone marrow and peripheral blood. Cell 122:303–315
9. Rossignoli F, Casellli A, Grisendi G, Piccinno S, Burns JS, Murgia A et al (2013) Isolation, characterization and transduction of endometrial decidual tissue multipotent mesenchymal/stem cells from menstrual blood. BioMed Res Int 2013:Article ID 901821, 14 pages
10. Hida N, Nishiyama N, Miyoshi S, Kira S, Segawa K, Uyama T et al (2008) Novel cardiac precursor-like cells from human menstrual blood derived mesenchymal stem cells. Stem Cells 26:1695–1704
11. Yang XY, Wang W, Li X (2013) In vitro hepatic differentiation of human endometrial stromal stem cells. In Vitro Cell Dev Biol Anim. doi:10.1007/s11626-013-9688-z
12. Li H-Y, Chen Y-J, Chen S-J, Kao C-L, Tseng L-M, Lo W-L et al (2010) Induction of insulin-producing cells derived from endometrial mesenchymal stem-like cells. J Pharmacol Exp Ther 335:817–829
13. Santamaria X, Massasa EE, Feng Y, Wolff E, Taylor HS (2011) Derivation of insulin producing cells from human endometrial stromal cells and use in the treatment of murine diabetes. Mol Ther 11:2065–2071
14. Tamagawa T, Ishiwata I, Sato K, Nakamura Y (2009) Induced in vitro differentiation of pancreatic-like cells from human amnion-derived fibroblast-like cells. Hum Cell 22:55–63
15. Mobarakeh ZT, Ai J, Yazdani F, Sorkhabadi SMR, Ghanbari Z, Javidan AN et al (2012) Human endometrial stem cells as a new source for programming to neural cells. Cell Biol Int Rep 19:art:e00015. doi:10.1042/CBR20110009
16. Wolff EF, Gao X-B, Yao KV, Andrews ZB, Du H, Elsworth JD et al (2011) Endometrial stem cell transplantation restores dopamine production in Parkinson's disease model. Cell Mol Med 15: 747–755
17. Peron JPS, Jazedje T, Brandao WN, Perin PM, Maluf M, Evangelista LP et al (2012) Human endometrial-derived mesenchymal stem cells suppress inflammation in the central nervous system of EAE mice. Stem Cell Rev Rep 8:940–952
18. Murphy MP, Wang H, Patel AN, Kambhampatia S, Angle N, Chan K et al (2008) Allogeneic endometrial regenerative cells: an "off the shelf solution" for critical limb ischemia. J Transl Med 6:45. doi:10.1186/1479-5876-6-45
19. Shinozuka K, Dailey T, Tajiri N, Ishikawa H, Kaneko YK, Cesar V (2013) Stem cell transplantation for neuroprotection in stroke. Brain Sci 3:239–261
20. Borlongan CV, Kaneko Y, Maki M, Yu S-J, Ali M, Allickson JG et al (2010) Menstrual blood cells display stem cell-like phenotypic markers and exert neuroprotection following transplantation in experimental stroke. Stem Cells Dev 19:439–451
21. Patel AN, Park E, Kuzman M, Benetti F, Silva FJ, Allickson JG (2008) Multipotent menstrual blood stromal stem cells: isolation, characterization and differentiation. Cell Transplant 17:303–311
22. Wang H, Jin P, Sabatino M, Ren J, Civini S, Bogin V et al (2012) Comparison of endometrial regenerative cells and bone marrow stromal cells. J Transl Med 10:207. http://www.translational-medicine.com/content/10/1/207
23. Bockeria L, Bogin V, Bockeria O, Le T, Alekyan B, Woods EJ et al (2013) Endometrial regenerative cells for treatment of heart failure: a new stem cell enters the clinic. J Transl Med 11:56. Http://www.translational-medicine.com/content/11/1/56

24. Jazedje T, Perin PM, Maluf M, Halpern S, Secco M, Bueno DF et al (2009) Human fallopian tube: a new source of multipotent adult mesenchymal stem cells discarded in surgical procedures. J Transl Med 18:46. doi:10.1186/1479-5876-7-46
25. Jazedje T, Bueno DF, Almeda BV, Caetano H, Czeresnia CE, Perin PM et al (2012) Human fallopian tube mesenchymal stromal cells enhance bone regeneration in a xenotransplanted model. Stem Cell Rev 8:355–362
26. Stimpfel M, Skutella T, Cvjeticanin B, Meznarik M, Dove P, Novakovic S et al (2013) Isolation, characterization and differentiation of cells expressing pluripotent/multipotent markers from adult human ovaries. Cell Tissue Res. doi:10.1007/s00441-013-1677-8
27. Bukovsky A, Svetllllikova M, Caudle MR (2005) Oogenesis in cultures derived from adult human ovaries. Reprod Biol Endocrinol 3:17
28. Honda A, Hirose M, Hara K, Matoba S, Inoue K, Miki H et al (2007) Isolation, characterization, and in vitro and in vivo differentiation of putative thecal stem cells. Proc Natl Acad Sci U S A 104:12389–12394
29. White YAR, Woods DC, Takai Y, Ishihara O, Seki H, Tilly JL (2012) Oocyte formation by mitotically-active germ cells purified from ovaries of reproductive age women. Nat Med 18:413–421
30. Zou K, Yuan Z, Yang Z, Luo H, Sun K, Zhou L et al (2009) Production of offspring from a germline stem cell derived from neonatal ovaries. Nat Cell Biol 11:631–636
31. Ratajczak MZ, Kucia M, Majka RR, Ratajczak J (2004) Heterogeneous populations of bone marrow stem cells-are we spotting on the same cells from different angles. Folia Histochem Cytobiol 42: 139–146
32. Ratajczak MZ, Zuba-Surma E, Kucia M, Poniewierska A, Suszynskas M, Rarajczak J (2012) Pluripotent and multipotent stem cells in adult tissues. Adv Med Sci 57:1–17
33. Bhartiya D, Unni S, Parte S, Anand S (2013) Very small embryonic-like stem cells: Implications in reproductive Biology. BioMed Res Intern 2013:Article ID 682326, 10 pages. http://dx.doi.org/10.1155/2013/682326
34. Dunlop CE, Telfer EE, Anderson RA (2013) Ovarian stem cells-potential roles in infertility treatment and fertility. Maturitas. http://dx.doi.org/10.1016/j.maturitas.2013.04.017
35. Regaud C (1901) Etudes sur la structure des tube seminifieres et sur la structures chez le mammieres. Arch Anat Microscop 4:101–156; 231–380
36. Clermont Y (1972) Kinetics of spermatogenesis in mammals: seminiferous epithelium cycle and spermatogonial renewal. Physiol Rev 52:198–236
37. Aman RP (2008) The cycle of the seminiferous epithelium in humans: a need to revisit? J Androl 29:469–487
38. Meng X, Lindahl M, HJyvonen ME, Parvinen M, de Rooij DG, Hess MW et al (2000) Regulation of cell fate decision of undifferentiated spermatogonia by GDNF. Science 287:1489–1493
39. Hofmann MC, Braydich-Stolle L, Dym M (2005) Isolation of male germ-line stem cells: influence of GDNF. Dev Biol 279:114–124
40. Kanatsu-Shinohara M, Miki H, Inoue K, Ogonuki N, Toyokuni S, Ogura A et al (2005) Long term culture of mouse male germline stem cells under serum or feeder-free conditions. Biol Reprod 72: 985–991
41. He Z, Kokkinaki M, Jian J, Dobrinski I, Dym M (2010) Isolation, characterization and culture of human spermatogonia. Biol Reprod 82:363–372
42. Piravar Z, Jeddi-Tehrani M, Sadeghi MR, Arash M, Eidi A, Akhondi MM (2013) In vitro culture of human testicular stem cells on feed-free condition. J Reprod Infertil 14:17–22
43. Conrad S, Renninger M, Hennenlotter J, Wiesner T, Just L, Bonin M et al (2008) Generation of pluripotent stem cells from adult human testis. Nature 456: 344–349
44. Kossack N, Meneses J, Shefl S, Nguyen N, Chavez S, Nicholas C et al (2009) Isolation and characterization of pluripotent human spermatogonial stem cell-derived cells. Stem Cells 27:138–149
45. Golestaneh N, Kokkinaki M, Pant D, Jiang J, DeStefano D, Fernandez-Bueno C et al (2009) Pluripotent stem cells derived from adult human testes. Stem Cells Dev 18:1115–1125
46. Schmidt KT, Andersen CY (2012) Recommendations for fertility preservation in patients with lymphomas. J Assist Reprod Genet 29:473–477
47. Jahnukainen K, Stukenborg J-B (2012) Present and future prospects of male fertility preservation for children and adolescents. J Clin Endocrinol Metab 97:4341–4351
48. Oatley JM, Brinster RL (2012) The germline cell niche unit in mammalian testes. Physiol Rev 92: 577–595
49. Koruji M, Shahverdi A, Arghavan J, Piryaei A, Lakpour MR, Sedighi MAG (2012) Proliferation of small number of human spermatogonial stem cells obtained from azoospermic patients. J Assist Reprod Genet 29:957–967
50. Bojnordi MN, Movahedin M, Tiraihi T, Javan M, Hamidabadi HG (2013) Oligoprogenitor cells derived from spermatogonia stem cells improve remyelination in demyelination mode. Mol Biotechnol. doi:10.1007/s12033-013-9722-0
51. Iwasa T, Baba S, Doi H, Kaichi S, Yokoo N, Mima T et al (2010) Neonatal mouse testis-derived multipotent germ line stem cells improve the cardiac function of acute ischemic heart mouse model. Biochem Biophys Res Commun 400:27–33
52. Shen MM, Shen-Abate C (2010) Molecular genetics of prostate cancer: new prospects for old challenges. Genes Dev 24:1967–2000

53. Collins AT, Habib FK, Maitland NJ, Neal DE (2001) Identification and isolation of human prostate, epithelial stem cells based on alpha(2)beta(1)-integrin expression. J Cell Sci 114(Pt 21):3865–3872
54. Richardson GD (2004) CD133, a novel marker for human prostate epithelial stem cells. J Cell Sci 117:3539–3545
55. Lapidot T, Sirard C, Vormoor J, Murdoch B, Hoang T, Caceres-Cortes J et al (1994) A cell initiating human acute myeloid leukemia after transplantation into SCID mice. Nature 367: 645–648
56. Podberezin M, Wen J, Chang C-C (2013) Cancer stem cells: a review of potential clinical applications. Arch Pathol Lab Med 137:1111–1116
57. Lopez J, Valdez-Morales FJ, Benitez-Bribiesca L, Cerbon M, Garcia Carranca A (2013) Normal and cancer stem cells of the human female reproductive system. Reprod Biol Endocrinol 11:53
58. Tomao F, Papa A, Rossi L, Strudel M, Vici MP, Tomao S (2013) Emerging role of cancer stem cells in the biology and treatment of ovarian cancer: basic knowledge and therapeutic possibilities for an innovative approach. J Exp Clin Cancer Res 32:48
59. Palk DY, Janzen DM, Schafenacker AM, Velasco VS, Sheng MS, Chung D et al (2012) Stem-like epithelial cells are concentrated in the distal end of the fallopian tube: a site for injury and serous cancer initiation. Stem Cells 30:2487–2497
60. Karst AM, Levanon K, Drapkin R (2011) Modeling high-grade serous ovarian carcinogenesis from the fallopian tube. Proc Natl Acad Sci U S A 108:7547–7552
61. Collins AT, Berry PA, Hyde C, Stower MJ, Maitland NJ (2005) Prospective identification of tumorigenic prostate cancer stem cells. Cancer Res 65:10946–10951
62. Wang G, Wang Z, Sarkar FH, Wei W (2012) Targeting prostate cancer stem cells for cancer therapy. Discov Med 13:135–142
63. Clark AT (2007) The stem cell identity of testicular cancer. Stem Cell Rev 3:49–59
64. Costa FF, Le Blanc K, Brodin B (2007) Concise review: cancer/testis antigens, stem cells and cancer. Stem Cells 25:707–711

Stem Cells of Human Endometrium: Trash to Treasure

Indumathi Somasundaram, Dhanasekaran Marappagounder, Vijayalakshmi Venkatesan, Padmanav Behera, and Ramesh R. Bhonde

Background

Reproduction can be defined as the process by which an organism continues its species. The development of the normal female reproductive tract is a complex process. The paramesonephric ducts arise from the intermediate mesoderm, which are the precursors of the female reproductive organs that include uterus, fallopian tubes, cervix, and upper vagina [1]. The female reproductive system is designed to carry out several functions. It produces the female egg cells necessary for reproduction, called the ova or oocytes. The system is designed to transport the ova to the site of fertilization. Conception, the fertilization of an egg by a sperm, normally occurs in the fallopian tubes. The fertilized egg then gets implanted into the walls of the uterus, beginning the initial stages of pregnancy. If fertilization and/or implantation do not take place, the system is designed to menstruate (the monthly shedding of the uterine lining). In addition, the female reproductive system produces female sex hormones that maintain the reproductive cycle. Fallopian tubes are the passageways that egg cells enter after release from the ovaries. The Fallopian tubes lead to the uterus a muscular organ in the pelvic cavity. The inner lining, called the *endometrium*, thickens with blood and tissue in anticipation of a fertilized egg cell. If fertilization fails to occur, the endometrium degenerates and is shed in the process of menstruation. Based on its dynamic tissue remodeling during the menstrual cycle and pregnancy, it has been suggested that stem cells of the endometrium must possess a high regenerative potential [2, 3]. In this regard, the chapter provides an overview on endometrial stem cells with a special emphasis on its proliferation and multilineage differentiation potentials coupled with its in vivo therapeutic applications.

Human Endometrium and Its Stem Cell Derivatives

The uterus is the largest female reproductive organ that plays a pivotal role in implantation and in the absence of pregnancy, menstruation.

I. Somasundaram (✉)
Assistant Professor, Department of Stem Cell and Regenerative Medicine, Centre for Interdisciplinary Research, D.Y. Patil University, Kolhapur, India
e-mail: indu.stemcell@gmail.com

D. Marappagounder
Stem Cell Banking & Research, Ree Laboratories Private Limited, Mumbai 400053, India
e-mail: dhanasekarbio@gmail.com

V. Venkatesan • P. Behera
Department of Stem Cell Research, National Institute of Nutrition (ICMR), Secunderabad, Hyderabad 500007, India
e-mail: v.venkateshan@gmail.com; saipadmanav@gmail.com

R.R. Bhonde
School of Regenerative Medicine, Manipal Institute of Regenerative Medicine, Manipal University, GKVK Post, Bellary Road, Allalasandra, Near Hotel Royal Orchid, Yelahanka, Bangalore 560065, India
e-mail: rr.bhonde@manipal.edu

I. Somasundaram (ed.), *Stem Cell Therapy for Organ Failure*,
DOI 10.1007/978-81-322-2110-4_19, © Springer India 2014

The innermost layer of the uterus is known as tunica mucosa, popularly termed as endometrium, opposed to the outer perimetrium and median myometrium. The uterus is the only organ whose lining is almost entirely expelled and reconstructed periodically, both phenomena taking place at each ovarian cycle. With the purpose of facilitating the periodic elimination of the endometrium that undergoes regression, shrinkage, and necrosis at the end of each cycle, the uterus also exhibits the unique peculiarity of physiological bleeding. The endometrium begins to reach full development at puberty and thereafter exhibits dramatic changes during each menstrual cycle. It undergoes further changes before, during, and after pregnancy, during the menopause, and in old age [3].

The endometrium is a simple columnar epithelium. It is divided into two zones: the inner functionalis which is adjacent to the uterine cavity and a deeper basalis layer which overlies the myometrium. The functionalis layer is shed each month with menstruation and is then regenerated from the basalis layer. The functionalis, comprising the upper two-third of the endometrium, is divided into stratum compactum and stratum spongiosum. The stratum compactum is a superficial thin layer nearest to the uterine cavity and contains the lining cells, necks of the uterine gland, and relatively dense stroma. The stratum spongiosum is the deeper part of functionalis composed of main portions of the uterine glands and accompanying blood vessels; the stromal cells are more loosely arranged and larger than in the stratum compactum. The lower basalis contains the basal region of the uterine glands, dense stroma (that remains relatively unaltered during the menstrual cycle), large blood vessel remains, and lymphoid aggregates. It serves as the germinal compartment for generating new functionalis each month [3–5]. It has been postulated that the niche of these adult stem or progenitor cells of the endometrium is the lower basalis. These stem or progenitor cells were also identified to be in the trophic endometrium of postmenopausal women [7, 8].

Accumulating evidence from the literature on the existence of epithelial and stromal/stem cells in endometrial tissue [6–9] has substantiated that it possesses a remarkable capacity for regeneration. Although endometrial tissue-derived stem cells are being reported, menstrual blood has become the most convenient source in the search for endometrial stem cells because collecting menstrual blood is easy and noninvasive and endometrial stem/progenitor cells are shed in menstrual blood [10–13]. For these reasons, reliable studies on menstrual blood-derived stem cells are in process. Furthermore, menstrual blood-derived stem cells demonstrate great promise for use in tissue repair and treatment of diseases, due to the plasticity and longevity of the cells. Menstrual blood has proven to be a unique and novel source of stromal stem cells from the endometrial functionalis. They have been identified through several in vitro and in vivo studies [12].

However, putative adult stem or progenitor cells that are responsible for the cyclical regeneration of the endometrial functionalis, every month, are thought to reside in the basalis region of the endometrium, as described earlier [4–8]. The study of these stem cells from the basalis is still in its infancy. Based on the dynamic tissue remodeling of the uterus, it has been suggested that adult stem cells must play a role in uterine tissue maintenance and function. Hence, a thorough characterization of the uterine/endometrial stem cells derived from the endometrial tissue biopsy of the inner lining of the uterus or from the intact uterus surgically removed in the treatment is equally important as that of studies on menstrual blood stem cells. Once a mechanical or functional characteristic platform has been constructed, it then becomes easier to understand the complex mechanisms underlying the morphogenesis and physiological generation of the female reproductive tract.

Marker Profiles

The concept that endometrial regeneration is mediated by endometrial stem/progenitor cells was proposed many years ago [5, 14]. Since then many evidences do arise for the existence of stem cells from endometrium. The first published evidence for the existence of adult stem/progenitor cells in the human uterus identified in the endometrium is the clonogenic epithelial and stromal cells suggesting the presence of two types of adult stem/progenitor cells [15, 16].

Schwab and Gargett demonstrated the existence of endometrial stem cell identification through the characterization of perivascular markers CD146 and PDGF-Rβ. This study also reveals that CD146 and PDGF-R β cells were co-localized in both the functional and basal layer [17, 18]. They demonstrated that these perivascular markers enabled isolation of stromal cells from human endometrium which exhibit phenotypic and functional properties of MSC. They hypothesized that these endometrial MSC like cells may contribute to the cyclic regeneration of the endometrium and might play a vital role in cell-based therapies.

Various studies have examined the expression of stem cell markers in the human endometrium including our earlier published data [9, 19–22]. Dimitrov and his coworkers [3] analyzed the cell surface markers for the cultured endometrial stem cells derived from both functionalis and basalis layer of the endometrium for phenotypic expression. Hematopoietic stem cell markers including CD45, CD14, CD19, CD56/16, CD34, and CD3 showed a negative expression, whereas markers like CD29, CD73, and CD90 were stained positive, strongly suggesting the mesenchymal nature of the cells. Despite these citations, the identifications of biomarkers of endometrium are uncertain and hence research is underway. The discovery of those markers highlights the importance of the stem cell system in human reproduction and also demonstrates its therapeutic implications.

Besides the MSC-specific markers, Oct-4, a transcription factor and marker of human embryonic stem cells, and more recently of adult stem cells, is expressed in almost half of tested endometrial samples [9, 23]. The expression of OCT-4 suggests the existence of endometrial stem cells, lending further support to the hypothesis of endometrial regeneration by local stem cells in endometrial tissue. More Oct-4+ cells are observed during the proliferative stage; however, the identity and location of the Oct-4+ cells have not been reported [19]. Several general adult stem cell markers, including bcl-2, c-kit (CD117), and CD34, have also been identified in endometrial tissues [9]. The importance of these markers, however, cannot be determined since they are expressed in many more endometrial cells than the numbers of clonogenic or side population cells identified in functional studies [16, 20]. Apart from the markers specified above, our team could identify multitude of markers that are specific to heterogenous endometrial cells [21]. The wide specificity toward multitude of marker characteristics similar to bone marrow stem cells favors its application in therapeutics.

Multi-differentiation Potential

Extensive literatures exist to support endometrial stem cells as an additional source of curative stem cell therapeutics. To quote a few, its dynamics in coordinated functions of proliferation, differentiation, and menstrual shedding has been reported [5, 23]. Endometrial stem cells, unlike other postnatal adult stem cells, could retain its stemness and multi-differentiation potential even under extensive culture condition [unpublished data]. Several researchers demonstrated the ability of CD146+PDGFRb+ MSC-like cells [18] or clonogenic human endometrial stromal cells [2, 3] to differentiate into mesodermal origin such as adipocytes, osteocytes, smooth muscle cells, and chondrocytes [2, 3, 16, 24]. Masuda et al. demonstrated that the endometrial tissue-reconstituting cells also possess the ability to differentiate into endothelial cells [25]. It has been demonstrated that not only the endometrial stem cells but also the SP cells of the endometrium have the ability to differentiate into endothelial and smooth muscle cells [20, 26, 27]. It was reported that endometrial stem cells could effectively differentiate into muscular cells of urinary bladder using myogenic growth factors, thereby making endometrial stem cells ideal for bladder cell replacement therapies [24]. Besides, endometrial stem cells have also been differentiated into megakaryocytes that produced functional platelets, thereby demonstrating its wide plasticity [28].

It is a well-known fact that angiogenesis plays a key role in the reproductive processes such as embryo implantation and endometrial regeneration after menstruation. Evidence is reviewed for the hypothesis that the endometrium in women has a high capacity of cell proliferation and angiogenesis [29, 30]. Evidences show the angiogenic phenotype in human endometrium by their establishment of their ability to implant [31]. 3D cultures of human endometrial cells demonstrate its high capacity of cell proliferation and angiogenesis [32].

With its built-in angiogenic role throughout reproductive phase of women coupled with proven experimental records, it is clear that endometrial stem cells could serve a better tool for treating vascular disorders. Apart from its mesodermal and ectodermal differentiation potency, endometrial cells were also shown to possess endodermal differentiation via differentiating into insulin-producing cells [33]. The differentiated islets that form endometrium were able to produce insulin both in vitro and in vivo in a murine model [34], thus serving its further applicability in treating diabetes.

In Vivo Applications of Endometrial Stem Cells

Endometrial stromal cells possess wide range of advantages as opposed to other postnatal stem cells, to prove themselves as a valuable tool in cell-based therapies. They are as follows—easy to isolate, high accessibility, trash source, immunogenic, longer preservation, highly clonogenic with a higher multi-differentiation, and angiogenic potential—thereby serving as a better autologous/allogenic therapeutic tool in regenerative medicine. This is evidenced by several preclinical and clinical trials on endometrial stem cells in autologous/ allogenic transplantations. Some of its in vivo preclinical and clinical applications are discussed below.

With the dynamic cyclical regenerative and angiogenic potency of endometrial stem cells, there are proven record tracks on its efficacy in treating vascular disorders. The applications of bone marrow-derived cells for heart failures and its related diseases are enticing. This is due to its directed cardiomyocyte differentiation [35] and its ability to secrete angiogenic and trophic factors. However, angiogenic potency of bone marrow in patients with coronary artery disease is impaired, in part due to its deficiencies in the CXCR4 migration activity [36]. The relationship between angiogenesis and post myocardial infarct healing is well known. Endometrial stem cells could outweigh this obstacle and many other such hindrances, thereby entering the clinic in an efficient manner.

As stated above, endometrium undergoes rapid angiogenesis in a controlled manner every month. With this built-in potency, upregulated production of angiogenic factors including PDGF, EGF, and VEGF has been described both in the mouse and human endometrium [29, 30, 37]. Besides, administration of endometrial regenerative cells into a post myocardial infarct model showed recovery as compared to bone marrow cells. Furthermore, the cells were capable of functionally integrating with existing cardiomyocytes and exerted effects through direct differentiation [13]. The possibility of using endometrial cells for treating critical limb ischemia has been demonstrated in mouse model due to its high levels of growth factors and MMPs. Besides, as it possesses superior immunomodulatory potential, its off-the-shelf allogenic therapeutic application is well demonstrated [38]. The first report of clinical use of ERC involved four patients with multiple sclerosis who received both intrathecal and intravenous injections. No adverse events were reported at the time of last follow-up [38]. Similar lines of clinical trial were also demonstrated with muscular dystrophies and heart failures [39–41]. Studies reported no adverse events at time of last follow-ups of these cases. Besides these applications, insulin-secreting capacity of endometrial cells to functionally recover insulin inefficiency in vivo has also been demonstrated. Preclinical murine model system treated for diabetes has restored its normoglycemic condition and produced insulin [31, 32].

Conclusion

Undoubtedly, it can be concluded that endometrial stem cells may become key players in treating various disorders because of their noninvasive mode of collection, ease of isolation, its enhanced clonogenecity, and multi-differentiation potentials. Furthermore, its off-the-shelf storage capacity and superior immunomodulatory property allow greater applicability for allogenic cell therapeutics. However, further investigations are warranted on use of endometrial tissue in cell-based therapies. Besides, studies on wide characteristics of putative endometrial stem cells that contribute to gynecological disorders might explore its applicability in targeted cell-based therapies for such reproductive disorders.

References

1. Wray S (2007) Insights into the uterus. Exp Physiol 92(4):621
2. Gargett CE, Schwab KE, Zillwood RM et al (2009) Isolation and culture of epithelial progenitors and mesenchymal stem cells from human endometrium. Biol Reprod 80:1136–1145
3. Dimitrov R, Timeva T, Kyurchiev D (2008) Characterization of clonogenic stromal cells isolated from human endometrium. Reproduction 135: 551–558
4. Spencer TE, Hayashi K, Hu J et al (2005) Comparative developmental biology of the mammalian uterus. Curr Top Dev Biol 68:85–122 (26)
5. (Padykula HA 1991). Padykula HA, Coles LG, McCracken JA et al (1984) A zonal pattern of cell proliferation and differentiation in the rhesus endometrium during the estrogen surge. Biol Reprod 31(5):1103–1118. (27)
6. Gargett CE, Chan RWS (2006) Label retaining cells in Estrogen- induced endometrial Regeneration. In: 4th International Society for Stem Cell Research, Toronto, Canada (28)
7. Gargett CE (2007) Uterine stem cells: what is the evidence? Hum Reprod Update 13(1):87–101. (29)
8. Marin Figueria PG, Abrao MS, Krikun G et al (2011) Stem cells in endometrium and pathogenesis of endometrium. Ann N Y Acad Sci 1221(1):10–17 (30)
9. Cho NH, Park YK, Kim YT, Yang H, Kim SK et al (2004) Lifetime expression of stem cell markers in the uterine endometrium. Fertil Steril 81:403–407 (31)
10. Cui CH, Uyama T, Miyado K et al (2007) Menstrual Blood-derived cells confer human dystrophin expression in the murine model of Duchenne molecular dystrophy via cell fusion and myogenic transdifferentiation. Mol Biol Cell 18:1586–1594. (32)
11. Xiaolong M, Ichim TE, Zhong J et al (2007) Endometrial regenerative cells: a novel stem cell population. J Transl Med 5:57. 33
12. Patel AN, Park E, Kuzman M et al (2008) Multipotent menstrual blood stromal stem cells: isolation, characterization, and differentiation. Cell Transplant 17:303–311 34
13. Hida M, Nishiyama N, Miyoshi S et al (2008) Novel cardiac precursor- like cells from human menstrual blood- derived mesenchymal cells. Stem Cells 26:1695–1704. 35
14. Prianishnikov VA (1978) On the concept of stem cell and a model of functional-morphological structure of the endometrium. Contraception 18(3):213–223
15. Gargett CE (2004) Stem cells in gynaecology. Aust N Z J Obstet Gynaecol 44:380–386. 36
16. Chan RWS, Schwab KE, Gargett CE (2004) Clonogenecity of human endometrial epithelial and stromal cells. Biol Reprod 70(6):1738–1750 37
17. Gargett CE, Chan RWS, Schwab KE (2007) Endometrial stem cells. Reprod Endocrinol 19(4):377–383
18. Schwab KE, Gargett CE (2007) Co-expression of two perivascular cell markers isolates mesenchymal stem-like cells from human endometrium. Hum Reprod 22:2903–2911 48
19. Matthai C, Horvat R, Noe M et al (2006) Oct-4 expression in human endometrium. Mol Hum Reprod 12(1):7–10
20. Kato K, Yoshimoto M, Kato K et al (2007) Characterization of side population cells in human normal endometrium. Hum Reprod 22:1214–1223
21. Indumathi S, Harikrishnan R, Rajkumar JS, Sudarsanam D, Dhanasekaran M (2013) Prospective biomarkers of stem cells of human endometrium and fallopian tube in comparison to bone marrow. Cell Tissue Res 352(3):537–549
22. Schwab KE, Hutchinson P, Gargett CE (2008) Identification of surface markers for prospective isolation of human endometrial stromal colony forming cells. Hum Reprod 23:934–943 49
23. Cervello I (2007) Identification, characterization and co-localization of label-retaining cell population in mouse endometrium with typical undifferentiated markers. Hum Reprod 22(1):45–51
24. Shoae-Hassani A et al (2013) Endometrial stem cell differentiation into smooth muscle cell: a novel approach for bladder tissue engineering in women. BJU Int 112:854–863

25. Masuda H, Maruyana T, Hiratsu E et al (2007) Non-invasive and real time assessment of reconstructed functional human endometrium in NOD/SCID/γc null immunodeficient mice. Proc Natl Acad Sci U S A 104(6):1925–1930 57
26. Tsuji S, Yoshimoto M, Kato K et al (2008) Side population cells contribute to the genesis of human endometrium. Fertil Steril 90:1528–1537. 42
27. Masuda H, Matsuzaki Y, Hiratsu E et al (2010) Stem cell- like properties of the endometrial side population: implication in endometrial regeneration. PLoS One 5(4):43
28. Wang J, Chen S, Zhang C, Stegeman S, Pfaff-Amesse T et al (2012) Human endometrial stromal stem cells differentiate into megakaryocytes with the ability to produce functional platelets. PLoS One 7(8):e44300. doi:10.1371/journal.pone.0044300
29. McLaren J (2000) Vascular endothelial growth factor and endometriotic angiogenesis. Hum Reprod Update 6:45–55. doi:10.1093/humupd/6.1.45 [10711829]
30. Girling JE, Rogers PA (2005) Recent advances in endometrial angiogenesis research. Angiogenesis 8:89–99
31. Lebovic DI, Bentzien F, Chao VA, Garrett EN, Meng YG, Taylor RN (2000) Induction of an angiogenic phenotype in endometriotic stromal cell cultures by interleukin-1beta. Mol Hum Reprod 6:269–275. doi:10.1093/molehr/6.3.269 [1069 4276]
32. Esfandiari N, Khazaei M, Ai J, Nazemian Z, Jolly A, Casper RF (2008) Angiogenesis following three-dimensional culture of isolated human endometrial stromal cells. Iran J Fertil Steril 2:19–22
33. Li H-Y et al (2010) Induction of insulin-producing cells derived from endometrial mesenchymal stem-like cells. JPET 335:817–829
34. Santamaria X et al (2011) Derivation of insulin producing cells from human endometrial stromal stem cells and use in the treatment of murine diabetes. Mol Ther 19(11):2065–2071
35. Hatzistergos KE et al (2010) Bone marrow mesenchymal stem cells stimulate cardiac stem cell proliferation and differentiation. Circ Res 107(7):913–922
36. Walter DH et al (2005) Impaired CXCR4 signaling contributes to the reduced neovascularization capacity of endothelial progenitor cells from patients with coronary artery disease. Circ Res 97(11): 1142–1151
37. Chegini N, Rossi MJ, Masterson BJ (1992) Platelet-derived growth factor (PDGF), epidermal growth factor (EGF), and EGF and PDGF beta-receptors in human endometrial tissue: localization and in vitro action. Endocrinology 130:2373–2385
38. Murphy MP et al (2008) Allogeneic endometrial regenerative cells: an "Off the shelf solution" for critical limb ischemia? J Transl Med 6:45
39. Zhong Z et al (2009) Feasibility investigation of allogeneic endometrial regenerative cells. J Transl Med 7:15
40. Ichim TE et al (2010) Mesenchymal stem cells as anti-inflammatories: implications for treatment of Duchenne muscular dystrophy. Cell Immunol 260(2):75–82
41. Ichim TE et al (2010) Combination stem cell therapy for heart failure. Int Arch Med 3(1):5

Stem Cell Therapy in Premature Ovarian Failure

Susinder Sundaram, Paramasivam Nithyanand, Shaik Mohammad Naushad, Nikhita Kutala, and Vijay Kumar Kutala

Premature ovarian failure (POF) occurs in women under the age of 40 years with primary or secondary hypergonadotropic amenorrhea and is accompanied by oestrogen deficiency in 75 % of cases. None of the women with primary amenorrhea have been reported to ovulate or conceive with their own oocytes, but more than a third of the women were pregnant at least once before developing hypergonadotropic POF.

Causes for POF

It is speculated that lack of follicular renewal may be caused by age-associated exhaustion of specific memory cells in the lymphoid system and bone marrow that are necessary to generate effector cells migrating to ovaries and stimulating transformation of the ovarian surface epithelium (OSE) cells into primitive granulose and germ cells. POF may be caused due to the following causes:

1. Delayed ovarian development during immunological adaptation
2. Earlier termination of immunological adaptation
3. Cytostatic chemotherapy affecting both the existing pool of primary follicles and committed ovary-specific mesenchymal cells (O-SMCs) required for the follicular renewal [1]

In hypergonadotropic amenorrhea, ovaries contain normal primary or even antral follicles not responding to gonadotropins by production of oestrogens [1].

Types of POF

Studies with laboratory animal models have shown that there are two types of POF with primary and antral follicles within the ovaries. The first persistent ovarian immaturity can be induced by inhibition of ovarian development (i.e. inhibition of androgen receptor expression with oestrogens) during immunological adaptation. The second premature ovarian ageing can be induced by the acceleration of ovarian development with androgens [1, 2].

S. Sundaram • P. Nithyanand • S.M. Naushad
School of Chemical & Biotechnology, SASTRA University, Tirumalaisamudram, Thanjavur, Tamil Nadu, India
e-mail: ssusinder@gmail.com; pnithyanand@gmail.com; naushadsm@gmail.com

N. Kutala
SVS Institute of Dental Sciences, Mahabubnagar, AP, India
e-mail: nikhitakutala@yahoo.co.in

V.K. Kutala (✉)
Department of Clinical Pharmacology and Therapeutics, Nizam's Institute of Medical Sciences, Panjagutta, Hyderabad 500082, India
e-mail: vijaykutala@gmail.com

I. Somasundaram (ed.), *Stem Cell Therapy for Organ Failure*,
DOI 10.1007/978-81-322-2110-4_20, © Springer India 2014

Ovarian Surface Epithelium (OSE) Cells

Experiments done with cultures of OSE produced cells with oocyte-like characteristics in 4–5 days [3]. A well-defined nuclear envelope of the oocyte was seen, and isolated oocyte-type cells exhibited germinal vesicle breakdown. Some of the cells showed characteristics of secondary oocytes with surface zona pellucida (ZP) expression, expulsion of polar body and poor nuclear/cytoplasmic separation [3]. The ability of OSE cultures to produce new oocytes from ovaries lacking follicular renewal suggests that OSE cells retain the potential to differentiate but fail to do so in vivo. As the putative stimulatory potential of O-SMCs is absent, uncommitted mesenchymal cells (u-MSCs) may exhibit a suppressive effect. For example: CD8 T-lymphocytes that are the predominant type of T-cells present in the epithelial and other non-lymphoid tissues [4] can function as effector or suppressor cells [5]. Marked changes in circulating subsets of T-cells are described in postmenopausal women and women with POF [6].

In order to test the effect of female germ line stem cells, these cells were isolated and cultured for more than 6 months. These cells were later transfected with a GFP virus and transplanted into ovaries of infertile mice that were sterilised by treatment with cyclophosphamide and busulfan to destroy the existing pre- and postmitotic cells [7–9]. Subsequently, transplanted cells underwent oogenesis and the mice produced offspring that expressed a GFP transgene [10].

It is thought that production of oocytes ceases before birth in most mammalian species. At 5 months of foetal age, the ovary contains the maximum population of germ cells. Between the sixth and seventh month of intrauterine life, the last oogonia enter meiosis and the formation of new follicles is terminated. But it has been seen that Drosophila melanogaster has an ability to produce a large number of eggs throughout its life due to the presence of permanent germ line stem cells [11]. Similarly, germ line stem cells of all stages of oogenesis have been observed in adult teleost fish [12].

Premeiotic Germ Line Cells Facilitate Oogenesis

The in vivo studies have shown that gonads of female mice may have regenerative capacity during juvenile and adult stages of life [7]. In yet another study, it was seen that injection of genetically modified GFP expressing primordial germ cells into ovaries of sterilised mice leads to the birth of pups expressing GFP [9]. A culture of female germ line cells from neonatal mice and adult mice showed the ability to produce normal oocytes as well as offspring after transplantation into ovaries [9]. Transgenic mouse lines with GFP expressing germ line cells were used to isolate OSCs for in vitro culture. The mouse OSCs spontaneously formed oocyte-like structures, and when these structures are combined with granulose cells of neonatal mouse ovaries, structures closely resembling primordial follicles were formed [13]. All the above findings supported the fact that mammalian ovaries contain a population of premeiotic germ line cells that can support postnatal oogenesis.

Experimental studies have shown an association between the mtDNA copy number and the ability of oocytes to become fertilised [14]. Oocyte of women with ovarian insufficiency has been found to have a lower mtDNA copy number in comparison to that in women with normal ovarian profile [15]. Further, studies in mouse models have shown that dramatic reduction in mitochondrial DNA did not compromise ovulation and fertilisation, but postimplantation development was impeded when mtDNA copy number was less than 50,000 [16].

Injecting 5–15 % ooplasm from a young fertile donor oocyte into a putative defective recipient oocyte at the time of intracytoplasmic sperm injection has resulted in the birth of several children worldwide [17–19]. However, children born after cytoplasmic transfer have demonstrated persistent mitochondrial heteroplasmy [20], and mouse model of neutral heteroplasmy has displayed a phenotype which is consistent with the symptoms of early onset metabolic syndrome [21]. Cortical tissues obtained from 20- to 30-year-old women and OSCs isolated using fluorescence-activated cell sorting method spontaneously generated immature oocytes in vitro. Co-culture of GFP-positive humans' OSCs with adult human ovarian cortex for a period of 72 h leads to the formation of large ovoid GFP-positive cells surrounded by somatic cells resembling immature follicles. These human OSCs were injected into human ovarian cortical biopsies and xenotransplanted in adult female mice. Few weeks after the transplantation of immature primordial, follicles containing GFP-positive oocytes were found [22]. However, the use of ovarian stem cells to produce normal fertilisable eggs in humans is an idea that remains to be tested.

Parthenogenesis

The parthenogenetic activation of oocytes has been tried by some groups. This method overcomes the ethical constraints associated with the use of embryonic stem cells (ESCs) from human embryos [23]. The parthenogenetic activation of oocytes may be used to generate isogenic cell lines relative to the donors and/or autologous transplantation [24]. Methods like electrical stimulation of oocytes, exposure to $Ca^{2+} - Mg^{2+}$-free medium, and exposure to medium containing hyaluronidase, Ca^{2+} ionophore chelators, or inhibitors of protein synthesis such as cycloheximide are used for parthenogenetic activation [25]. A method pioneered by Taupin has a higher efficiency of generating parthenogenetic oocytes than the above mentioned methods. Oocytes were obtained from voluntary donors and involved standard hormonal stimulation for superovulation. Human oocytes were parthenogenetically activated by cultivating the oocytes with an ionophore at high oxygen tension and a serine threonine kinase inhibitor at low oxygen tension until the blastocyst formation. The blastocysts were transferred to feeder layers and cultured under high oxygen tension. However, diploid parthenogenetic embryos lack extraembryonic tissues and develop only till late somite stages in mice models [26], and human oocytes when parthenogenetically activated have also been shown to have a limited developmental potential [27]. The limited developmental potential is due to lack of/or overexpression of imprinted genes that are expressed only from one parental allele [28]. If the hurdles associated with the development of such embryos can be addressed then, this method could be used as a treatment option for infertility.

Treatment Strategies

Bone Marrow Transplantation for the Treatment of Infertility

It has been observed that gonadotropin-releasing hormone (GnRH) agonist minimises and prevents gonadotoxic effects of chemotherapy in humans. Patients with GnRH-a/chemotherapy co-treatment resumed spontaneous ovulation and menses or conception [29, 30]. Studies have shown that infertility can be treated using bone marrow stem cells. Bone marrow transplantation restored oocyte production in wild-type mice sterilised by chemotherapy as well as in ataxia telangiectasia-mutated (ATM) gene-deficient mice which are otherwise incapable of making oocytes [8]. In a different study by the same group, it was reported that bone marrow transplantation generated immature oocytes and restored long-term fertility in a mouse model of chemotherapy-induced premature ovarian failure [31].

Potential of Human Amniotic Epithelial Cells in Treating Infertility

Fangyuang Wang et al. [32] have shown that cultured human amniotic epithelial cells (hAECs) expressed markers like OCT4 and NANOG. The cells also expressed markers of surface antigens like c-KIT. Wild-type female mice were sterilised by pretreatment with cyclophosphamide or busulfan to destroy the pre- and post meiotic germ cell pools of the mice that were to be used as recipients of hAECs. The hAECs were then infected with GFP lentivirus. The postinfection and cultured hAECs were transplanted by tail vein injection into recipient females 7 days after chemotherapy. In the analysis of ovaries 14–21 days post hAEC transplantation, it was seen that they possessed several primary and developing follicles in which GFP-positive follicle cells were found. The ovarian follicles were double positive for GFP and human nuclei antigen expression. Anti-müllerian hormone (AMH) levels are strongly correlated with the follicular pool and are used as a marker of ovarian ageing, ovarian responsiveness and pathophysiology [33]. The AMH expression was not observed in ovaries within 7 days post hAEC transplantation, but was seen in ovarian follicles 14, 21 and 60 days post hAEC transplantation. The hAECs which do not have any karyotypic abnormalities or transformation potential in vitro and no tumorigenic potential in vivo may have some advantages over adult bone marrow stem cells.

Role of Somatic Granulosa Cells in Treatment Outcomes

The in vitro differentiation of putative ESC-derived oocytes and follicle-like cells has been reported in mice [34, 35]. A recent work has shown that primordial germ cell-like cells (PGCLCs) derived from embryonic stem cells or induced pluripotent stem cells which resemble endogenous primordial germ cells in foetal gonads require interaction with developmentally matched embryonic ovary somatic cells to realise their full potential in vivo [36]. This study used fox12 as a marker which is expressed in the granulosa cells [37] and involved a fox12 promoter DsRed-driven reporter system. When ESCs generated from transgenic TgOG2 mice that differentiated into GFP-positive oocyte-like cells in vitro [34, 38] are transplanted into wild-type mouse ovaries, immature follicles containing GFP-positive oocytes can be detected [38]. Woods et al. [36] observed that only those cultures containing cells expressing DsRed by day 5 of differentiation eventually generated follicle-like structures suggesting that both granulose cell and germ cell need to be specified and be present in these cultures. DsRed-positive cells were observed to associate with oocytes within a week after injection and become incorporated exclusively localised into granulose cell layer of immature growing follicles 2 weeks after transplantation [36].

Adipocyte-Derived Stem Cells in Treatment of Premature Ovarian Failure

To study the effects of adipocyte-derived stem cells in the treatment POF, CB57/L6 mice were administered cyclophosphamide (15 mg/kg) continuously for 15 consecutive days via intraperitoneal injections. Subcutaneous adipose tissue was derived from inguinal region of mice [39]. POF mice were injected intravenously with 1×10^6 autologous ADSCs. This was later followed by injection of 1×10^5 ADSCs in bilateral ovaries in the POF mice with a microinjector device [40]. One week after the transplantation, the number of follicles in the female mice that received ADSCs via intravenous and in situ routes was increased. Several hundred follicles were seen at all stages of development. In Contrast, the control POF mice (without ADSC therapy) did not recover to the normal level within 1 month, thus suggesting that intravenous and in situ ADSC therapy does significantly improve the ovary injury caused by the chemical treatment [40]. In addition to this, quantification of TUNEL-positive cells in control groups and 1-week-old ADSC transplantation group showed that the number of apoptotic granulosa cells in the intravenous ($42.8 \pm 13.7/\text{mm}^2$) and in situ ($50.4 \pm 13.1/\text{mm}^2$) ADSC-treated groups was lower than that of control POF group ($150.8 \pm 48.4/\text{mm}^2$) indicating that both intravenous and in situ methods rescue apoptosis of granulosa cells in ovary sections [40].

In Vivo Studies with iPSCs and miRNAs

In yet another study induced pluripotent stem cells (iPSCs) were generated using CD34+ human amniotic fluid cells using the method mentioned in Liu et al. [41]. This method also involved treatment of iPSCs with miR-17-3p, a micro RNA that suppresses vimentin expression. MiR-17-3p was found to bind to vimentin mRNA 3′ UTR. The downregulation of vimentin expression with miR-17-3p was to limit the differentiation of iPSCs to epithelial cells. It was seen that transfection of miR-17-3p in iPSCs leads to the expression of epithelial markers like cytokeratin7, AE3, E-cadherin, and the oestrogen receptor ERβ. The levels of these markers were higher than miR-mutant-transfected group or miR-mutant-untransfected control group. It was also observed that expression levels of vimentin are reduced due to miR-17-3p-transfected iPS group cells after differentiation. This result suggested that when endogenous levels of vimentin are reduced by miR-17-3p, the iPS cells can differentiate into oestrogen-sensitive OSE-like cells. To create a POF model, mouse was given a single intraperitoneal injection of cyclophosphamide (70 mg/kg) at 6 weeks of age. To investigate the effect of transplanting OSE-like cells derived from iPS cells, red fluorescent label iPS cell groups (miR-17-3p-transfected and miR-mutant-transfected groups) and PBS vehicle control were injected into the ovaries of POF mice. Fourteen days after injection, the ovarian tissues were analysed by haematoxylin and eosin (H&E) staining. H & E staining of the OSE-like cell (miR-17-3p-transfected iPS group) transplantation group showed a significant reduction in the number of atretic follicles and an increase in the number of mature follicles. In contrast, analysis of cells from miR-mutant-transfected cell group and vehicle group showed a larger number of atretic follicles in the ovarian tissues. The ovarian weight of POF mice transplanted with OSE-like cells was also increased in comparison to other two transplantation groups.

Application of Stem Cell Therapy to Treat Infertility in Humans

Although it has been possible to develop parthenogenetic embryos with limited developmental potential in mice and human models, many aspects and factors involved in genomic imprinting are not yet known. However, providing growth factors like fibroblast growth factor 2 (FGF2), insulin-like growth factor 2 (IGF2) and transforming growth factor-α (TGF-α) to a developing parthenogenetic embryo has increased the developmental potential of parthenogenetic embryos. IGF2 is maternally imprinted and has a role in regulating embryonic growth [42]. Parthenogenetic embryos when treated initially with FGF2 and explanted in culture for further development with IGF2 at the stage of 18–21 somites develop much longer (up to 50 somites) in comparison to embryos not treated with IGF2 [43] and provide the hope that it is just a matter of time before going into large-scale human clinical trials and implementation of the technology for infertility treatment. It was observed that in the presence of TGF-α in the nutrient medium with PG embryos, the number of somites developed in utero to the stage of 30–45 had increased, while more than a third of the embryos had a well-formed placenta and resembled the placenta of 11-day-old normal fertilised mouse embryos [44].

Studies done by Woods et al. [36] had shown the primordial germ cell-like cells require developmentally matched ovary somatic cells to realise their full potential in vivo. This study in addition to work done by Lei et al. [45] shows that germ cell and granulosa precursor synchronisation during foetal life is required for normal folliculogenesis. In addition, these studies have identified factors like Fox12 and FIGalpha to play a role in normal folliculogenesis. These studies have laid the foundation to identify many other factors that may have a role in normal folliculogenesis. Such factors could have a pivotal role in future infertility treatments.

Application of human amniotic epithelial cells in infertile nude-mice has shown revival of folliculogenesis [32] in these models. Advantages like lack of karyotypic abnormalities and absence of tumorigenesis after in vivo transplantation could pave the way for human clinical trials using hEACs in the near future. To conclude, application of various types of stem cells in the treatment of premature ovarian failure has shown promising results in animal models. To address the side effects of this therapy such as development of heteroplasmy, premature metabolic syndrome, and the mitochondrial copy number variations, alternative therapeutic modalities that preserve genomic integrity should be developed. Future studies are warranted to explore the long-term effects of such a treatment in humans.

References

1. Bukovsky A (2006) Oogenesis from human somatic stem cells and a role of immune adaptation in premature ovarian failure. Curr Stem Cell Res Ther 1(3):289–303
2. Bukovsky A, Ayala ME, Dominguez R, Keenan JA, Wimalasena J, Elder RF, Caudle MR (2002) Changes of ovarian interstitial cell hormone receptors and behavior of resident mesenchymal cells in developing and adult rats with steroid-induced sterility. Steroids 67(3–4):277–289
3. Bukovsky A, Svetlikova M, Caudle MR (2005) Oogenesis in cultures derived from adult human ovaries. Reprod Biol Endocrinol 3:17
4. Beagley KW, Husband AJ (1998) Intraepithelial lymphocytes: origins, distribution, and function. Crit Rev Immunol 18:237–254
5. Doherty PC, Topham DJ, Tripp RA, Cardin RD, Brooks JW, Stevenson PG (1997) Effector CD4+ and CD8+ T-cell mechanisms in the control of respiratory virus infections. Immunol Rev 159:105–117
6. Giglio T, Imro MA, Filaci G et al (1994) Immune cell circulating subsets are affected by gonadal function. Life Sci 54:1305–1312
7. Johnson J, Canning J, Kaneko T, Pru JK, Tilly JL (2004) Germline stem cells and follicular renewal in the postnatal mammalian ovary. Nature 428(6979):145–150
8. Johnson J, Bagley J, Skaznik-Wikiel M, Lee HJ, Adams GB, Niikura Y, Tschudy KS, Tilly JC, Cortes ML, Forkert R, Spitzer T, Iacomini J, Scadden DT, Tilly JL (2005) Oocyte generation in adult mammalian ovaries by putative germ cells in bone marrow and peripheral blood. Cell 122(2):303–315
9. Shiromizu K, Thorgeirsson SS, Mattison DR (1984) Effect of cyclophosphamide on oocyte and follicle number in Sprague-Dawley rats, C57BL/6 N and DBA/2 N mice. Pediatr Pharmacol (New York) 4(4):213–221
10. Zou K, Yuan Z, Yang Z, Luo H, Sun K, Zhou L, Xiang J, Shi L, Yu Q, Zhang Y, Hou R, Wu J (2009) Production of offspring from a germline stem cell line derived from neonatal ovaries. Nat Cell Biol 11(5):631–636
11. Kirilly D, Xie T (2007) The Drosophila ovary: an active stem cell community. Cell Res 17(1):15–25
12. Nakamura S, Kobayashi K, Nishimura T, Tanaka M (2011) Ovarian germline stem cells in the teleost fish, medaka (Oryzias latipes). Int J Biol Sci 7(4):403–409
13. Pacchiarotti J, Maki C, Ramos T, Marh J, Howerton K, Wong J, Pham J, Anorve S, Chow YC, Izadyar F (1984) Differentiation potential of germ line stem cells derived from the postnatal mouse ovary. Pediatr Pharmacol (New York) 4(4):213–221
14. Reynier P, May-Panloup P, Chrétien MF, Morgan CJ, Jean M, Savagner F, Barrière P, Malthièry Y (2001) Mitochondrial DNA content affects the fertilizability of human oocytes. Mol Hum Reprod 7(5):425–429
15. May-Panloup P, Chrétien MF, Jacques C, Vasseur C, Malthièry Y, Reynier P (2005) Low oocyte mitochondrial DNA content in ovarian insufficiency. Hum Reprod 20(3):593–597
16. Larsson NG, Wang J, Wilhelmsson H, Oldfors A, Rustin P, Lewandoski M, Barsh GS, Clayton DA (1998) Mitochondrial transcription factor A is necessary for mtDNA maintenance and embryogenesis in mice. Nat Genet 18(3):231–236
17. Cohen J, Scott R, Alikani M, Schimmel T, Munné S, Levron J, Wu L, Brenner C, Warner C, Willadsen S (1998) Ooplasmic transfer in mature human oocytes. Mol Hum Reprod 4(3):269–280
18. Barritt JA, Brenner CA, Malter HE, Cohen J (2001) Mitochondria in human offspring derived from ooplasmic transplantation. Hum Reprod 16(3):513–516
19. Barritt J, Willadsen S, Brenner C, Cohen J (2001) Cytoplasmic transfer in assisted reproduction. Hum Reprod Update 7(4):428–435
20. Harvey AJ, Gibson TC, Quebedeaux TM, Brenner CA (2007) Impact of assisted reproductive technologies: a mitochondrial perspective of cytoplasmic transplantation. Curr Top Dev Biol 77:229–249
21. Acton BM, Lai I, Shang X, Jurisicova A, Casper RF (2007) Neutral mitochondrial heteroplasmy alters physiological function in mice. Biol Reprod 77(3):569–576
22. White YA, Woods DC, Takai Y, Ishihara O, Seki H, Tilly JL (2012) Oocyte formation by mitotically active germ cells purified from ovaries of reproductive-age women. Nat Med 18(3):413–421
23. Hao J, Zhu W, Sheng C et al (2009) Human parthenogenetic embryonic stem cells: one potential resource for cell therapy. Sci China C Life Sci 52:599–602

24. Taupin P (2006) Autologous transplantation in the central nervous system. Indian J Med Res 124:613–618
25. Pashaiasl M, Khodadadi K, Holland MK et al (2010) The efficient generation of cell lines from bovine parthenotes. Cell Reprogram 12:571–579
26. Surani MA, Barton SC (1983) Development of gynogenetic eggs in the mouse: implications for parthenogenetic embryos. Science 222(4627):1034–1036
27. Imesch P, Scheiner D, Xie M, Fink D, Macas E, Dubey R, Imthurn B (2013) Developmental potential of human oocytes matured in vitro followed by vitrification and activation. J Ovarian Res 6(1):30
28. McGrath J, Solter D (1984) Completion of mouse embryogenesis requires both the maternal and paternal genomes. Cell 37(1):179–183
29. Blumenfeld Z, Eckman A (2005) Preservation of fertility and ovarian function and minimization of chemotherapy-induced gonadotoxicity in young women by GnRH-a. J Natl Cancer Inst Monogr 34:40–43
30. Del Mastro L, Boni L, Michelotti A, Gamucci T, Olmeo N, Gori S, Giordano M, Garrone O, Pronzato P, Bighin C, Levaggi A, Giraudi S, Cresti N, Magnolfi E, Scotto T, Vecchio C, Venturini M (2011) Effect of the gonadotropin-releasing hormone analogue triptorelin on the occurrence of chemotherapy-induced early menopause in premenopausal women with breast cancer: a randomized trial. JAMA 306(3):269–276
31. Lee HJ, Selesniemi K, Niikura Y, Niikura T, Klein R, Dombkowski DM, Tilly JL (2007) Bone marrow transplantation generates immature oocytes and rescues long-term fertility in a preclinical mouse model of chemotherapy-induced premature ovarian failure. J Clin Oncol 25(22):3198–3204
32. Wang F, Wang L, Yao X, Lai D, Guo L (2013) Human amniotic epithelial cells can differentiate into granulosa cells and restore folliculogenesis in a mouse model of chemotherapy-induced premature ovarian failure. Stem Cell Res Ther 4(5):124
33. Visser JA, de Jong FH, Laven JS, Themmen AP (2006) Anti-Müllerian hormone: a new marker for ovarian function. Reproduction 131(1):1–9
34. Hübner K, Fuhrmann G, Christenson LK, Kehler J, Reinbold R, De La Fuente R, Wood J, Strauss JF 3rd, Boiani M, Schöler HR (2003) Derivation of oocytes from mouse embryonic stem cells. Science 300(5623):1251–1256
35. Qing T, Shi Y, Qin H, Ye X, Wei W, Liu H, Ding M, Deng H (2007) Induction of oocyte-like cells from mouse embryonic stem cells by co-culture with ovarian granulosa cells. Differentiation 75(10):902–911
36. Woods DC, White YA, Niikura Y, Kiatpongsan S, Lee HJ, Tilly JL (2013) Embryonic stem cell-derived granulosa cells participate in ovarian follicle formation in vitro and in vivo. Reprod Sci 20(5): 524–535
37. Schmidt D, Ovitt CE, Anlag K, Fehsenfeld S, Gredsted L, Treier AC, Treier M (2004) The murine winged-helix transcription factor Foxl2 is required for granulosa cell differentiation and ovary maintenance. Development 131(4):933–942
38. Nicholas CR, Haston KM, Grewall AK, Longacre TA, Reijo Pera RA (2009) Transplantation directs oocyte maturation from embryonic stem cells and provides a therapeutic strategy for female infertility. Hum Mol Genet 18(22):4376–4389
39. Gimble JM, Katz AJ, Bunnell BA (2007) Adipose-derived stem cells for regenerative medicine. Circ Res 100(9):1249–1260
40. Sun M, Wang S, Li Y, Yu L, Gu F, Wang C, Yao Y (2013) Adipose-derived stem cells improved mouse ovary function after chemotherapy-induced ovary failure. Stem Cell Res Ther 4(4):80
41. Liu T, Zou G, Gao Y, Zhao X, Wang H, Huang Q, Jiang L, Guo L, Cheng W (2012) High efficiency of reprogramming $CD34^{+}$ cells derived from human amniotic fluid into induced pluripotent stem cells with Oct4. Stem Cells Dev 21(12):2322–2332
42. Rappolee DA, Sturm KS, Behrendtsen O, Schultz GA, Pedersen RA, Werb Z (1992) Insulin-like growth factor II acts through an endogenous growth pathway regulated by imprinting in early mouse embryos. Genes Dev 6(6):939–952
43. Penkov LI, Platonov ES, New DA (2001) Effects of fibroblast growth factor 2 and insulin-like growth factor II on the development of parthenogenetic mouse embryos in vitro. In Vitro Cell Dev Biol Anim 37(7):440–444
44. Platonov ES (2005) Genomic imprinting and problem of parthenogenesis in mammals. Ontogenez 36(4):300–309
45. Lei L, Zhang H, Jin S, Wang F, Fu M, Wang H, Xia G (2006) Stage-specific germ-somatic cell interaction directs the primordial folliculogenesis in mouse fetal ovaries. J Cell Physiol 208(3):640–647

Breast Milk Cells: Bliss to Neonates

Indumathi Somasundaram, Namitha Haridoss, and Phillip G. Stubblefield

Introduction

Breast milk is a complex mixture of interacting compounds including carbohydrates, proteins, antibodies, vitamins, growth factors, hormones, cytokines, and several immunizing factors for the newborn. Breast milk also encompasses epithelial cells, colostral corpuscles, polymorphonuclear leukocytes, mononuclear phagocytes and lymphocytes, with those of epithelial lineage forming the main bulk of cells within 2 weeks of establishing lactation. These complex mixtures of interacting components of breast milk contribute to the beneficial effects of breast feeding. This may extend well beyond weaning and has been shown to prevent or mitigate several diseases later in life. The cellular constituents of breast milk and its significance to the neonate for growth, immunity, and regeneration have been highlighted.

I. Somasundaram (✉)
Assistant Professor, Department of Stem Cell and Regenerative Medicine, Centre for Interdisciplinary Research, D.Y. Patil University, Kolhapur, India
e-mail: indu.stemcell@gmail.com

N. Haridoss
Department of Stem Cells Research, National Institute of Nutrition (ICMR), Jamai Osmania, Tarnaka, Secunderabad, Hyderabad 500007, India
e-mail: haridasnamitha86@gmail.com

P.G. Stubblefield
Department of Obstetrics and Gynecology, School of Medicine, Boston University, Boston, MA, USA
e-mail: phillip.stubblefield@gmail.com

Breast and Its Components

The breast is an accessory sex organ of the female body that contains mammary glands, milk ducts, and adipose tissue. Mammary glands are unique to mammals, with the specific function of synthesizing, secreting, and delivering milk to the newborn upon demand for its optimal nourishment, protection, and development [1]. Mammary glands arise from the terminal end buds (TEB) which consist of two cell lineages: the cap cells that form a single outer layer and the body cells that make up multiple inner layers. The current dogma is that mammary stem cells exist in the TEBs and give rise to mammary epithelial cells [2]. The special function of the breast includes the production of milk for lactation. The mammary gland is composed of two types of epithelial cells. The inner or luminal cells, which are potential milk-secreting cells, are surrounded by an outer basal layer of contractile myoepithelial cells. The epithelial component of the tissue consists of lobules, where milk is made, which connect to ducts that lead out to the nipple. These lobules and ducts are located spread throughout the background fibrous tissue and adipose tissue that makes up the main mass of the breast. It has been hypothesized that the epithelial cells are shed from the ductal and luminal epithelial layers

I. Somasundaram (ed.), *Stem Cell Therapy for Organ Failure*, DOI 10.1007/978-81-322-2110-4_21, © Springer India 2014

through either a heightened turnover of the secretory tissue or as a consequence of the mechanical shear forces associated with the continued filling and emptying cycle associated with breast milk synthesis and lactation [3].

The milk is produced when the hormone called prolactin binds with the luminal epithelial cells. Oxytocin acts on the myoepithelial cells of the breast to produce milk ejection or "milk letdown." Breast milk is an undeniably unique, natural source of nutrition for the human infant. Human milk is a food that evolved to ensure optimal growth, development, and survival of human infants and young children. All female mammals are uniquely equipped to provide species-specific nourishment and immunity through the provision of milk to their newborns [4]. Thus breastfeeding is an important factor helping the newborn adapt to the environment.

Nutrients in Breast Milk

Breast milk contains the right balance of nutrients to help infant grow into a strong and healthy toddler. The beneficial effect of breastfeeding may extend well beyond weaning and has been shown to prevent or mitigate several diseases later in life [5] due to the presence of various nutrients and complex mixture of interacting compounds. The WHO recognizes breastfeeding as the normal way of providing nutrition to infants requiring special care. Colostrum is high in protein, fat-soluble vitamins, minerals, immunoglobulins and a strong antiviral property, strengthens the newborn's immune system, and acts as a laxative to remove meconium (first feces) from the digestive tract. Later colostrum will be replaced by transitional milk which appears creamy. The content of transitional milk includes high levels of fat, lactose, and water-soluble vitamins and contains more calories than colostrum. Mature milk begins to appear near the end of the second week after childbirth. Mature milk is produced in as great a volume as transitional milk but is thinner and more watery. 90 % is water, which is necessary to maintain hydration of the infant. The other 10 % is comprised of carbohydrates, proteins, and fats which are necessary for both growth and energy [6–8].

Cellular Constituents of Breast Milk

The immense potency of nutritional components of human breast milk and the importance of breastfeeding is known worldwide. It is imperative to explore the cellular constituents of human breast milk, including of stem cells, to open new avenue in child's development and regeneration. Accumulating evidences suggest that breast milk comprises epithelial cells, colostral corpuscles, polymorphonuclear leukocytes, mononuclear phagocytes, and lymphocytes, with those of epithelial lineage forming the main bulk of cells within two weeks of establishing lactation [8–10]. Besides these cellular components, human breast milk was also identified to possess heterogeneous stem cells and other cellular components [11, 12]. Hence, it was believed that breast milk-derived stem cells might possess considerable potency for the treatment of a wide horizon of neonatal diseases.

Special interest lies on identification of stem cells and other cellular constituents of breast milk. Breast milk encompasses various mammary stem cell populations that have been exfoliated from the mammary glands during ejection. Putative breast milk stem cells were identified by Cregan and colleagues using the markers CK5, CK14, CK19, and nestin [13]. In addition, Fan and his coworkers demonstrated the presence of cytokeratin (CK) 5, CK14, and CK18. They identified CK5 as a marker for mammary progenitor cells, CK14 as a marker for both mammary progenitor cells and mature myoepithelial cells, and CK18 as an established marker for mature luminal epithelial cells [11]. These mammary stem cells are hypothesized to be of epithelial origin and are considered to be present in milk through the suckling force from the infant. Our group studied a panel of epithelial markers and found high expression of CK19, MUC1, CK14, CD24, and CD10.

Thus the results denote the higher content of heterogenous epithelial cell population in fresh and cultured milk samples. This indicates the fact that probability of getting these cells in breast milk is directly from the breast tissue epithelium [14]. Cultured milk-derived cells were found to be more of epithelial in nature from its higher expression of epithelial markers and lower MSC marker. This confirms these cultured breast milk cell population as myoepithelial cells in accordance to its significant expression of other myoepithelial markers CD29 and CD44, as discussed later, which is also coherent with other literatures [15, 16]. The cells were also found to express a slight positivity for the adhesion-related universal mesenchymal marker, vimentin, and other MSC markers, though insignificant along with other lineage markers including markers of pluripotency (Fig. 1). This suggests that breast milk cells are epithelial in nature which gets amplified through passages in culture medium (Fig. 2). Hence, it substantiates the fact that breast milk components are predominantly epithelial with a small mesenchymal component. Hence, investigating the potential of these epithelial cells from breast milk in mammary epithelial culture medium or its equivalence might unravel its potential therapeutic applications.

These aforesaid features of the preponderance of epithelial cells in culture with very few mesenchymal stem cells might be explained by epithelial-mesenchymal transition. Epithelial-to-mesenchymal transition (EMT) is a developmental process quite important for cell fate determination. It is defined by the loss of epithelial characteristics and the acquisition of the mesenchymal phenotype. It is presumed that the breast milk cells are predominantly epithelial in nature which upon culture might undergo EMT and give rise to little fibroblast-like mesenchymal cells. Nevertheless, it is just a speculation, and further investigation is warranted to substantiate this hypothesis. Breast milk cells might be a noninvasive model for understanding the physiological and pathological role of EMT in milk samples from normal and cancerous breasts. Breast milk is a readily accessible source to

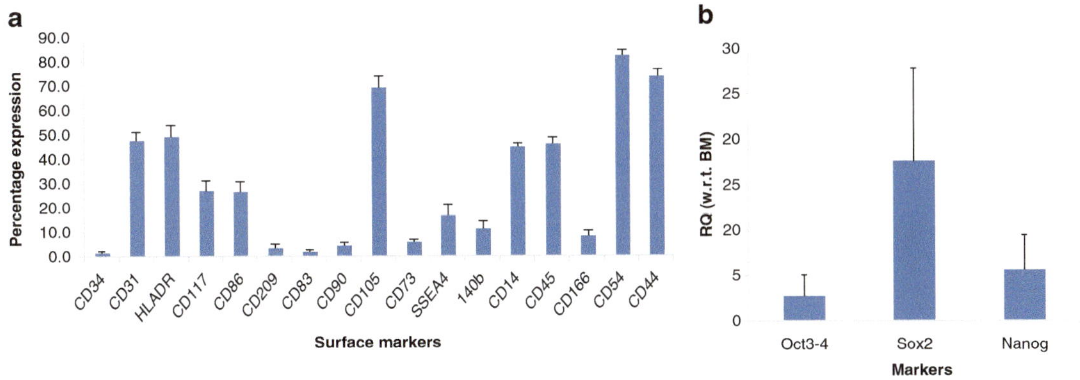

Fig. 1 Characterization of breast milk cells

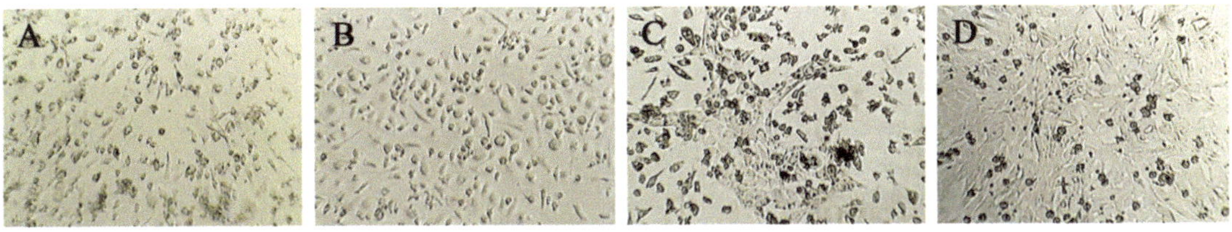

Fig. 2 Cultures of breast milk cells. Culture of breast milk cells: primary culture at Day 2 (**a**), Day 10 (**b**), Day 15 (**c**), and Day 21 (**d**)

understand the role of EMT and its implications in response/resistance to the current and/or new approaching drugs in the clinical management of breast cancer.

Besides these mammary stem cell/epithelial cell populations, breast milk is believed to possess other heterogenous cellular components including hematopoietic, side population (SP), and endothelial cells as well with the expression of its lineage-specific markers. By flow cytometric analysis, Patki and his coworkers demonstrated the breast milk stem cells exhibit the markers CD44, CD29, and SCA-1 and are negative for CD33, CD34, CD45, and CD73 confirming their identity as mesenchymal stem cells. The breast milk stem cells were also found to show positivity for cytoskeletal protein markers such as vimentin and smooth muscle actin and also manifest the presence of e-cadherin, an epithelial-mesenchymal transition marker, thus proving the aforesaid hypothesis [12], thereby substantiating our aforesaid report.

Similar to his studies, Fan and his coworkers also showed the positive expression of various primitive stem/progenitor cell markers like CD133, Stro-1, and nestin and the presence of SP within the whole cell population of breast milk. This suggests that expressed human breast milk may be a novel source of putative stem/progenitor cells. Since CD133 is a primitive cell marker present on hemopoietic and neuroectodermal stem/progenitor cells, they hypothesized that the putative stem cells in breast milk may be found in the CD133 fraction [11]. In addition, they reported the presence of other markers like osteonectin, alkaline phosphatase, and other neuroepithelial lineage markers. In addition, the breast milk was also demonstrated to possess a subpopulation of cells that shows positivity for the markers of pluripotency such as oct4, SOX2, and Nanog [17]. These findings suggest the existence of pluripotency in human lactating glands, which is also found in breast milk.

Although there are studies pertaining to the characterization of these constituents, we hypothesized that the wide characterization of cellular constituents of human breast milk in a single roof might contribute to research and therapeutics. From our extensive comparative biomarker characteristic studies of breast milk, we found that breast milk possesses more hematopoietic population cells and a scanty population of mesenchymal stem cells [18]. However, the higher expressions of CD90 and CD105 could be correlated with other functions such as immunity, immunomodulation, host-defense mechanism, and anti-inflammatory actions required for normal growth and development and other endothelial functions over the lactation period [7]. Higher expressions of certain cell adhesion and myoepithelial markers were also identified. Various adhesion molecules in different proportions suggest that homing ability of these cells is required for proper growth necessary for neonatal development and homeostasis. It is predicted that high expression of CD44 is required for firm adhesion of cell to endothelium. Besides CD44 being a myoepithelial mammary stem cell population [13], data show that CD44 is activated by PDGF [16, 17], which plays an important role in exogenous migration to injured sites by interaction with hyaluronate. Besides other immune markers, markers of pluripotency and growth factors were also identified (Fig. 1). Although the stem cell population of breast milk is less as compared to other adult sources, we speculate that cells positive for immunological components, progenitors, cell adhesion molecules, and growth factors identified so far can probably be explained by the higher needs of neonates for immunological protection, growth, leukocyte-transendothelial migration, protein synthesis, neurocognitive development, and angiogenesis.

Differentiation Potential of Breast Milk Cells

Breast milk was found to encompass several cellular constituents, and as a result recent research focus lies on investigating the multipotent differentiation potential of breast milk cells. The differentiation potential of breast milk cells is limited at its infancy with confounding aspects of results. Patki et al. have confirmed the mesenchymal nature of isolated breast milk cells by differentiating them into mesodermal lineages. The isolated cells showed an epithelial-like cell population and fibroblast-like cell phenotype within 2 weeks of culture. Furthermore, incubation of human breast milk MSCs in adipogenic induction medium led to differentiation into adipocyte- like cells after almost 21 days. Similarly, the breast milk cells were identified to differentiate into osteogenic and chondrogenic cells as well in the appropriate media [12].

Other workers studied the multilineage differentiation potency of breast milk pluripotent cell population. Culture of CD49f human breast milk-derived stem cells (hBSCs) in 2D conditions and 3D biomatrices revealed their ability to differentiate toward myoepithelial (CK14þ) and luminal (CK18þ) mammary cells, demonstrating properties of MaSCs. In keeping with the presence of markers of pluripotency identified in breast milk, the study demonstrated the multilineage differentiation potential of these pluripotent cells in breast milk and identified its differentiation ability into the cells of ectodermal, mesodermal, and endodermal origin, including bone, liver, and neuronal cells [17]. This opens new avenues for the use of breast milk as a noninvasive source of pluripotent cells for stem cell therapies. With such in vitro differentiation potential, it was believed that breast milk-derived stem cells might possess colossal potency for the treatment of wide horizon of neonatal diseases. Nonetheless, there is little evidence on the stem cell constituents of breast milk and its proliferation and differentiation potential. Furthermore, the full spectrum of cellular constituents of human breast milk is yet to be defined.

Conclusion

It has been postulated that the reproductive organs including the accessory sex organ, breast tissue, and breast milk possess several key attributes that call themselves as "pluripotent." Upon conclusion, we advocate the significance of cellular components of breast milk, thereby supporting the breastfeeding policy of WHO and UNICEF. It is hypothesized that these integral components might migrate to the bloodstreams of neonates and potentially favor cell fate development. Besides, these cellular components of breast milk could be used as a natural, oral, cellular, first remedy for the treatment of neonatal disorders due to its nonantigenic property. Although we draw focus to the breast milk components, the present study is limited to in vitro analysis and warrants further investigation in its application toward treatment of neonatal disorder.

References

1. Medina D (1996) The mammary gland: a unique organ for the study of development and tumorigenesis. J Mammary Gland Biol Neoplasia 1:5–19
2. Esmailpour T, Huang T (2008) Advancement in mammary stem cell research. J Cancer Mol 4(5):131–138
3. Russo J, Russo IR (2004) Development of the human breast. Maturitas 49:2–15
4. Michie CA (1998) The long term effects of breastfeeding: a role for the cells in breast milk? J Trop Pediatr 44:2–3
5. Hanson LA (1999) Human milk and host defense: immediate and long-term effects. Acta Paediatr Supply 88:42–46
6. Hamosh M (2001) Bioactive factors in human milk. Pediatr Clin North Am 48:69–86
7. Pabst HF (1997) Immunomodulation by breastfeeding. Pediatr Infect Dis J 16:991–995
8. Labbok MH, Clark D, Goldman AS (2004) Breastfeeding: maintaining an irreplaceable immunological resource. Nat Rev Immunol 4:565–572
9. Boutinaud M, Jammes H (2002) Potential uses of milk epithelial cells: a review. Reprod Nutr Dev 42:133–147
10. Hanson L, Silfverdal SA, Stromback L, Erling V, Zaman S, Olcén P, Telemo E (2001) The immunological role of breast feeding. Pediatr Allergy Immunol 12:15–19
11. Fan Y, Chong YS, Choolani MA, Cregan MD, Chan JKY (2010) Unravelling the mystery of stem/progenitor cells in human breast milk. PLoS One 5:e14421
12. Patki S, Kadam S, Chandra V, Bhonde R (2010) Human breast milk is a rich source of multipotent mesenchymal stem cells. Hum Cell 23:35–40
13. Cregan MD, Fan Y, Appelbee A, Brown ML, Klopcic B, Koppen J, Mitoulas LR, Piper KM, Choolani MA, Chong YS, Hartmann PE (2007) Identification of nestin-positive putative mammary stem cells in human breast milk. Cell Tissue Res 329:129–136

14. Sleeman KE, Kendrick H, Ashworth A, Isacke CM, Smalley MJ (2006) CD24 staining of mouse mammary gland cells defines luminal epithelial, myoepithelial/basal and non-epithelial cells. Breast Cancer Res 8:R7
15. Reis-Filho JS, Milanezi F, Paredes J, Silva P, Pereira EM, Maeda SA, de Carvalho LV, Schmitt FC (2003) Novel and classic myoepithelial/stem cell markers in metaplastic carcinomas of the breast. Appl Immunohistochem Mol Morphol 11(1):1–8
16. Chu PG, Lyda MH, Weiss LM (2001) Cytokeratin 14 expression in epithelial neoplasms: a survey of 435 cases with emphasis on its value in differentiating squamous cell carcinomas from other epithelial tumors. Histopathology 39:9–16
17. Hassiotou F, Beltran A, Chetwynd E, Stuebe AM, Twigger A- J, Metzger P, Trengove N, Lai CT, Filgueira L, Blancafort P, Hartmann PE (2012) Breast milk is a novel source of stem cells with multilineage differentiation potential. Stem Cells 30:2164–2174
18. Indumathi S, Dhanasekaran M, Rajkumar JS, Sudarsanam D (2013) Exploring the stem cell and non-stem cell constituents of human breast milk. Cytotechnology 65(3):385–393